Automotive Tune-Up and Test Equipment

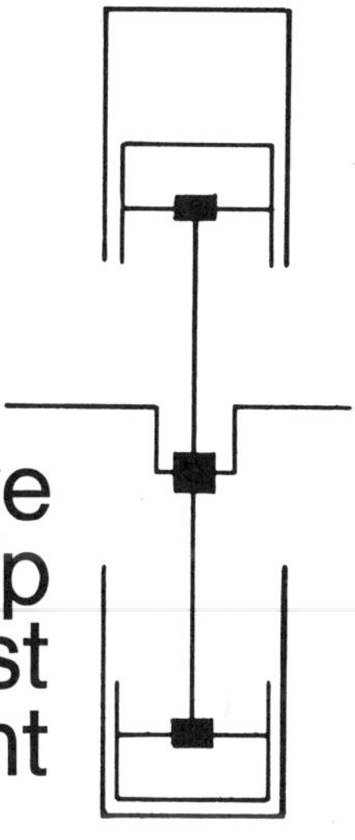

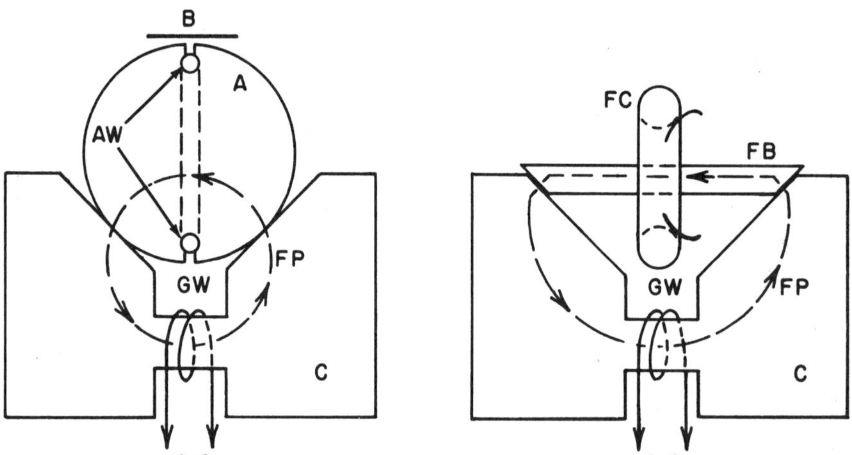

(a) (b)

Automotive Tune-Up and Test Equipment

Charles R. Cantonwine, E.E.

CHILTON BOOK COMPANY
Philadelphia New York

PATENT WARNING

ALL DISCLOSURES OF CIRCUITS AND IDEAS HEREIN ARE FOR THE SOLE PURPOSE OF INSTRUCTION TO INDIVIDUALS. ANY COMMERCIAL USE OF SAID CIRCUITS AND IDEAS SHOULD BE PRECEDED BY A PATENT SEARCH AND ENTERED INTO WITH FULL EXPECTATION THAT THEY MAY BE PATENTED OR THE SUBJECT OF A PATENT APPLICATION PENDING. SAID DISCLOSURES ARE NOT INTENDED TO IMPLY OR CONVEY ANY RIGHTS HELD BY THE INVENTORS.

First Edition

Published in Philadelphia by Chilton Book Company
and simultaneously in Ontario, Canada,
by Thomas Nelson & Sons, Ltd.

ISBN: 0-8019-5672-2

Library of Congress Catalog Card Number 78-180201
Designed by Cypher Associates, Inc.
Manufactured in the United States of America

Preface

The purpose of this book is to explain and illustrate the principles of operation, adjustments, and calibrating procedures for representative models of tune-up and test devices. The book should be of interest as a general reference to anyone concerned with the maintenance of automotive equipment—salesmen, servicemen, mechanics, technicians, engineers, and manufacturers of associated equipment.

This book covers the fundamentals and principles of operation of all electrical equipment likely to be encountered in the automotive service field. It is written so that the beginner can understand the subject, yet it provides useful information for even the most advanced engineer. It is assumed that the reader has a basic knowledge of electronics. If he does not, he should benefit by reading the author's book, *Battery Chargers and Testers,* which treats the elements of this subject.

Until fairly recently the only automotive tune-up and test equipment in general use consisted of a plain neon timing light and a feeler gauge for adjusting distributor points and spark plug gap. Today, however, there is available an array of electronic test equipment that is even more elaborate and accurate than could be found in the most advanced engine development laboratory of only a few years ago.

The industry has grown so rapidly that most of the knowledgeable writers on the subject have been pressed to get out operating instructions and service manuals, and thus have had little time to provide information for students and technicians interested in learning the special applications of electronics to this broad category of equipment.

All the circuits in this book have been actually traced from equipment serviced by the author, and are up to date. The author has developed several pieces of special test equipment to service automotive engines. In this book are many original ideas not found anywhere else in print. The author has invented, and has applied for a patent on, the "Cantonwine timing light to measure dwell angle." The details of this invention are disclosed and explained here. Also included are discussions of the scope type of engine analyzers; the "Hall effect" clamp-around dc ammeter; exhaust gas analyzers; transistorized, condenser-discharge, and magnetic-pulse ignition systems; the Cantonwine precision dc voltmeter; the diode and transistor tester; ignition simulators and calibrators, and many more.

A listing of sources of supply and brand names is provided, and these are referred to in the text by source number. Appreciation and credit is given to all sources mentioned for material prepared from operating instructions, service manuals, bulletins, and the like that have been made available.

Charles R. Cantonwine E.E.

Contents

Automotive Tune-Up and Test Equipment

1
Automotive Engines

1-1 Engine Types

The ability to use, repair, and calibrate modern automotive tune-up and test equipment properly requires a thorough knowledge of the principles of operation of the internal-combustion engine.

Internal-combustion engines are so-called because the pressure is produced by gas or vapor exploding inside the cylinder and against the piston. Internal-combustion engines include the gasoline (butane, propane, etc.) engine, diesel engine, and electronic fuel injection engine. Other automotive engines not considered internal-combustion engines include, the gas turbine, jet engine, steam turbine, steam piston engine, and electric engine.

The gasoline internal-combustion engine has remained basically unchanged since the invention of high voltage spark plug ignition.

Increased demand for performance and anti-pollution requirements have, however, required higher standards in the design, manufacture, and adjustment of the modern engine. These requirements have caused a rapid evolution of tune-up and test equipment. The electronics industry has made it possible to bring to every garage and filling station an array of sophisticated equipment that enables the trained mechanic to diagnose quickly and efficiently any malfunction of an internal-combustion engine.

1-2 Engine Cylinder Arrangements

An internal-combustion engine may have any number of cylinders, which can be arranged in several ways, and it may have four cycles or two cycles.

Some of the more common cylinder arrangements are shown in Figure 1-1. Figure 1-1a shows a single-cylinder engine viewed from the end of the crankshaft CS and Figure 1-1b shows the same engine from the side of the crankshaft. The cylinder CY is stationary, and the piston P, moving up and down in the cylinder, is connected to the crankshaft by a connecting rod CR. The crankshaft rotates about the axis AX.

Figure 1-1c shows a two-cylinder, or twin-cylinder, engine from the side and "in line," with the two cylinders side by side in the same plane. The crankshaft has the "throw" in one direction so that the relative positions of pistons P1 and P2 are always the same in cylinders CY1 and CY2.

Figure 1-1d shows a two-cylinder arrangement with the cranks 180° apart. Piston P1 will be at the top dead center (TDC) when piston P2 is at the bottom of the stroke. This arrangement may have any number of cylinders in line.

Figure 1-1e shows a two-cylinder engine with the two cylinders on opposite sides of the crankshaft. It is called an "opposed cylinder" engine.

Figure 1-1f shows the rotary type engine, commonly used in aircraft. Only three cylinders are shown but any number can be used.

Figure 1-1g shows the V-type engine; it may have any even number of cylinders, such as, V-2, V-4, V-6, or V-8. Not shown is the "slant" engine, which has the same appearance as either half of a "V" in Figure 1-1g; the cylinders are in line but slant at an angle.

1-3 The Four-Cycle Engine

The four-cycle engine is so-called because the piston must make four excursions, two up and two down, inside the cylinder to complete a cycle in the following order: suction stroke downward, compression stroke upward, power stroke downward, and exhaust stroke upward. It then repeats the cycle.

The crankshaft makes two complete revolutions per cycle for one cylinder. A one-cylinder engine will fire once and have one power stroke for every two revolutions of the crankshaft. A two-cylinder engine will

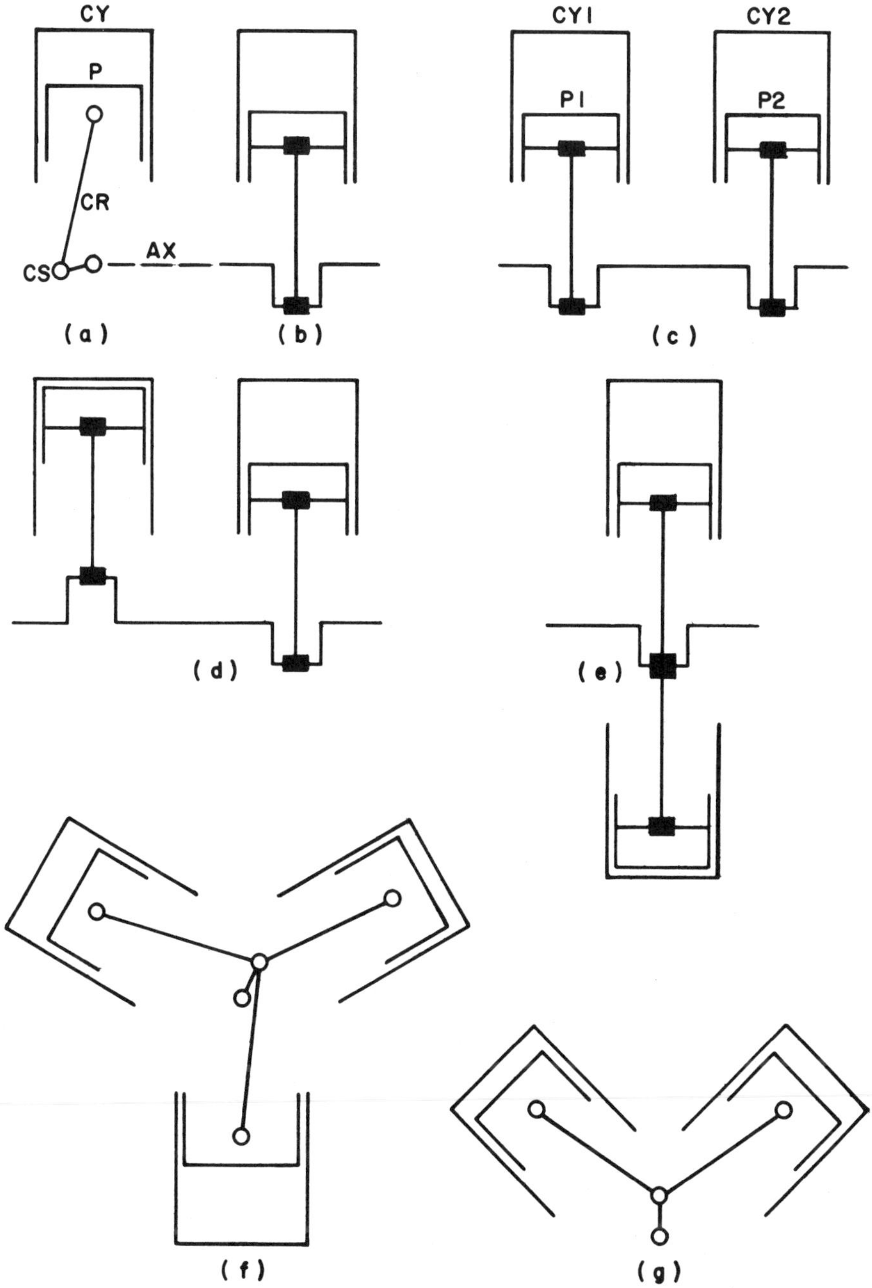

Figure 1-1 Engine cylinder arrangements

fire twice and have two power strokes for every two revolutions of the crankshaft, or one power stroke every revolution. A four-cylinder engine will have four power strokes for every two revolutions, or two per revolution. Therefore, the number of power strokes per revolution will always be one-half the number of cylinders. For example, an eight-cylinder engine will fire four times and have four power strokes per revolution, or eight for every two revolutions. The distributor, which selects the cylinder to fire, must make one revolution in every two revolutions of the crankshaft in order to fire all cylinders in one cycle. The distributor is, therefore, geared down to turn at one-half the speed of the crankshaft.

The principle of operation of the four-cycle engine can best be understood by studying the action of a single-cylinder engine. Engines having more than one cylinder are simply arranged so that all of the cylinders fire at equally spaced intervals in two revolutions. The order in which they fire (firing order), dependent on the crankshaft arrangement, is important to engine balance and smooth running.

A single-cylinder four-cycle engine is shown in schematic form in Figures 1-2, 1-3, 1-4, and 1-5. In practice, only one camshaft is used and the valves are placed in line; also, the distributor is driven from the camshaft at right angles.

In Figure 1-2, piston P slides within cylinder CY and is connected to a rotating crankshaft CS by connecting rod CR. The crankshaft, in rotating, carries a gear that is meshed with three other gears of twice the diameter of the crankshaft gear to turn an intake valve camshaft CMSI, an exhaust valve camshaft CMSE, and a distributor camshaft DC. Arrows indicate the directions of rotation.

The intake camshaft CMSI causes an intake valve INV to open and close between the combustion chamber and the carburetor at specified intervals. The exhaust camshaft CSME causes an exhaust valve EXV to open and close between the combustion chamber and the exhaust pipe EX at specified intervals.

Spark plug SP is mounted so that the spark gap is located within the combustion chamber.

The distributor points PTS are actuated by distributor cam DC so that they open and close at specified intervals.

Ignition coil IC consists of a low-resistance primary winding with relatively few turns of wire, connected in series with a battery BAT,

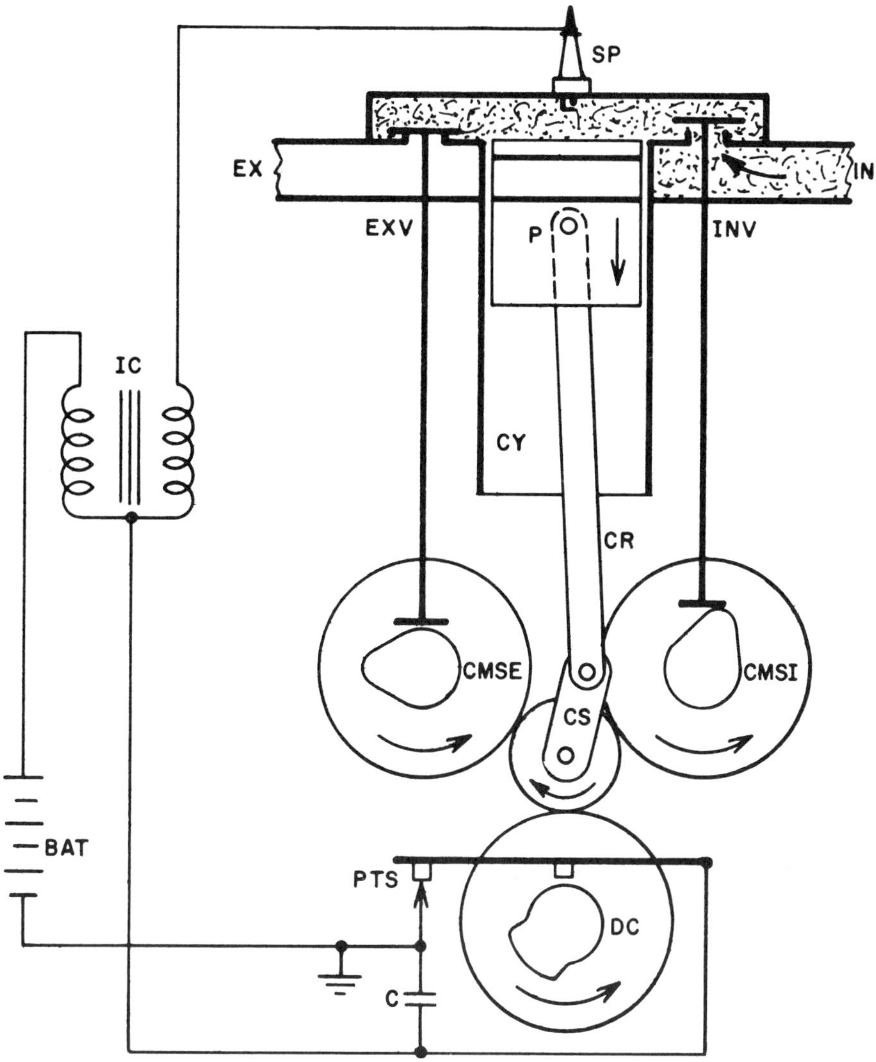

Figure 1-2 Four-cycle engine suction stroke

and the distributor points. A high-resistance secondary winding with a relatively large number of turns is connected between the ground, or frame, and the spark plug SP center electrode.

Figure 1-2 shows the engine at the beginning of the suction stroke.

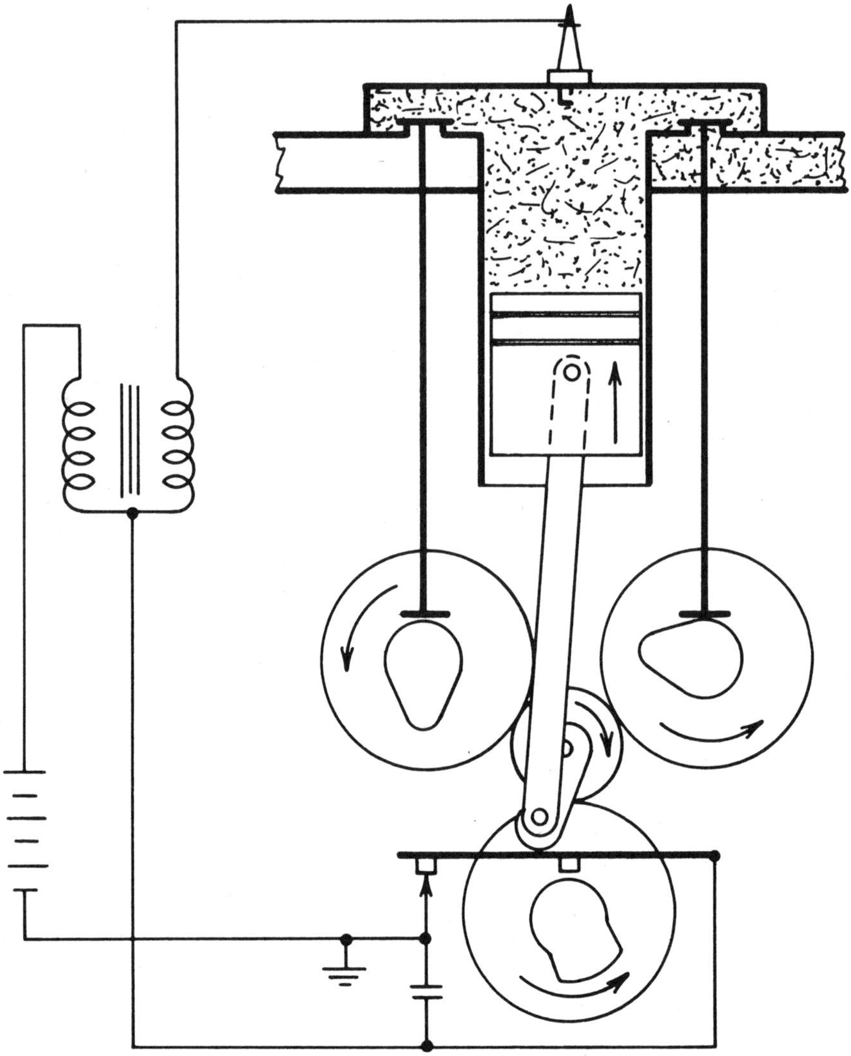

Figure 1-3 Four-cycle engine compression stroke

The intake valve INV is starting to open (the exhaust valve EXV is closed), and as the piston P moves downward in the cylinder CY it creates a suction that pulls in a mixture of gasoline vapor and air from the carburetor through intake port IN. As the piston reaches the bottom of the stroke the intake valve closes, trapping the combustible gas.

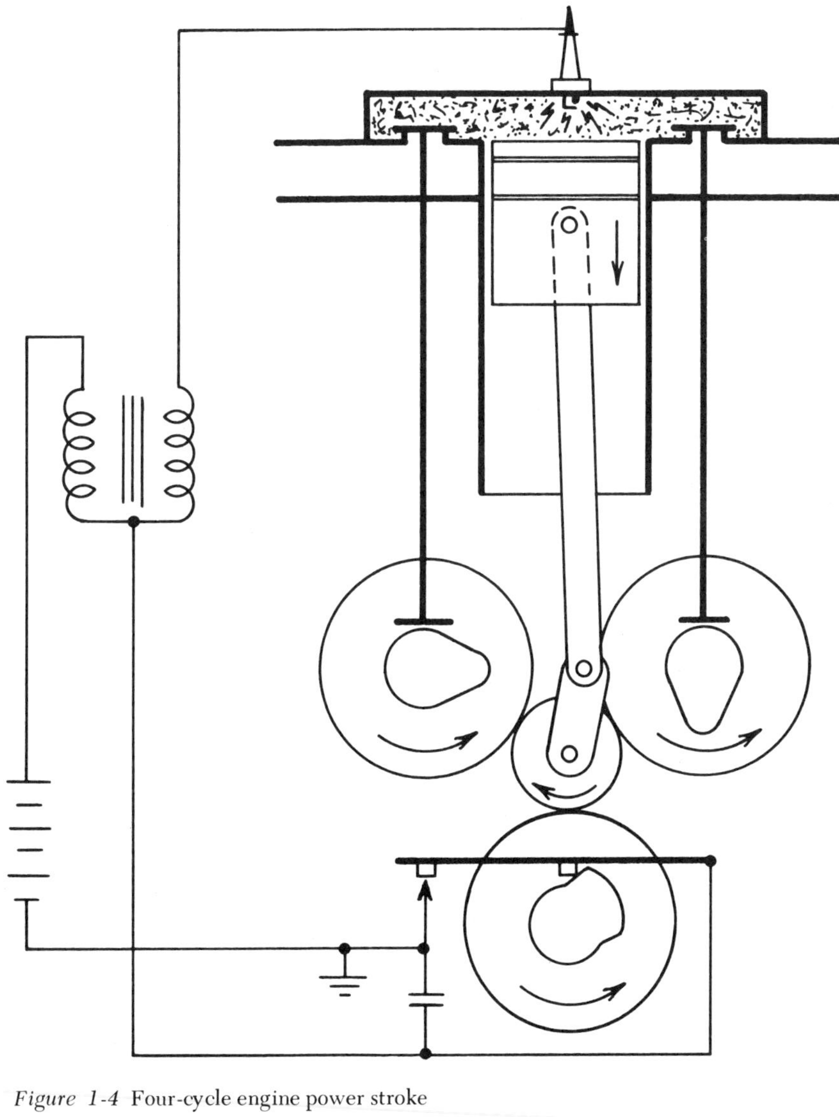

Figure 1-4 Four-cycle engine power stroke

Figure 1-3 shows the piston on the upward compression stroke. Since both valves are closed, the gases are compressed to a fraction of their original volume, which causes an increase in pressure and temperature. This temperature must not be high enough to ignite the combustible gases; this is the job of the spark plug at just the right time.

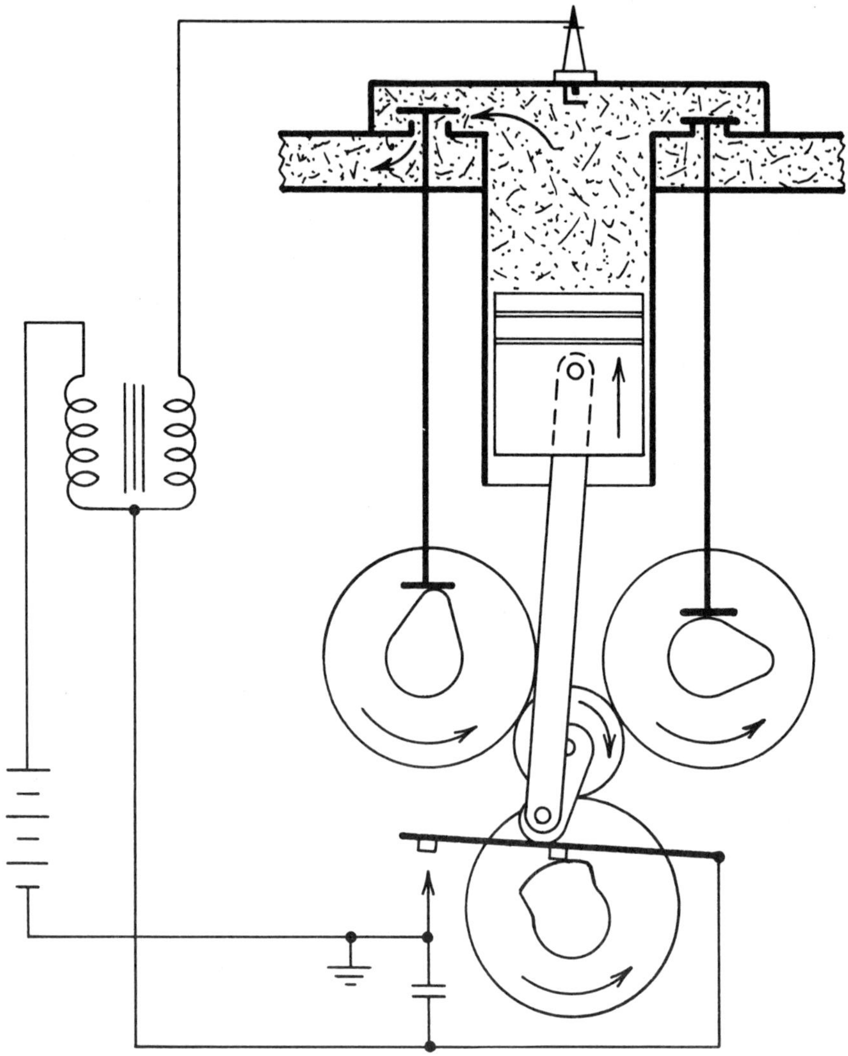

Figure 1-5 Four-cycle engine exhaust stroke

In Figures 1-2 and 1-3 the distributor cam has caused the points PTS to remain closed. This permits a flow of current from the battery to flow through the ignition coil IC primary. This flow of current builds up a flux in the iron core during the suction and compression strokes.

As the piston reaches the top, or when it is near the top of the stroke (Figure 1-3), the distributor cam starts to open the points PTS, as shown in Figure 1-4. When the points open, the magnetic field which was built up in the ignition coil collapses and generates two voltages. One voltage in the primary circuit charges capacitor C across the points; then the capacitor discharges back into the coil primary. This is repeated several times until the current dies down. This is called a damped wave oscillation. At the same time, a second damped wave of oscillating high voltage is developed across the secondary of the ignition coil, and it has voltage high enough to jump the air gap of the spark plug. This causes the compressed gases to explode, which increases the pressure above the piston, forcing it downward on the power stroke.

As the piston reaches the bottom of the power stroke or just before reaching it, the exhaust valve EXV is opened, permitting the burned gases to escape through the exhaust pipe. The piston then moves upward on the exhaust stroke, as shown in Figure 1-5, expelling the remainder of the burned gases out exhaust pipe EX. As the piston approaches the top of the stroke, the exhaust valve closes and the intake valve starts to open for the beginning of a new cycle, which starts again, as shown in Figure 1-2. The points have closed during the exhaust stroke, at a time that depends on their adjustment. The points are usually closed about two-thirds of the cycle and open about one-third of the time. The percentage of time they are in the closed position is expressed in degrees and is called the "dwell angle."

1-4 The Two-Cycle Engine

The two-cycle engine differs from the four-cycle in that there is a power stroke for every revolution of the crankshaft. In a four-cycle engine the cylinder fires (power stroke) one time on every second revolution of the crankshaft. In the two-cycle engine there is a power stroke and an intake stroke every revolution of the crankshaft; hence the name "two-cycle."

The main difference between the operation of the four-cycle and the two-cycle engine is in the method of conducting the air-fuel mixture from the carburetor to the cylinder. In a four-cycle engine, a manifold conducts the air-fuel mixture externally to the top of the cylinder through the intake valve. In a two-cycle engine, the air-fuel mixture is drawn into the cylinder through the crankcase. The two-cycle engine has an arrangement of ports, rather than valves.

The principle of operation of the two-cycle engine is illustrated in Figures 1-6, 1-7, 1-8, and 1-9. In Figure 1-6 piston P is at the bottom of the stroke, which leaves the intake port IP and the exhaust port EXP fully open. On its downward stroke, piston P has compressed the air-fuel mixture in the crankcase CC, and this mixture is then at a higher pressure than the gases between the top of the piston and the cylinder CY. The reed valve RV is closed under pressure at this stage. The air-fuel mixture from the crankcase then flows through the intake port and strikes the vertical step at the top of the piston. This step deflects the

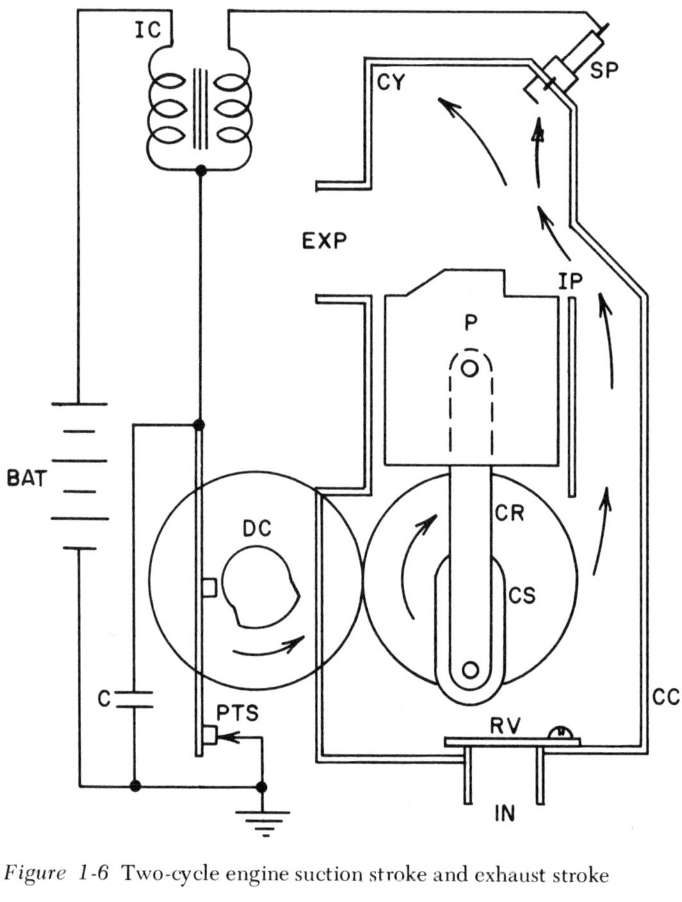

Figure 1-6 Two-cycle engine suction stroke and exhaust stroke

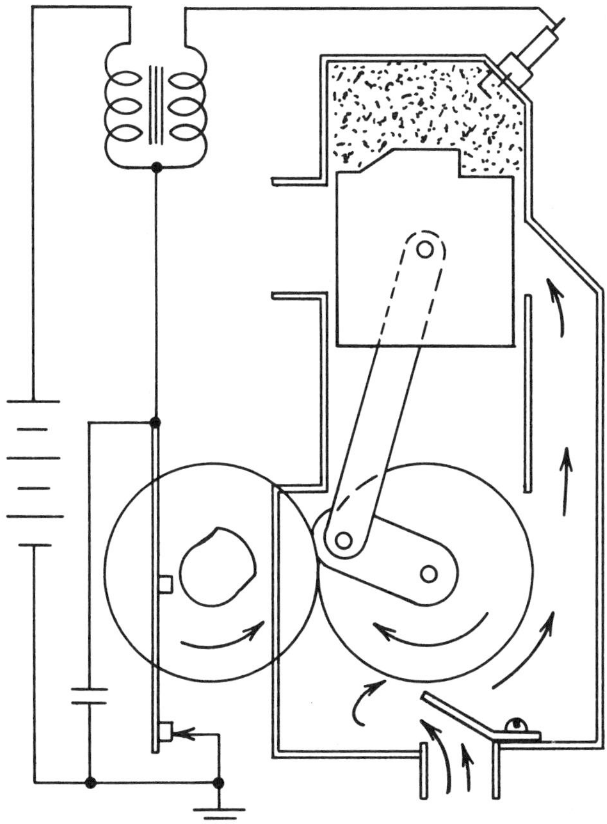

Figure 1-7 Two-cycle engine compression stroke

air-fuel mixture along the cylinder wall and away from the exhaust port.

In Figure 1-7 the piston has moved upward and past the intake and exhaust ports and has trapped the air-fuel mixture inside the cylinder. This is the beginning of the compression stroke, and it continues until the piston is at the top of the stroke, where the space within the cylinder is at a minimum. During this stroke, additional air-fuel mixture has been drawn from the carburetor, through the intake manifold IN and the reed valve RV (shown open), into the crankcase.

Figure 1-8 shows the piston at the top of the compression stroke, where the air-fuel mixture is ignited by spark plug SP. The distributor contact points PTS have been closed, or on dwell, up to now. At this point the

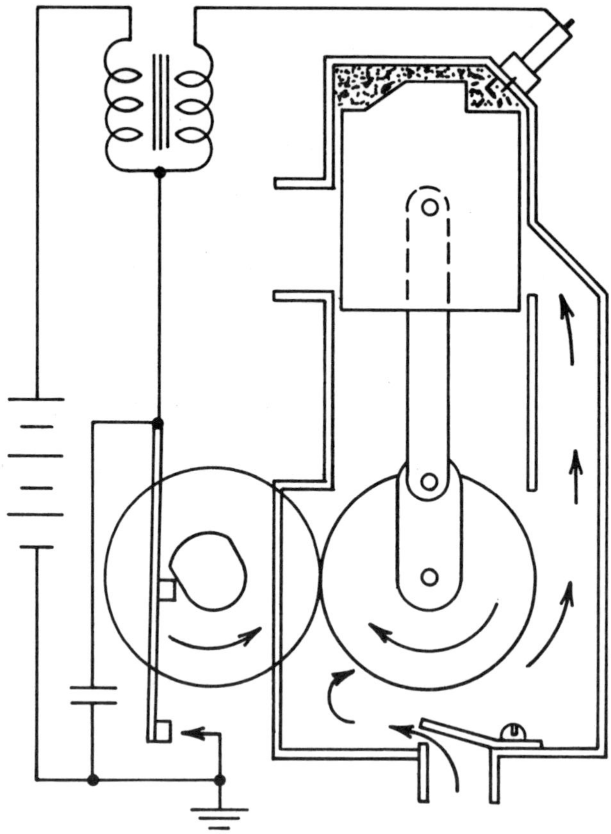

Figure 1-8 Two-cycle engine power and suction stroke

contact points open by the action of the distributor cam DC, which fires the spark plug. The explosive action forces the piston downward, turning the crankshaft CS through connecting rod CR until the exhaust port (first) and the intake port IP (second) are again open. However, during this power stroke the exhaust port is uncovered first. The burned gases are blown through the exhaust port, and the pressure in the cylinder drops to about zero. Then, the fresh, compressed air-fuel mixture forces the remaining burned gases out the exhaust port, as in Figure 1-9, and a new compression stroke starts. The contact points PTS are now closed. The ignition coil IC, capacitor C, and battery BAT function the same as for a four-cycle engine.

The four-cycle engine is lubricated by oil in the crankcase, whereas in the two-cycle engine the lubrication is provided by mixing oil with the fuel in the ratio of about 0.5–1 pint of oil per gallon of gasoline.

In Figures 1-6, 1-7, 1-8, and 1-9, the distributor cam DC is shown driven by a 1-to-1 gear ratio arrangement to simplify the drawing. In practice, the distributor cam is usually mounted on the crankshaft and turns at the same speed as the crankshaft on a two-cycle engine.

The two-cycle engine usually has a magneto for ignition instead of a battery-powered coil. This is also true of small one-cylinder four-cycle engines.

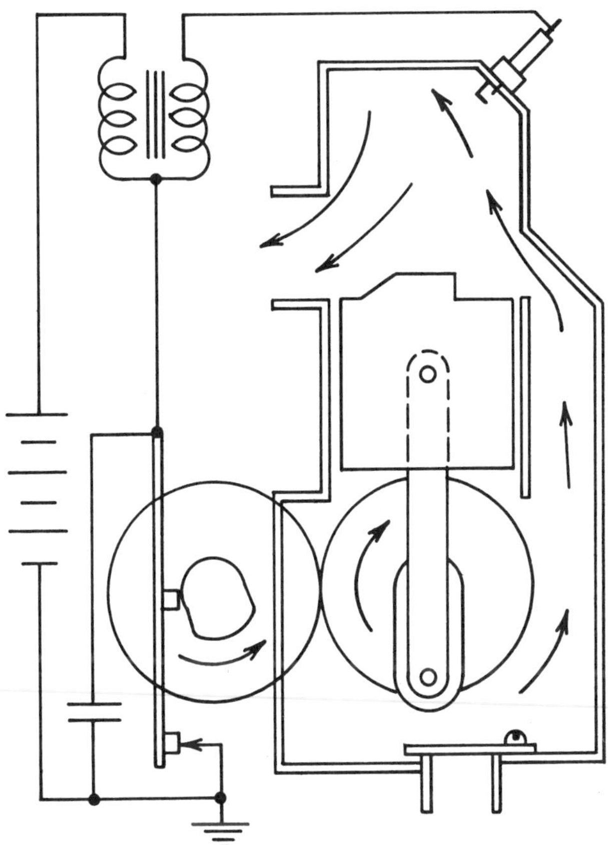

Figure 1-9 Two-cycle engine exhaust and suction stroke

1-5 Electronic Fuel Injection System of Carburetion

The electronic fuel injection system is generally thought to be a relatively new method of introducing the fuel into the cylinders of an internal-combustion engine. On the contrary, fuel injection systems have been used in experimental automobile models for a number of years but have never become popular in the United States. Interest has been renewed, however, since anti-pollution requirements have been imposed, and at least one foreign car is using this system.

The engine is the same as any other internal-combustion engine using high voltage spark ignition, except in the method of carburetion. The electronic fuel injection system introduces the fuel into the air intake manifold close to the engine intake valves. The fuel is under a pressure of less than 30 pounds per square inch and is atomized to mix with the intake air.

The amount of fuel, the timing, and duration of the injection is controlled by a rather complicated electronic computer. The computer takes information from several sensing devices and translates it into the correct amount of fuel for the driving conditions. These sensing devices measure such things as temperature, pressure, vacuum, speed, acceleration, and deceleration. The electronic computers are not considered, at the present time, to be serviced in the field because of their complicated nature and also because of the elaborate equipment and specialized knowledge required. They must be replaced as a unit and are simply plugged in.

In addition to the customary tune-up and test equipment required for internal-combustion engines, the fuel injection system can be serviced with a volt-ohm-milliammeter (VOM) and a 0–30 pounds per square inch pressure gauge.

1-6 Diesel Engines

The diesel engine is an internal-combustion engine similar to a gasoline engine except that it does not require a carburetor or an electric spark to ignite the fuel in the cylinder. The four-cycle diesel engine can be compared with the gasoline engine as follows.

1. *Suction Stroke.* The exhaust valve is closed, but the intake valve opens and the piston, on its downward stroke, draws in *air only.*

2. *Compression Stroke.* At the bottom of the suction stroke, both valves close as the piston starts upward on the compression stroke. The compression ratio is more than twice that of a gasoline engine, and the air is compressed to a very high pressure and temperature. Compression ratio is the ratio of the volume of air before compression to that after compression. For a gasoline engine, the compression ratio might be 8 to 1; for a diesel, 16 to 1.

3. *Power Stroke.* At the top of the compression stroke, or near the top, diesel oil under pressure atomizes into the hot compressed air and self-ignites from the heat of compression. The resulting explosion forces the piston downward on the power stroke.

4. *Exhaust Stroke.* At the end of the power stroke, or near the end, the exhaust valve opens to enable the burned gases to escape from the cylinder. The piston then starts upward on the exhaust stroke and pushes the rest of the burned gases out to the atmosphere. At the end of the exhaust stroke, the exhaust valve closes and the intake valve opens, and the piston starts again on a suction stroke. The cycle then repeats itself.

We have described a four-cycle diesel engine, but it can also be constructed in two-cycle design.

The tune-up and test equipment required for the diesel engine consists essentially of pressure, temperature, and mechanical measuring devices. The electrical test equipment for starters, generators, alternators, voltage regulators, and batteries are the same as for gasoline engines.

2
Ignition Systems

2-1 Early Ignition Systems

In the first gasoline engines, the air-fuel mixture in the cylinder was ignited by an open flame or by a very hot metallic surface in the shape of a tube or ball. It was discovered that the air-fuel mixture could be ignited by the heat of compression. The early engine using this principle was the forerunner of the diesel engine.

The air-fuel mixture is fired by creating an electric spark inside the cylinder head of the engine at exactly the right time. A battery, or magneto, and an induction coil are used to create a spark when the contacts are opened. This spark is carried through a pair of electrical contacts located in the cylinder head and operated so as to open and close by a cam-operated lever, or electromagnetically. This system has been used for many years on slow-speed single-cylinder stationary engines.

In modern engines, the air-fuel mixture is ignited by a high-voltage spark across the spark plug electrode gap located inside the combustion chamber. The voltage can be generated with a high- or low-tension magneto and one or more high-voltage transformers (coils), or by a battery and one or more high-voltage coils.

The high-voltage transformer may provide a single spark, or it may have a vibrator to give a continuous spark while the distributor contacts are closed. In the vibrator system, which is now obsolete, there can be a separate vibrator coil for each cylinder, or a single master vibrator feeding several separate coils (one for each cylinder), or one vibrator coil can supply the high voltage to all cylinders using a high-voltage distributor cap and rotor. These early systems of ignition have been mentioned because of the renewed interest in antique car restoring.

2-2 Kettering System of Ignition

Modern automotive engines use the Kettering system of ignition, although many high-tension magnetos are still used for special applications, and the magneto system is still in common use for small gasoline engines, outboard motors, and the like.

The Kettering system of ignition is shown schematically in Figure 2-1 for a six-cylinder engine with a six-lobe cam spaced at 60° and having six spark plug wires leading to the distributor cap DC. For a four-cylin-

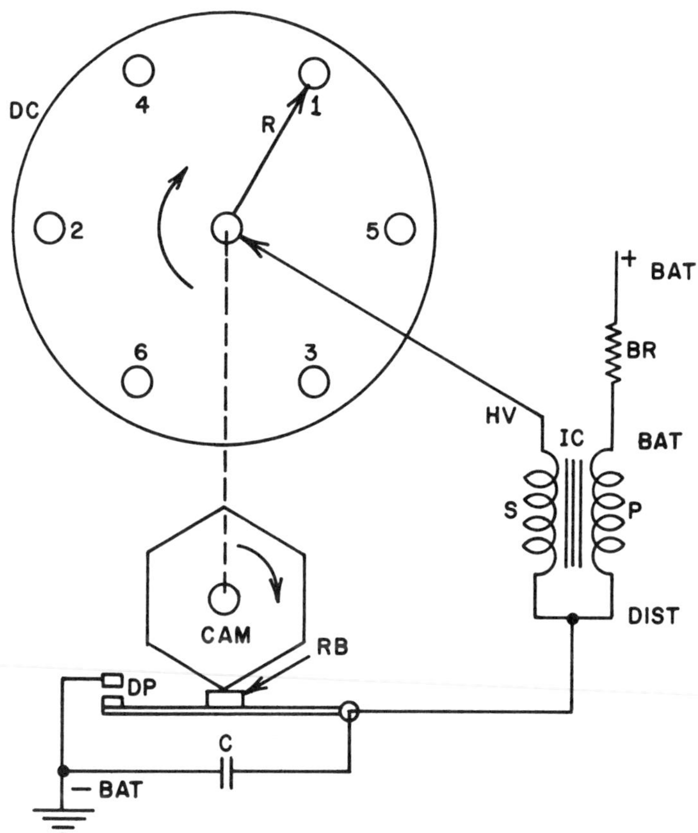

Figure 2-1 Kettering system of ignition

der engine there would be only four cam lobes and four wires to the distributor cap spaced 90°. Similarly, for an eight-cylinder engine there would be eight cam lobes spaced at 45° and eight wires leading to the distributor.

The cam is driven at one-half the crankshaft speed on a four-cycle engine. The cam drives the distributor rotor R inside the distributor cap. The numbers 1, 5, 3, 6, 2, and 4 indicate the firing order and the spark plug wires going to the corresponding cylinder numbers. The cam actuates the distributor points DP through the insulated rubbing block RB. Capacitor C is connected across the distributor points. Ignition coil IC has an iron core wound with a low-resistance low-voltage primary P winding and a high-resistance high-voltage secondary S winding, which is connected to the center of the distributor cap DC. The primary has two terminals, BAT (or DIST), or it may have no marking at all. It is very important that the primary be connected properly as marked. If it is connected in reverse, the high voltage may be approximately 30% below normal.

With a 12-volt battery connected across the positive and negative battery connections, and when the cam has turned to permit the distributor points DP to close, positive current flows through ballast resistor BR, through ignition coil IC and primary P, through the distributor points, and back to the battery negative terminal. This causes a flux to build up in the iron core of the coil. Capacitor C is shorted by the distributor points. When the distributor points open, the current is interrupted in the coil primary, but the distributor points do not spark because the capacitor tends to hold the voltage across the points at zero volts for a split second until the points are sufficiently open. Then the flux in the iron core collapses, inducing a high voltage HV in the secondary S. At the same time, a relatively high voltage is induced in the primary P. With the primary now in series with capacitor C, a rapid charge and discharge takes place, setting up a damped oscillation that dies down after a few cycles, at which time the distributor points close again and the cycle is repeated. The distributor rotor R moves into position to conduct the spark to the next cylinder spark plug in the firing order.

The distributor points are closed approximately two-thirds of the time and are open about one-third of the time. For a six-cylinder engine with six cam lobes spaced at 60° the distributor points would be closed for 40° and open for 20°. This 40 degrees of closing time is called the dwell angle. For a four-cylinder engine with four cam lobes spaced at 90°, they would be closed 60° (dwell angle) and open for 30° of cam rotation. For an eight-cylinder engine with eight cam lobes spaced at 45°, they would be closed for 30° (dwell angle) and open for 15° of cam rotation.

Newer ignition systems use the same distributor points but do not carry the battery current through the coil primary. Instead, only the distributor points, at very low current, trigger transistors or other solid state devices which carry the heavier current. This results in longer distributor point life and better engine performance. Newer ignition systems use no distributor points at all; the ignition coil is energized with a pulse generator and electronic means, and even better performance is claimed.

2-3 Transistorized Ignition System Using One Transistor, Typical of Source 16

In the conventional ignition system the distributor points interrupt the full current of several amperes in the circuit of the ignition coil primary, causing arcing, burning, and pitting of the points, thus shortening the life of the distributor. The transistorized ignition system overcomes this objectionable characteristic by requiring the distributor points to interrupt a current of only a small fraction of an ampere. The transistor, acting as an electronic relay, interrupts the heavy current of several amperes, leaving the distributor points only to trigger the transistor.

A transistorized ignition system typical of Source 16 is shown schematically in Figure 2-2, where only one transistor (Q1) is used. Others may use two or more transistors with a driver stage and a power stage, but they all work according to the same fundamental principles.

They are designed for a fixed ground polarity and voltage. For example, a unit may be marked 12 V neg, which must be observed, since they will not work on reverse polarity or other voltages. For best results, a special ignition coil is used. The coil is usually of conventional design but modified with a higher turns ratio that has fewer primary turns at higher current. This results in increasing the amount of coil current, reducing its inductance, and hence obtaining a smaller time constant that is more favorable for rapid build-up in the primary.

Some elaborate experimental models have been designed to maintain the ignition voltage at high speeds, when the voltage normally reduces

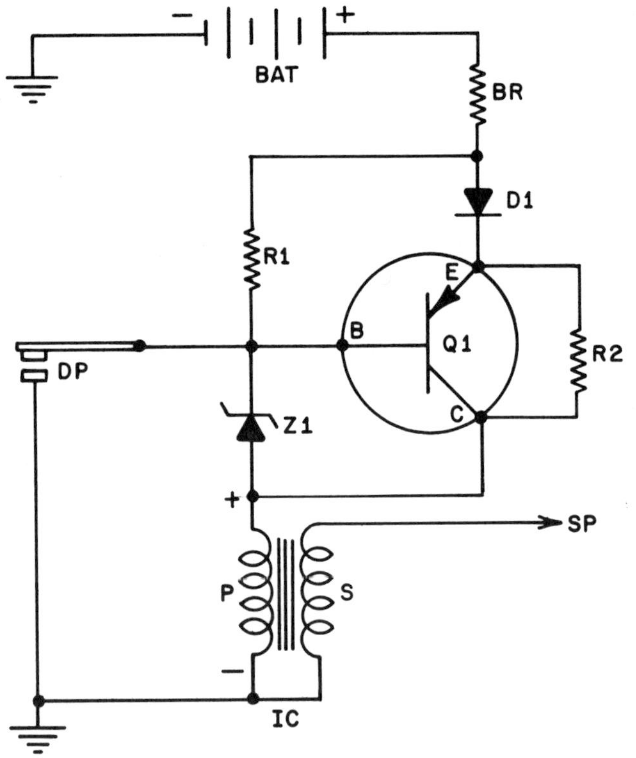

Figure 2-2 Transistorized ignition system using one transistor, typical, Source 16

to about half of its level at idle speed. It can be maintained with a transistorized tachometer circuit that increases the current of the ignition coil primary as the speed increases.

In Figure 2-2, when the distributor points DP are closed, a current flows from the positive battery BAT through ballast resistor BR, through resistor R1, and back to ground through the distributor points. This places a negative bias on the base B of transistor Q1, which causes it to conduct from the collector C to emitter E and through the ignition coil IC primary P, which builds up the coil voltages as usual. Zener diode Z1 serves to limit the bias on the base B to a safe level. Diode D1 and resistor R2 suppress high-voltage surges, which would damage the transistor Q1. When the distributor points open, the base of the transistor is restored to zero bias and the transistor ceases to conduct between C and

E, causing the coil to fire through the secondary S and the spark plugs SP.

The coil is sometimes connected between the transistor and battery with the collector C of the transistor grounded.

2-4 Transistorized Ignition System, Typical of Source 6

Figure 2-3 shows the schematic wiring diagram of a transistorized ignition system typical of Source 6 using one transistor.

When the distributor points DP are open, the transistor Q1 base B is held at the emitter potential (no bias) by resistor R3, and no current flows from the battery through the coil primary because the transistor is not conducting between collector C and emitter E.

When the distributor points close, the base of the transistor has a negative bias relative to the emitter from the voltage divider network R3

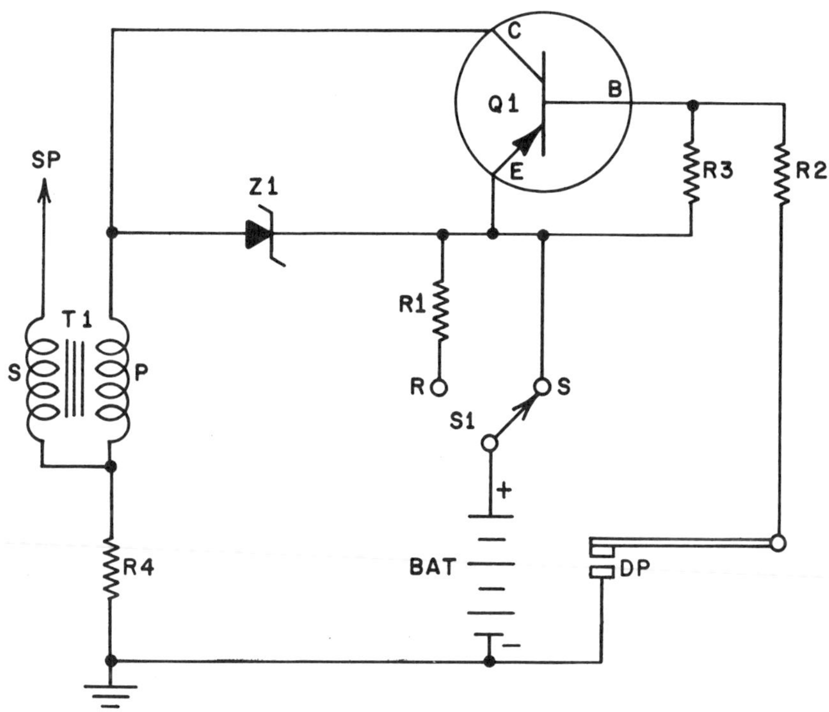

Figure 2-3 Transistorized ignition system, typical, Source 6

cuit. Because the control circuit triggers the power circuit, current stops in the power circuit also.

This causes the magnetic field produced by the transformer T1 coil primary P winding to collapse, generating high voltage in the primary circuit, and, in turn, inducing very high voltage in the secondary S of transformer T1, and the high voltage is fed to the distributor and spark plugs.

This would ordinarily be sufficient to destroy the transistor, inasmuch as the induced voltage instantly rises above the load level of the transistor until it reaches the zener voltage of Z1. For this reason a capacitor C2 (2 mfd) is incorporated into the power circuit to absorb the high voltage and keep it within the load limits until the protection level of the zener diode Z1 is reached. Discharge of the capacitor voltage from zener diode Z1 level to system voltage takes place by reverse flow through the power circuit, except that the transistor is bypassed by the zener diode, which freely passes this reverse current.

The closing of the ignition points completes the cycle by closing the "point" circuit again and causing current flow in the "power" circuit.

The zener diode Z1 is rated at 91 V zener voltage, and the transistor Q1 is rated cathode to emitter, at 120 V.

The capacitor C1 (40 mfd) is a safety device that prevents transistor damage caused by the opening of the battery circuit with the engine running.

A "tach block" TB is incorporated into the circuit for attaching service instruments only. In the conventional ignition system these instruments are usually connected across the distributor primary lead and ground, but in the transistorized circuit, connecting the instruments in this manner would jump the breaker point contacts, contributing to a current build-up in the base circuit and in the collector. Consequently, it would overheat and damage the transistor, or change its characteristics and cause improper operation and shorten its life. With the use of the tach block for tachometer and dwell meter, readings can now be taken without causing transistor damage.

A fuse F1 (2 amp) has been inserted into the collector C to the blue lead of the tach block preventing the transistor from being damaged by the application of external devices other than normal testing equipment.

A cold-start relay RL1 and resistor R1 are incorporated into the circuit at the starter relay RL2, interrupting the conventional battery-to-coil lead. The purpose of this is to furnish additional current to the coil primary winding during situations when the starter draw is excessive. The cold-start relay contacts are normally closed. They open only during the cranking cycle. However, when the available battery voltage drops below a specified value during the cranking cycle, they again close, by-passing the ignition resistor R2 and furnishing full battery current to the system. Thus, this relay is generally operative only during cold starting conditions with a low battery voltage or high starter current flow.

2-6 Transistorized Ignition Systems Precautions

On all ignition systems using transistors, there are certain precautions that must be observed. Refer to the manufacturer's instructions concerning the following procedures.

1. Use of measuring instruments, dwell meters, tachometers, and the like.
2. Installation of capacitors for suppression of radio interference.
3. Quick charging and battery boosting.
4. Removing the battery or its connections while the engine is running.
5. Remote control cranking switch for "jogging."
6. Cranking engine for compression tests and the like.
7. Working on the electrical system.

2-7 Capacitor Discharge Ignition System, Typical of Source 7

The capacitor discharge (C-D) system is not usually found as original equipment on automotive engines, but it is available from several sources as an add-on item.

Figure 2-5 shows a typical, Source 7, C-D ignition system. It functions as follows. The system consists of a dc-to-dc converter (to change the battery voltage to a higher dc voltage for storage), a storage element (capacitor), a switching element (silicon-controlled rectifier), and a high-voltage output transformer (coil) to transform the low-voltage stored dc to a level that will fire the spark plugs.

The applied battery voltage is converted from a nominal 12 V to approximately 400 V by the converter circuitry (transistors Q1, Q2, and

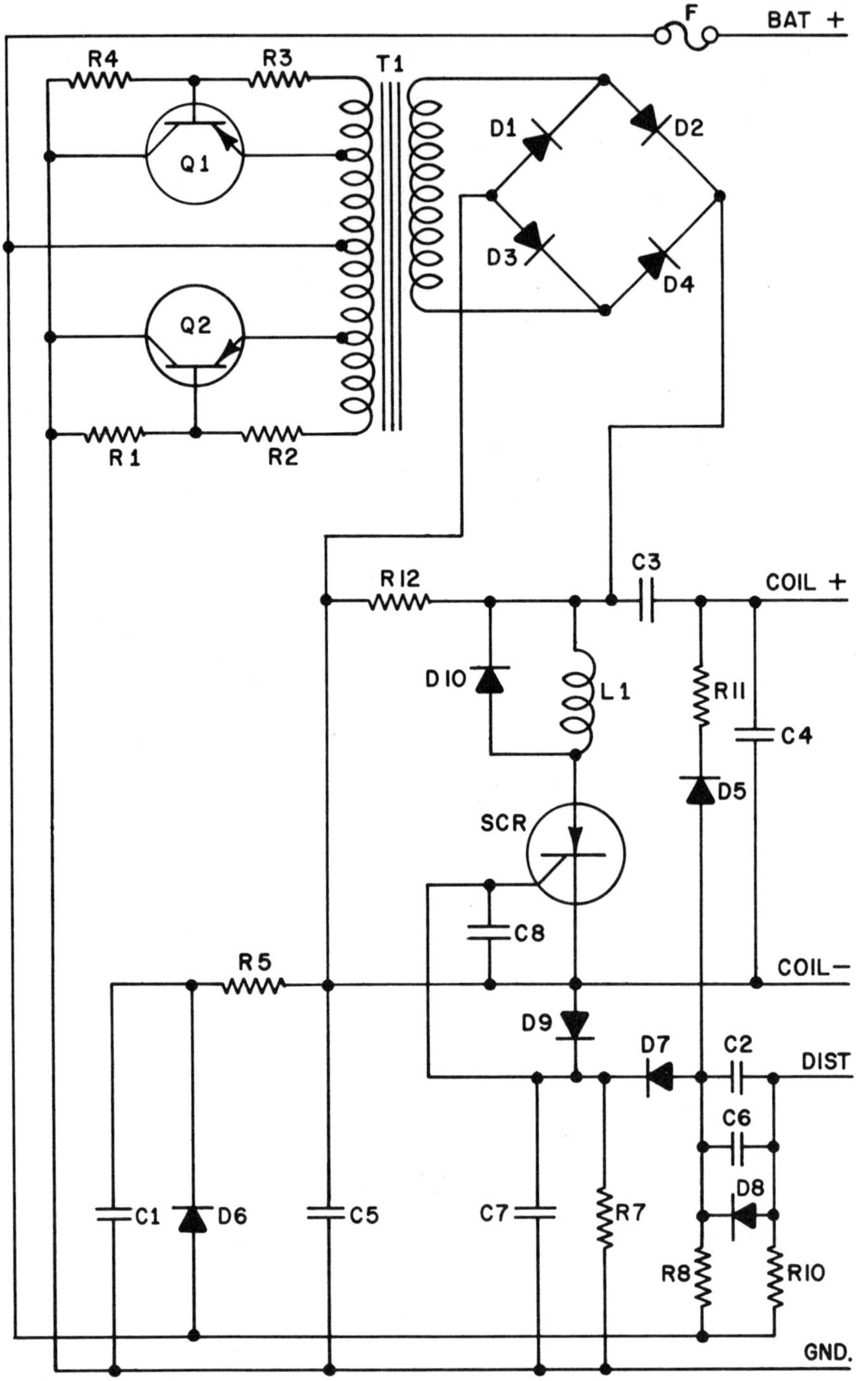

Figure 2-5 Capacitor-Discharge (C-D) ignition system, Source 7

transformer T1). This converter operates as follows. The battery voltage applied to transformer T1 causes current to flow through resistors R1 (680 ohms), R2 (15 ohms), R3 (15 ohms), and R4 (680 ohms). Since it is impossible for these two paths to be exactly equal in resistance, one-half of the primary winding will have a higher current flow.

Assuming that the upper half of the primary winding carries slightly higher current than the lower, the voltages developed in the two feedback windings (the ends connected to R3 and R2) tend to turn off Q2 and Q1. This increases the current through the upper half of the transformer winding. The increase in current further drives Q2 into conduction and Q1 into cutoff, simultaneously transferring energy to the secondary of T1.

When the current through the upper half of the primary of T1 reaches a point where it can no longer increase, because of resistance in the primary circuit and/or transformer core saturation, the signal applied to the transistor from the feedback winding drops to zero, turning Q2 off. The current in this winding immediately starts decreasing, causing a collapse of the magnetic field. This collapsing field, cutting across all the windings in the transformer, develops voltages in the transformer opposite in polarity to the voltage developed in the expanding field. This voltage now drives Q2 into cutoff and Q1 into conduction and simultaneously delivers power to the diode bridge (D1, D2, D3, and D4). Once started, this action converts the applied battery voltage into an alternating signal.

The voltage applied to the diode bridge is rectified to a dc potential of about 400 V, charging capacitor C3 (1.5 mfd), through the coil, connected through the negative and positive coil terminals to the negative terminal of the diode bridge. This action takes place as soon as power is applied by turning on the ignition key. Simultaneously, battery voltage is applied to the point circuit through resistor R10 (37.5 ohms), and to the SCR through diode D6 and resistor R5 (15 ohms), with capacitor C1 (0.47 mfd) serving as a filter to prevent false triggering caused by noise transients.

Assume that the points are closed and the ignition key is turned on. As the first cylinder comes up on the compression stroke and reaches the position where the spark plug should be fired, the points open. The current available at the junction of R10 and C2 (0.47 mfd) now flows through C2 and D7 to the silicon-controlled rectifier. This current switches the SCR on.

When the SCR turns on, two things happen simultaneously. The silicon-controlled rectifier short circuits the power supply, with the effect of the short reflected to the primary of T1, which removes the drive from transistors Q1 and Q2, stopping converter operation. The SCR also connects the positive side of C3 to the lower coil connection. This forms a closed circuit consisting of the capacitor, the silicon-controlled rectifier, and the coil primary. The energy stored in the capacitor is now delivered to the ignition coil. The coil primary voltage rises from zero to 400 V in approximately two microseconds.

In the circuit made up of the silicone-controlled rectifier, capacitor, and coil, a resonant circuit is formed between the primary coil inductance and capacitor C3. The flywheel effect of this circuit restores unused energy to the capacitor as follows: the capacitor discharge current flows through the SCR and coil primary, creating a magnetic field in the coil. This current continues to flow in the circuit until the capacitor is charged in a reverse direction to approximately 300 V. At this point the current attempts to reverse through the SCR, causing the SCR to return to its off condition. The reverse voltage now causes the diode bridge to conduct as a short circuit (all diodes simultaneously in a conduction mode), discharging the capacitor to zero from its reverse direction and recharging the capacitor toward its normal state. When the current supplied by the coil inductance again drops to zero, the bridge returns to a normal state, the load is removed from transformer T1, and normal converter operation resumes.

Diodes D5 and D8 in conjunction with R11 (1000 ohms) serve to discharge the triggering capacitor C2 completely when the SCR turns on. Diode D7 and resistor R8 (2200 ohms) are used to reverse bias the SCR to prevent erratic triggering caused by point bounce. Resistor R7 (1000 ohms) is used to reverse bias the SCR to prevent erratic triggering due to noise. Diode D10 and RF choke L1 are used to control the "turn on" characteristics of the SCR.

Other component values: resistor, R12, 10 megohms; capacitors (in microfarads), C4, 0.01; C5, 0.01; C6, 0.47; C7, 0.01; and C8, 0.02.

2-8 Magnetic Pulse Transistorized Ignition System, Typical of Source 6

A relatively new ignition system has been developed that completely eliminates the conventional distributor points. One type, primarily for large, stationary engines, even eliminates the distributor rotor and cap by providing a separate coil and transistorized control unit for each spark plug. When this breakerless ignition system is used on automotive engines, however, it has an electromagnetic pulse distributor and generator, a special ignition coil, and a transistorized control unit.

The transistor-controlled magnetic pulse ignition system is comprised of a specially designed magnetic pulse distributor, an ignition pulse amplifier, and a special coil. Although the external appearance of the distributor resembles a standard distributor, the internal construction is quite different.

As shown in Figure 2-6, an iron timer core TC replaces the conventional breaker cam. The timer core has the same number of equally spaced projections, or teeth, as engine cylinders (eight shown). The timer core rotates inside a magnetic pickup assembly, which replaces the conventional breaker plate, contact points set, and capacitor assembly.

The magnetic pickup assembly consists of a ceramic permanent magnet PM, a pole piece PP, and a pickup coil PC. The pole piece is a metal plate having equally spaced internal projections, or teeth, one tooth for each cylinder of the engine (eight shown). This assembly is mounted over the main bearing on the distributor housing and is rotated by the vacuum control unit, providing vacuum advance. The timer core TC is turned by the shaft and is made to rotate about the shaft by conventional advance weights, thus providing centrifugal advance.

Figure 2-6 shows the two separate ballast resistors used in this type of circuit. These resistors may be separate units or they may be in the form of resistance wires in the wiring harness. Resistor R5, connected directly to the switch SW, is bypassed during cranking, whereas the other resistor R6 is always in the circuit. On some applications, the bypass lead may be connected to the cranking motor solenoid instead of to the switch. The use of two resistors permits the required value of resistance to be bypassed during cranking.

In order to fire the spark plugs SP it is necessary to develop a high voltage in the ignition coil T1 secondary S winding by opening the circuit of the coil T1 primary P winding. This is accomplished in standard systems by opening the distributor contact points. In this ignition system, this is accomplished electronically.

When switch SW is closed, with the engine not running, current flows

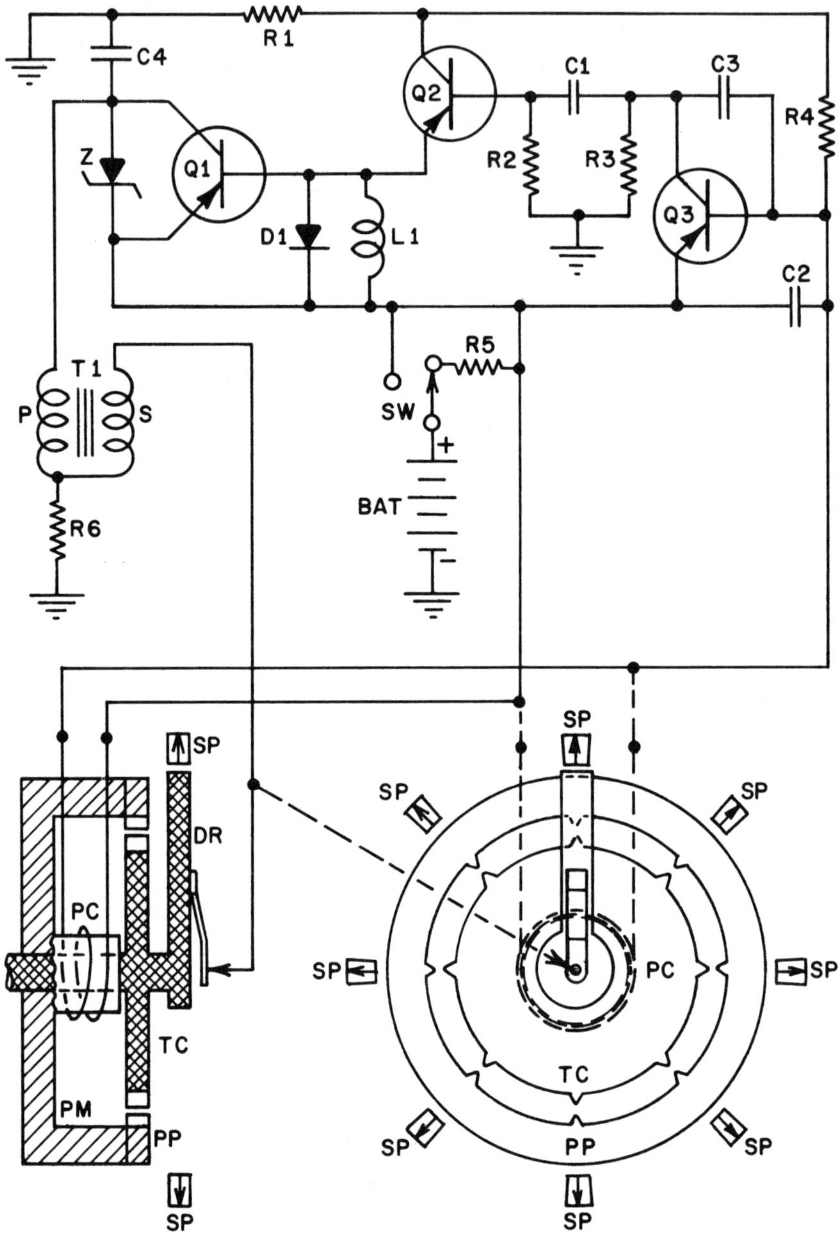

Figure 2-6 Magnetic pulse-type transistorized ignition system, Source 6

through a part of the circuit from the battery positive through the switch SW and resistor R5 to the amplifier. Current then flows through choke L1, transistors Q1 and Q2, resistors R1 and R2, the coil T1 primary P, and resistor R6 back to ground and battery negative. It is important to note that, under this condition, full current flows through the coil primary winding, and that capacitor C1 is charged with the positive voltage toward transistor Q2 base.

When the engine is running, the teeth on the rotating iron core TC in the distributor line up with the internal teeth of the pole piece PP. This establishes a magnetic path through the center of the pickup coil PC, causing a voltage to be induced in the pickup coil. This voltage causes transistor Q3 to conduct, resulting in current flow in the circuit.

The charge on capacitor C1 causes transistor Q2 to turn off, which, in turn, causes transistor Q1 to turn off. This interrupts the circuit to the ignition coil primary winding, and the high voltage needed to fire the spark plug is induced in the coil secondary winding. This high voltage is fed to the distributor rotor DR, where it is lined up with the proper spark plug SP electrode. This current flow condition exists until the charge on capacitor C1 has been dissipated through resistor R2. When this happens, the system reverts back and is then ready to fire the next spark plug in order of firing.

Resistor R4 is a feedback resistor, and its purpose is to turn transistor Q3 off when Q2 returns to the "on" condition. Zener diode Z protects transistor Q1 from high voltages which may be induced in the coil primary winding.

Capacitors C2 and C3 protect transistor Q3 from high voltages which appear in the system. Choke coil L1 and diode D1 operate to help turn off transistor Q1, and capacitor C4 reduces the amount of peak power that Q1 must dissipate.

2-9 Pulse Generator Transistorized Ignition System, Typical of Source 3

Figure 2-7 shows schematically the Pulse-Tronic ignition system typical of Source 3 and intended primarily for large, stationary engines.

This system uses one generator and timer unit with an induction coil for each cylinder. A separate transistorized control unit and a separate ignition coil is used for each cylinder. This has the advantage of short,

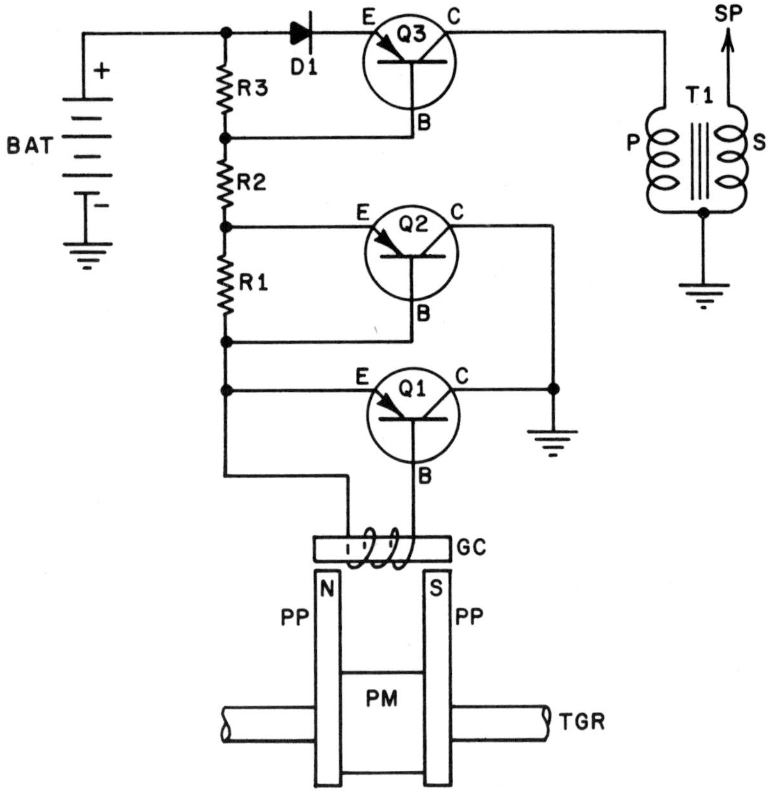

Figure 2-7 Pulse generator transistorized ignition system, Source 3

high-tension leads to the spark plugs. Further, a complete engine failure will not result if one of the transistorized switching units or coils fails.

In Figure 2-7, the timer-generator rotor TGR is driven by the engine. The rotor consists of a permanent magnet PM and two pole pieces PP, one a north pole N and the other a south magnetic pole S.

The timer-generator stator consists of a number of generator coils GC determined by the number of cylinders, one for each cylinder.

As the magnet rotor in the timer-generator sweeps past the generator coil, an ac control voltage with a sine wave form is generated. Each generator coil is connected to its corresponding transistor control unit and ignition coil. Figure 2-7 shows only one unit, but they are all identical and operate in the same way.

With the generator coil GC connected across the base B and emitter E of transistor Q1, during the negative portion of the ac cycle at the base of the transistor it causes Q1 to conduct from emitter E to collector C to ground. This permits a small current to flow from the battery BAT positive, down through the resistor network of R3, R2, and R1, and back to the battery negative and ground through transistor Q1, emitter E, and collector C. As this small current passes through the resistor network, it provides a strong negative bias to base B of transistor Q2, which now conducts from E to C and draws additional current from the battery. This action turns on transistor Q3 and causes current to flow from the battery positive, through diode D1, transistor Q3, emitter E to collector C, and through the primary P of the ignition coil T1 to ground and the battery negative.

A current flow is now set up in the primary circuit of the transformer T1, which produces a strong magnetic field that surrounds both the primary P and secondary S windings.

As the magnet rotor TGR continues on its sweep past the generator coil GC, the generated alternating current enters the positive half of the cycle at base B of transistor Q1, which switches off all three transistors abruptly. This action provides the interruption of the primary current, and the magnetic field collapses, causing a high voltage to be induced in the secondary, which is fed to the spark plug SP.

As the rotor continues to turn, it passes each generator coil in turn of firing order, and the associated transistor control, transformer, and spark plug of each cylinder is served in the same manner.

Full output voltage is attained at rotor speeds above 15 rpm. The maximum limit at the present time is 1000 rpm. This will serve any two-cycle engine at 1000 rpm or a four-cycle engine up to 2000 rpm.

Timing of the ignition system should be done with a standard timing light. With this ignition system, the timing will shift in the retarded direction about 2 degrees in the timer's first 100 rpm, and about 0.5° for each subsequent 100 rpm. Therefore, for a two-cycle 300-rpm engine, the flywheel should be set about 3° in advance of the running spark position. On a two-cycle 500-rpm engine the flywheel should be set about 4° in advance of the running spark position. Double these advance settings for a four-cycle engine.

3
Ignition Simulators

3-1 Complete Mechanical, Calibrated Ignition Simulator

An ignition simulator is an exact duplication of an automotive engine ignition system, but with additional timing calibrations, consisting of a variable-speed electric motor drive, a distributor, coil, spark plugs, and a calibrated dial.

The ignition simulator is required to properly test and calibrate scope-type engine analyzers, as well as other tune-up and test equipment.

These simulators are available, with various features, from all manufacturers of scope engine analyzers.

A simulator constructed from readily available materials has the advantage of being able to incorporate features to suit the needs of the builder. The following simulator was designed by the author for his own use.

In Figure 3-1 the basic requirements of a simulator are shown. Motor M1 is a two-speed 1725/1140 rpm constant-speed capacitor-start induction motor of ⅓ or ½ hp. Pulley P1, mounted on the motor M1 shaft, is 5 in. in diameter and is self-adjusting and spring-loaded. Pulley P2 is a fixed pulley 3½ in. in diameter driven by a V-belt from pulley P1. Pulley P3 is mounted on the same shaft and is keyed to turn with pulley P2. Pulley P3 is 5 in. in diameter, self-adjusting, and spring-loaded. Pulleys P2 and P3 are mounted on an idler base IB that is free to rock. Fixed pulley P4 has a diameter of 6 in. and is mounted on a drive shaft

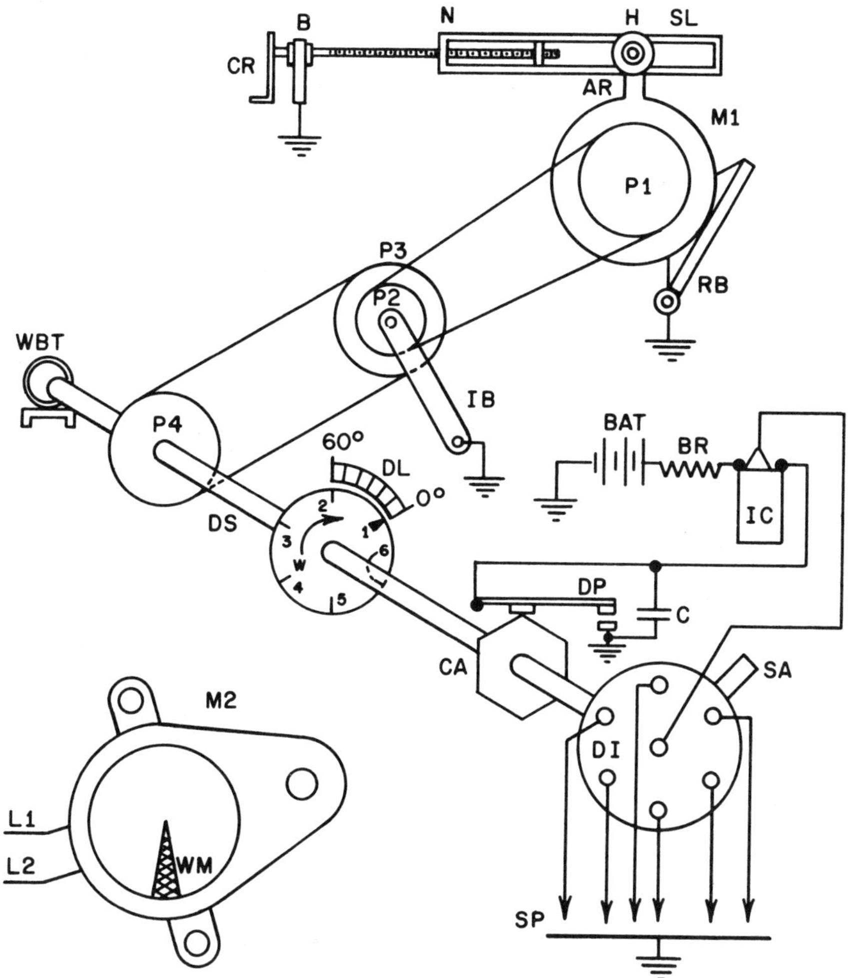

Figure 3-1 Ignition simulator construction

DS. The disc wheel W can be incorporated in the periphery of pulley P4, or it can be a larger separate disc mounted on shaft DS. Disc wheel W is divided by six equally spaced markers for a six-cylinder simulator. Dial DL, either a radial dial as shown, or a drum-type dial, is divided in degrees between zero and 60 to span precisely between any two divisions on disc wheel W.

The six-lobe cam CA, distributor points DP, and capacitor C are actually located inside the distributor housing. The distributor points should be adjustable externally. The six spark plugs SP are mounted on a metal plate connected to the common ground of the system. The mechanical governor advance mechanism of the distributor DI is left intact and must rotate in the proper direction. The vacuum advance lever SA is constructed so as to stay in any manually adjusted position. The battery BAT, ballast resistor BR, ignition coil IC, and capacitor C are standard components.

The drive motor M1 is mounted on a rocking base RB or a sliding base for adjusting the speed. Arm AR, fastened to the motor base, has a threaded handle H which locks the arm to slide SL for coarse speed adjustments. Fine speed adjustments are made by crank CR turning a threaded rod through nut N, with bearing B anchored stationary. With a two-speed switch in the motor circuit and the mechanical speed adjustments, the rpm of the distributor can be varied between 200 and 2000. This corresponds to a crankshaft or tachometer speed of 400 to 4000 rpm.

A wheel balancer tester cam WBT can be added easily to the protruding end of shaft DS. This cam WBT can be made from a standard ball bearing by pressing in a collar. The collar bore should be ⅛ in. larger than the diameter of the shaft DS. This will give a full throw of ¼ in. To adjust the throw, place three equally spaced setscrews around the collar. A U-shaped piece of iron is welded to the collar that is pressed on the outside of the ball bearing to engage with the magnet of the wheel balancer pickup head. A timing mark should be placed on the end of the shaft or collar to indicate if the flash tube firing is in phase with the pickup head. An elaborate but more flexible method of adjusting the throw of the pickup head would be to use a 12-in.-long fulcrum arm pivoted at one end, with the eccentric ball bearing running in a channel 6 in. from either end. The magnetic pickup head could then be attached at any position on the 12-in. arm to give a movement between zero and twice the throw of the ball bearing cam WBT. The speed of the simulator motor can be adjusted as desired.

A terminal board should be provided to bring out the following connections.

A. Distributor points
B. Ground
C. Hot battery

There should be a reversing switch for reversing the battery polarity.

Motor M2 is a 450-rpm (200 rpm preferred) 60-Hz synchronous timer, or clock, motor having an exposed rotor. A white marker WM is placed on the rotor to calibrate tachometers using the simulator and a timing light.

It will be shown under appropriate headings how the simulator is used for various functions, such as, the complete test and calibration of scope engine analyzers; the calibration of tachometers, dwell angle meters, engine ignition advance meters; the testing of timing lights, wheel balancers, and the like.

3-2 Electronic Trigger Supply for Source 15 Engine Analyzers

An inexpensive and portable electronic trigger supply and calibrator, or ignition simulator, can be built from readily available components as shown schematically in Figure 3-2. The component values are given in the parts list which follows.

Ref. No.	*Description and Rating*
C1, C2	Electrolytic capacitor 20 mfd, 250 V
C3	Paper capacitor, 0.02 mfd, 400 V
C4, C5	Paper capacitor, 1 mfd, 600 V
R1	Carbon resistor, 470 Kohms, 0.5 W, 10%
R2	Carbon resistor, 120 Kohms, 0.5 W, 10%
R3	Carbon resistor, 100 Kohms, 0.5 W, 10%
R4	Carbon potentiometer 250 Kohms, 2 W
R5	Carbon resistor, 10 Kohms, 1 W, 10%
R6	Carbon resistor, 220 Kohms, 0.5 W, 10%
CR1, CR2	Selenium rectifier, 50 ma, 400 PIV
F1	Fuse, 0.5 amp, 125 V, Slo-Blo
T1	Power transformer primary, 120 V; secondary, 125 V at 60 ma, 6.3 V at 0.6 amp (Stancor number PA8421 or equivalent)
T2	Ignition coil, 12 V (Delco number 087 or equivalent)
V1	Thyratron tube 2D21
V2	Neon lamp NE-2
S1	Switch SPST, 3 amp
S2	Switch SPDT, 3 amp

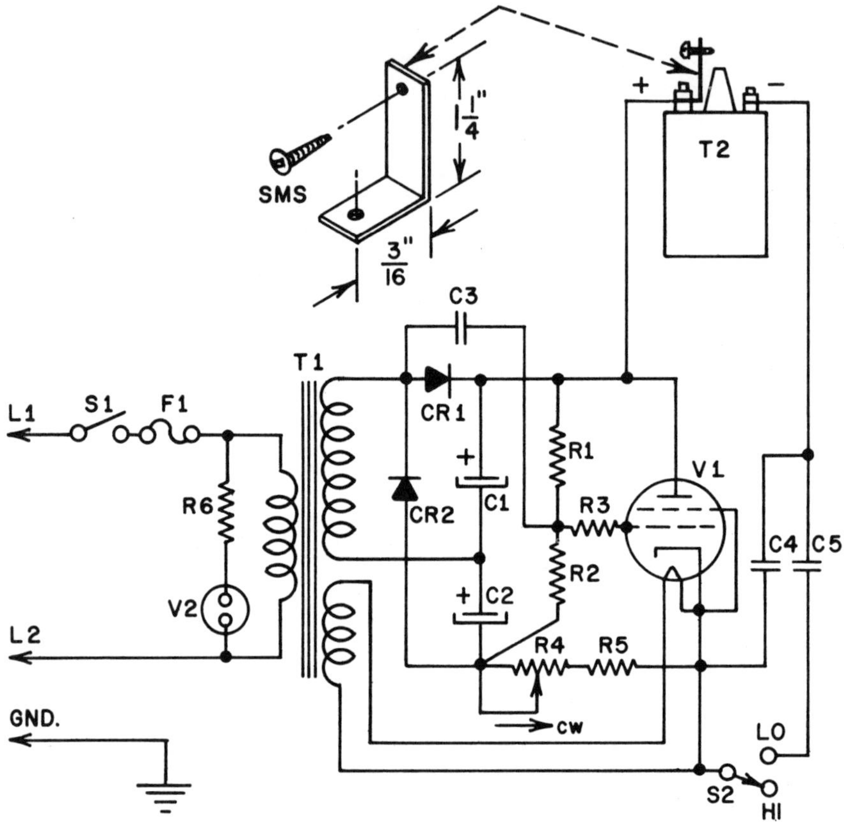

Figure 3-2 Electronic trigger supply for Source 15 engine analyzers

This self-contained electronic service instrument was designed by Source 15 for testing their scope engine analyzers, Model 1000, and others. It provides simulated ignition type pulses comparable to those obtained from a typical automobile ignition system, and it is designed to check out analyzers rather than to demonstrate ignition scope patterns.

The connections to the engine analyzer harness and the appearance of the assembled unit are shown in Figure 3-3. The HI TACH–LOW TACH switch S2 and the TACH ADJ control R4 are used to vary the frequency of the simulated ignition pulses (simulating engine rpm). Potentiometer R4 should be wired so that a clockwise (CW) rotation of the knob will move the rotor contact, or wiper, in the direction of the arrow. This will

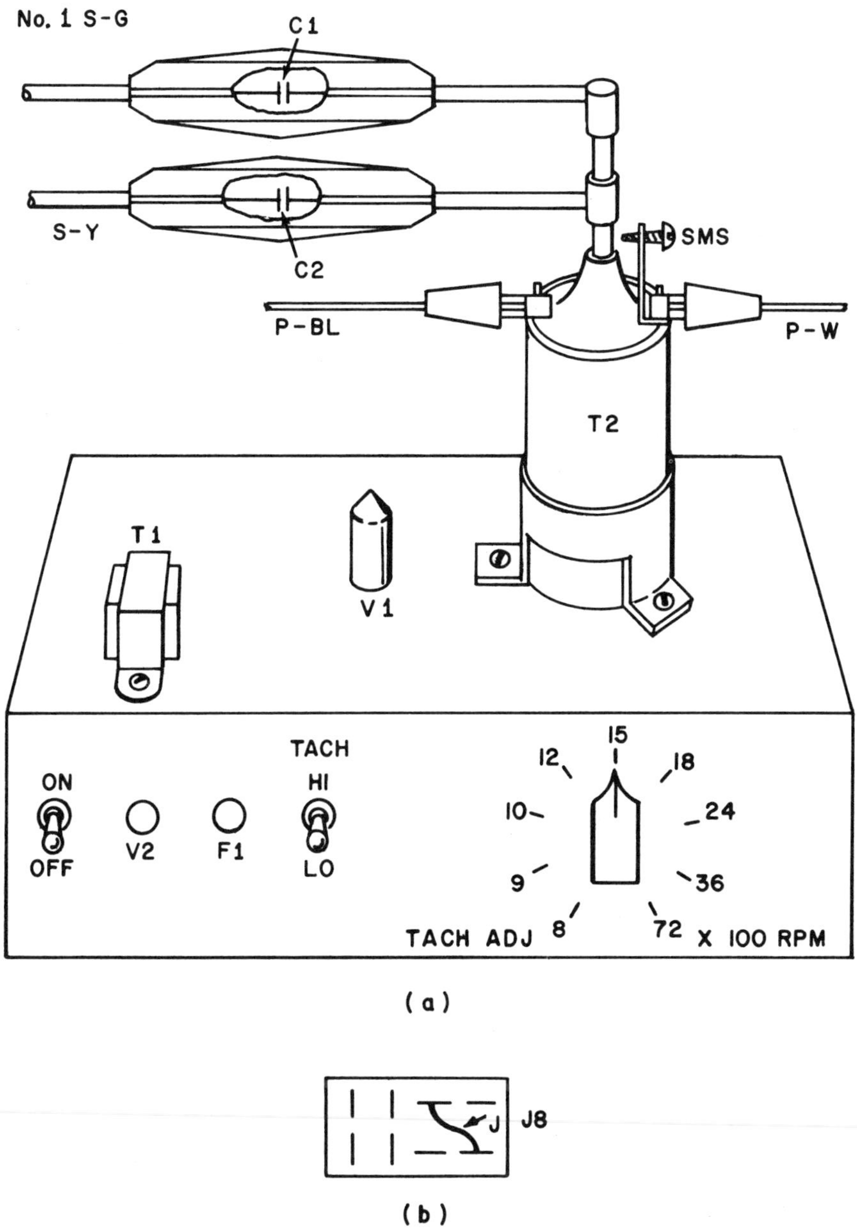

Figure 3-3 Electronic trigger supply assembled from Figure 3-2

result in an increase of frequency and rpm as the knob is turned clockwise. The sheet metal screw SMS is an adjustable spark gap to simulate the electrode air gap of the spark plug.

To calibrate the TACH ADJ control R4 in rpm, proceed as follows.

A. Connect trigger supply to Model 1000 analyzer as shown in Figure 3-3a. Adjust spark gap to approximately 3/32 in.

B. Insert jumper J into ACCESSORY jack J8 as shown in Figure 3-3b. Jack J8 is located on the rear of Model 1000 analyzer chassis directly below the CRT base. This jumper connects 6.3-V 60-Hz ac sine waves from the filament transformer to the vertical section of the CRT to give sine waves as shown in Figure 3-4.

C. Set controls of analyzer as follows. Put SPREAD control in AUTO position, and TEST SELECTOR in position 8 (accessory).

D. Set controls on trigger supply as follows: HI TACH–LOW TACH switch S2 in HI position; turn TACH ADJ control R4 clockwise until scope pattern on analyzer appears as shown in step 1 of the calibration chart,

STEP	ADJUSTABLE TRIGGER SUPPLY DATA	PULSES PER SEC.	ENGINE RPM
1	 1 SINE WAVE	60	7200
2	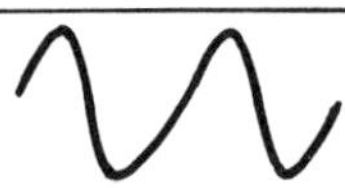 2 SINE WAVES	30	3600
3	 3 SINE WAVES	20	2400
4	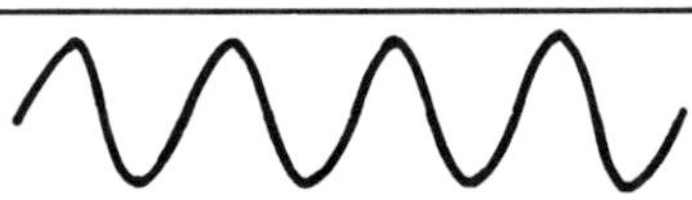 4 SINE WAVES	15	1800
5	5 SINE WAVES	12.5	1500
6	6 SINE WAVES	10	1200
7	7 SINE WAVES	8.5	1028
8	8 SINE WAVES	7.5	900
9	9 SINE WAVES	6.66	800

Figure 3-4 Calibration chart for trigger supply shown in Figures 3-2 and 3-3

Figure 3-4. If the pattern appears upside down, reverse the primary leads on the power supply transformer T1 in the trigger supply, or reverse the connections to the polarized grounded ac plug.

E. Make a reference mark on the chassis or a prepared dial, in line with the pointer of the R4 control knob and label the mark 72 or 7200.

F. Continue with steps 2 through 9 of the calibration chart, Figure 3-4, and make reference marks at 3600, 2400, 1800, 1500, 1200, 1028, 900, and 800 rpm. For rpm below 800 turn the HI TACH–LOW TACH switch S2 to LO.

This can be calibrated similarly on any standard scope. Once calibrated, it can be used on any brand or model of scope engine analyzer.

In Figure 3-3, the high-voltage pickup head No. 1 S-G (green) would normally go to the engine No. 1 spark plug and head S-Y (yellow) would go to the coil. Clip P-BL is the primary blue lead to the distributor, and P-W is the white lead to ground.

4
Diode and Transistor Tester

4-1 Diode and Transistor Tester Construction

A rectifier, diode, and transistor tester is a big time-saver and a necessity. No tester available will test both the heavy-current diodes and rectifiers as well as the smaller low-current diodes, or indicate polarity of rectifiers, diodes, transistors, and the like. For the heavy-current diodes, a unit is available from Source 4, or one can be purchased from an auto parts supply house as an alternator diode tester.

A good diode and transistor tester can be constructed from readily available material by following the details of Figure 4-1. This unit will test all rectifiers and diodes, as well as transistors, for opens, shorts, conductivity, and polarity, and it will test the polarity and condition of other solid state devices rated at 40 ma or over. Diodes and transistors may have to be disconnected for testing in some cases. The following tester was designed by the author for his own use.

In Figure 4-1, transformer T1 can be any battery charger transformer having a tapped secondary capable of charging at a 6-amp rate on 6 or 12 V. The primary 6-volt and the common taps are used; these are the two outside leads. Toggle switch SW1 is a SPST, 3–6 amp, 115 V, and it turns the tester off and on. Switch SW2 is a three-contact rotary switch rated at 6 amp, 115 V, and it is used to calibrate the two meter ranges: position H to calibrate for diodes over 5-amp rating; position L to calibrate for diodes under 5-amp rating; position T for testing conductivity through prods PR3 and PR4. Switch SW3 is a DPDT with no off position, rated at 6 amp, 115 V, and it is used to select the high- and low-

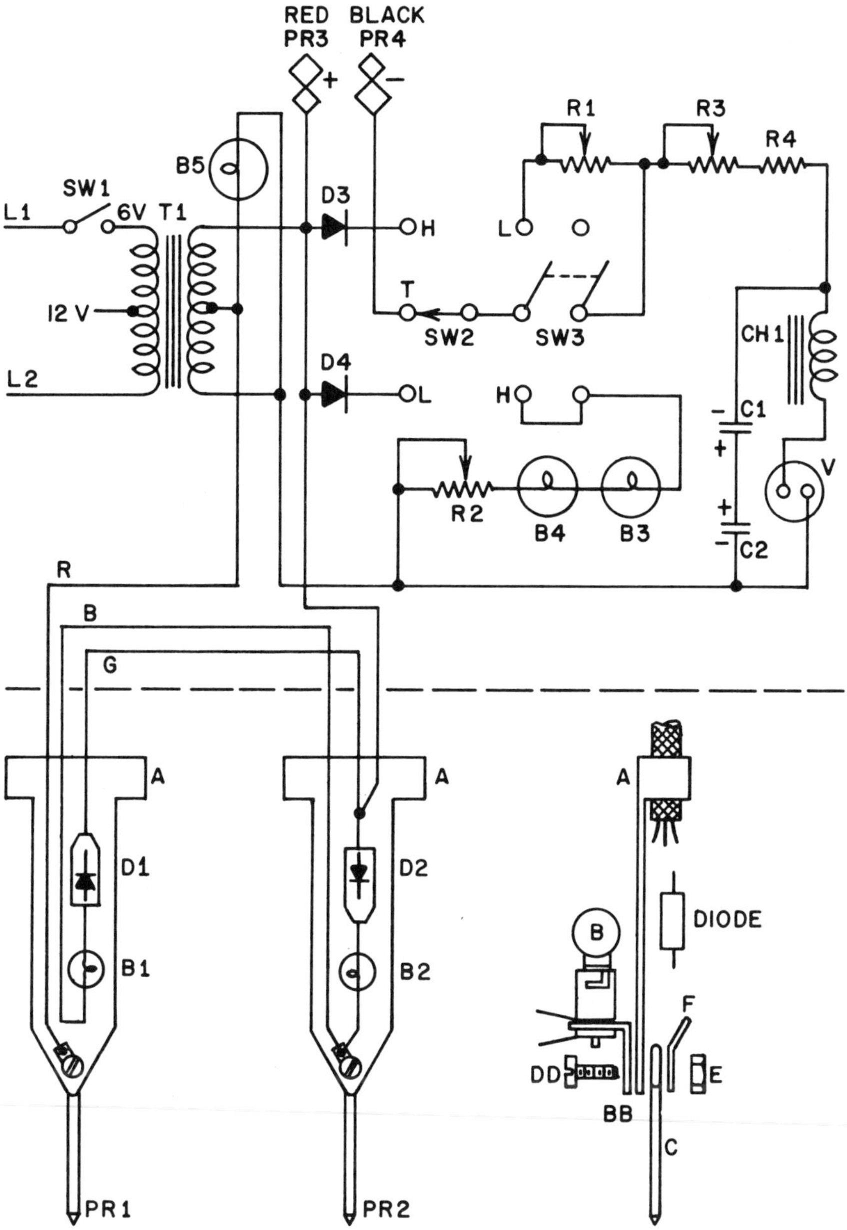

Figure 4-1 Diode and transistor tester construction

current ranges; L position is for diodes and transistors under 5-amp rating (it draws about 100 ma current); and position H is for diodes and power transistors rated over 5 amp (it draws about 5 amp).

Potentiometer R1 is a meter-calibrating resistor rated 100 ohms. It is a 2-W wire-wound linear-type taper for the low range under 5 amp. Potentiometer R2 is a 2-ohm, 50-W wire-wound, linear taper for calibrating the high range above 5 amp. Potentiometer R3 is a meter-calibrating resistor for both ranges and is a 25-ohms, 1-W, wire-wound linear taper. Resistor R4 is a 20-ohm, 1-W, fixed value.

Meter V is an expanded-range 6-V voltmeter. The one used for the resistance values shown is a No. B179305, Battery Condition Test Meter, from Source 10. Any 6-V expanded range battery tester voltmeter can be used, providing resistors R1, R3, and R4 are correct for the substitute meter. It should have a colored section scale. An expanded range voltmeter gives a more accurate reading because diodes having less than 80% conductivity are useless in alternators and other applications.

Capacitors C1 and C2 are each 100 mfd, 25-V electrolytics and are connected back-to-back to be nonpolarized, should a diode under test be connected for reverse polarity. Choke CH1 is a 7-ohm primary of a transformer taken from a 6-V 4-amp charger and is not critical; choke CH1 and capacitors C1 and C2 filter the rectifier output and hold the meter needle steady.

Bulbs B3 and B4 are the same, No. 2330, 6- to 8-V auto headlight bulbs connected in series, with the two filaments in each connected together. These bulbs have a flange which can be used for mounting and connecting them, and the two filaments in each bulb are connected by soldering the lead across both terminals.

Diode D4 is a calibrating silicon diode for under 5 amp and can be any small diode. The one used is a 750-ma, 200-PIV diode. Diode D3 is a calibrating diode for over 5 amp and can be any 35-amp 50-PIV silicone diode. The one used is a 1N1183. Pilot light B5 is a No. 53 bulb, or any 12- to 14-V bulb.

The leads to each of the test prods PR3 and PR4 should be about 4 to 5 ft long and may terminate in alligator clips or pointed prods. Test prods PR3 and PR4 are used for conductivity and quality tests on all diodes and transistors from the smallest to the largest. Although they will also indicate polarity, shorts, and opens, prods PR1 and PR2 are better for determining all of these conditions except quality and are used first in testing diodes and transistors. Prods PR1 and PR2 are mechanically the same but are wired differently as shown.

Diodes D1 and D2 are identical and are the same as D4. Bulbs B1 and B2 are the same, No. 53. Sockets BB are the same and are single-contact, minature bayonet, with offset bracket insulated from both terminals.

A copper sheet is cut as shown as part A, with two tabs on one end folded over to act as a cable clamp, and a hole drilled in the opposite end. Part C can be a number 12 copper wire, ground to a point on one end and an eye loop formed on the other end for mounting. Parts A, BB, C, and solder terminal F are held together by a 6-32 x ⅜-in. machine screw DD and a 6-32 nut E. After complete assembly and test, wrap the entire prod with plastic tape except the glass part of the bulb for viewing. The three-wire extra-flexible cable connecting prods PR1 and PR2 to each other and to the tester is color-coded with black B, green G, and a red R.

To use the diode and transistor tester, prods PR1 and PR2 determine the polarity and indicate shorts or opens in diodes and transistors. This will show whether or not they are working but not necessarily whether they are up to standard in conductivity. When prods PR1 and PR2 are placed across any diode or transistor section capable of carrying 40 ma or over (regardless of which prod is connected to which terminal of the unit being tested) only the bulb B1 or B2, associated with the prod on the anode of the unit, will burn, indicating rectifier action. If neither bulb burns, the unit is an open circuit. If both bulbs burn, the unit is shorted. If both bulbs burn but one is brighter than the other, there is shunt leakage, as may occur on "in-circuit testing" in some cases, or it may indicate a rectifier that has almost as low a backward resistance as forward resistance, which is reason enough to reject the unit. If the unit shows it is shorted or open, it is not necessary to continue to test for conductivity with prods PR3 and PR4.

Prods PR3 and PR4 are used for testing diodes and power transistors capable of carrying 100 ma or over; for testing conductivity and quality; and to confirm shorts or opens.

Initially, and from time to time, it is necessary to calibrate meter V to the built-in standard diodes D3 and D4. To calibrate, plug in line cord of tester and close switch SW1. Set switch SW2 to position L. Set switch SW3 to position L. Set potentiometer R3 in mid-range, and adjust potentiometer R1 until meter V reads at the far edge to the right in the green SAFE scale under CHARGE. Ignore the two upper scales, LOAD TEST and GEN REGUL. If unable to reach this point, adjust R3 and R1 until it is reached. Bulbs B3 and B4 do not glow on this setting. Next, turn switch SW2 to H position and turn switch SW3 to H position (bulbs B3 and B4 should glow), and adjust R2 until meter V reads at the same point as under the L range adjustment, that is, to the right edge of the green SAFE area. It may be necessary to repeat these two adjustments, re-adjusting R3, R2, and R1 accordingly, until the meter reads the same on both ranges. No diode or rectifier tested should exceed this point if diodes D3 and D4 are up to standard. All good rectifiers, even the older low-conductivity selenium rectifiers, should fall within the SAFE area.

4-2 Testing Diodes

After the meter is calibrated, the following procedure should be followed to test diodes. First, use prods PR1 and PR2 to test for opens, shorts, and polarity. If there are no opens or shorts, one of the bulbs in prod PR1 or PR2 will glow, indicating the anode. The red test prod PR3 should be connected to this terminal, the anode, and the black prod PR4 should be connected to the dark light terminal, the cathode. Set switch SW3 either to H for over 5-amp rating or to L for under 5-amp rating, and set switch SW2 to test position T. Turn on switch SW1. If the unit is acceptable, the meter should read in the green SAFE area. On the H position of switch SW3, a shorted unit will cause bulbs B3 and B4 to increase in brilliancy. Be sure of the polarity, since a reversed connection would show no reading on the meter.

The tester should not be left on or connected for calibration or testing for any longer than is necessary to obtain a reading.

This tester can be used to test large power transistors on the high H range. For other transistors, test only on the low range L, which draws 100 ma.

Prods PR3 and PR4 draw a current, through the diode or transistor, of 5 amp on the H range and 100 ma on the L range. Prods PR1 and PR2 draw a current of 40 ma through the diode or transistor section.

4-3 Testing Transistors

To test transistors, use the diode and transistor tester shown in Figure 4-1. Using prods PR1 and PR2, connect one of the prods to the base B of the transistor, as shown in Figure 4-2, and the other prod to the emitter E, and then to the collector C. If the lamp connected to the emitter E and collector C glows, and the B lamp does not glow, and neither lamp glows when connected between collector C and emitter E, then the transistor is neither open nor shorted, has good conductivity, and is a PNP type as shown in Figure 4-2a. If only the base B lamp glows when the other prod is touched to the emitter E and collector C, and neither lamp glows when connected between collector C and emitter E, then the transistor is neither open nor shorted, has good conductivity, and is a NPN type as shown in Figure 4-2b. Neither lamp should

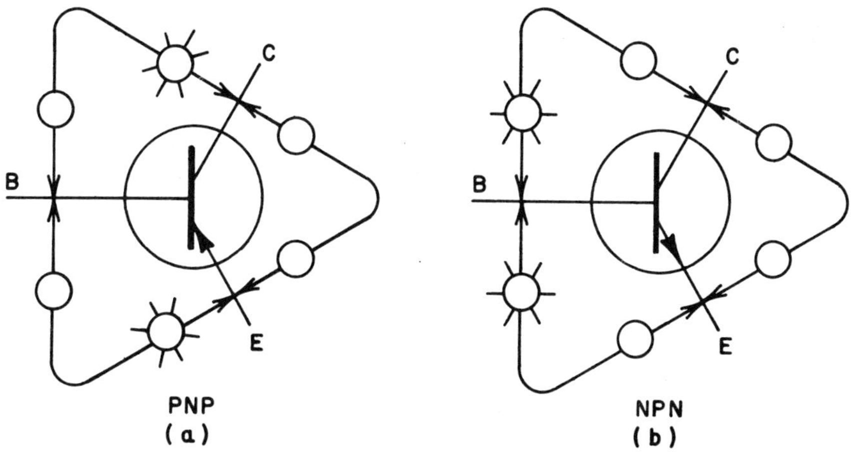

Figure 4-2 Testing transistors using diode and transistor tester

glow when connected across the emitter E and collector C for either the PNP or the NPN type. If neither lamp glows, when one should glow, the transistor is open. If both lamps glow at any position, the transistor is shorted. In either case it should be replaced without further testing.

This test is sufficient in most cases, but for power transistors, a conductivity test may also be made if there is doubt.

4-4 Power Transistor Conductivity Test

Power transistors used in transistorized ignition systems and voltage regulators can be further tested for conductivity by using prods PR3 and PR4 of the diode and transistor tester shown in Figure 4-1. Most of the smaller, low-power transistors used in other electronic circuits may also be tested for conductivity on this tester, but the test should be made as rapidly as possible to avoid overheating. Usually, if transistors pass the test with prods PR1 and PR2, they will work satisfactorily in most circuits. Silicon transistors and diodes are either good or bad, very seldom weak.

When first testing with prods PR1 and PR2, determine the polarity of the transistor, that is, whether it is a PNP or NPN type, and make a diagram showing which terminal caused the lamp to glow. This is the terminal which should be connected by the positive red lead PR3, and where the lamp was dark, connect the negative black lead PR4. Always

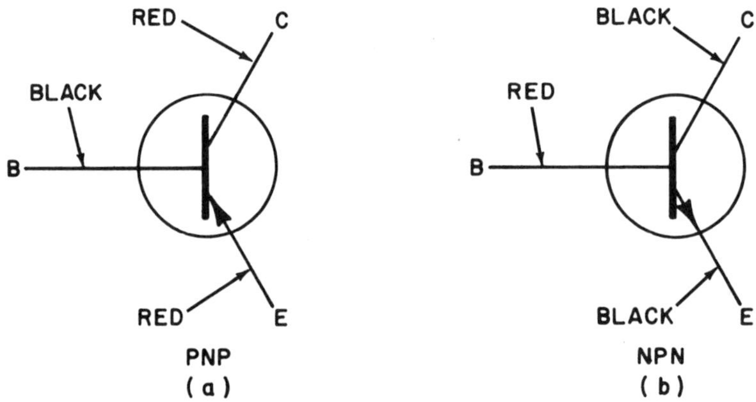

Figure 4-3 Power transistor conductivity test using diode and transistor tester

use the low range, under 5 amp, for testing transistors, unless it is known that the transistor will carry 5 amp safely. The current on the low range is 100 ma.

For a PNP transistor, Figure 4-3a, connect the black lead PR4 to the base B of the transistor and the red lead PR3 to the collector C, and the meter should read in SAFE. Next connect the red lead PR3 (with the black lead PR4 still on the base B) to the emitter E, and the meter should read in SAFE.

For a NPN transistor, Figure 4-3b, connect the red lead PR3 to the base B and leave it there while the black lead PR4 is connected, first to the emitter E and then to the collector C. In each case the meter should read in SAFE.

It is easy to remember the type designation by looking at the arrow in the symbol. If the arrow on emitter E points to base B, E is *P*ositive, B is *N*egative, and C is *P*ositive, thus PNP. Similarly, if the arrow on E points away from B, then E is *N*egative, B is *P*ositive, and C is *N*egative, thus NPN. By observing the polarities indicated above, transistors can also be tested with an ohmmeter or test lamp.

To test for conductivity between C and E, connect B to C and test as a diode between B-C and E.

5
Symbols and Circuit Tracing

5-1 Principles of Symbol Usage

Understanding the principles of operation and the functions of electrical automotive tune-up and test equipment requires a fundamental knowledge of electricity and electronics, the symbols used, and the ability to trace circuits through a chassis or a schematic wiring diagram.

Although it is beyond the scope of this book to teach the fundamentals of electricity and electronics, except for components and circuits peculiar to the equipment discussed, we will show all symbols used in schematic wiring diagrams, describe how to trace circuits through the chassis, and make schematic wiring diagrams. A symbol is a standard drawing that represents a particular component. It usually bears no physical resemblance to the component itself.

Some symbols are conventional, such as those for diodes, zener diodes, transistors, transformers, resistors, capacitors, and relays, and they will be defined in the process of explaining the various circuits. These can also be learned from any textbook on the subject.

The abbreviations used in terminal designations in the schematic wiring diagram usually give a clue as to what they designate and where to look for the source to the connection. For example, TB may stand for terminal board, TL for timing light, JH for ignition pickup harness input jack, and PH for ignition pickup harness input plug.

5-2 Vacuum Tube Symbols

Tube symbols used in automotive tune-up and test equipment are the same as used in all radio, TV, and electronic circuit schematics.

Vacuum tubes having two independent tubes in one envelope will have the same tube designation number but will have the first section designated with a suffix "A," such as V8A. The second section will be designated as V8B, for example, and if there is a third section, as V8C and so on. These will appear on the schematic as shown in Figure 5-1a and 5-1b, respectively. The small numbers designate the pin connection on the tube socket taken in a clockwise direction when looking at the bottom of the tube base. Number 1 starts at the first connection clockwise to the key or open space (missing terminal).

Vacuum tube V8 is a type 12AX7, with two independent triode sections V8A and V8B, where 6 and 1 are the plates, 7 and 2 the grids, and 8 and 3 the cathodes. The heater numbers are not usually shown on the

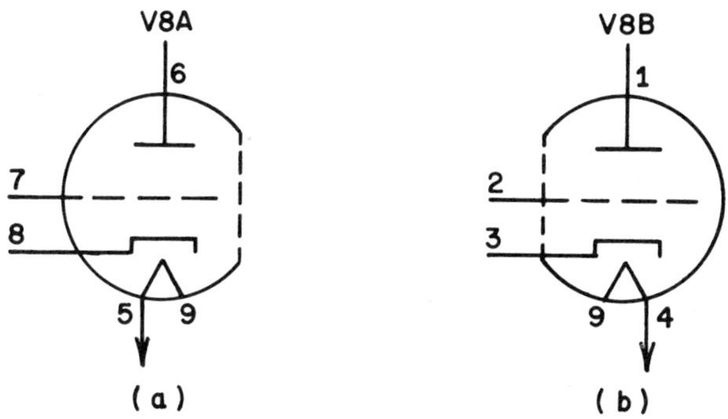

Figure 5-1 Vacuum tube symbols

schematic but will be found on a separate power supply drawing. Terminal 9 is common to both heaters. The other heater to V8A is pin 5, and the other heater connection to V8B is pin 4.

5-3 Gas-Filled Tube Symbols

Gas-filled tubes, such as thyratrons and voltage regulator tubes, use the same symbols as a corresponding vacuum tube except that there is a black dot in the circle of the tube symbol.

5-4 Symbols for Plugs and Jacks

A plug is the male socket, and a jack is the female socket. Either may be mounted on the chassis or on the end of a cable.

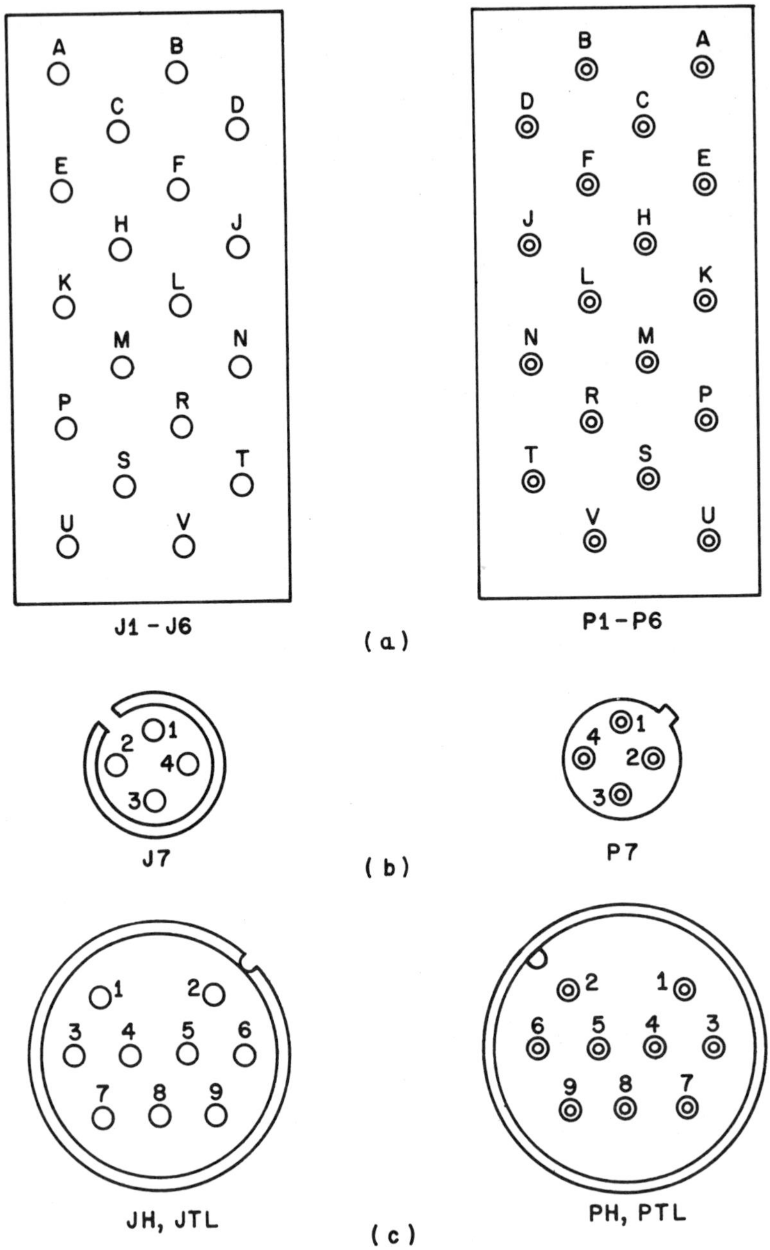

Figure 5-2 Plug and jack terminal guides

Switching arrangements, plugs, and jacks are used more generously in scope engine analyzers than they are in even the most elaborate television set. This makes the schematic wiring diagram more difficult to read and follow without some helpful hints and clues. Once learned, this system of schematic reference is much simpler than would at first be suspected.

In Figure 5-2 are shown some of the common plugs and jacks used. Figure 5-2a shows jacks 1 to 6 (J1–J6) on the left and plugs 1 to 6 (P1–P6) on the right.

A connection as shown in Figure 5-3a in the schematic wiring diagram will show a solid dot for the plug connection designated, P1A, for example, which would mean terminal A of plug No. 1. Similarly, the jack connection would be shown as a small open circle and designated

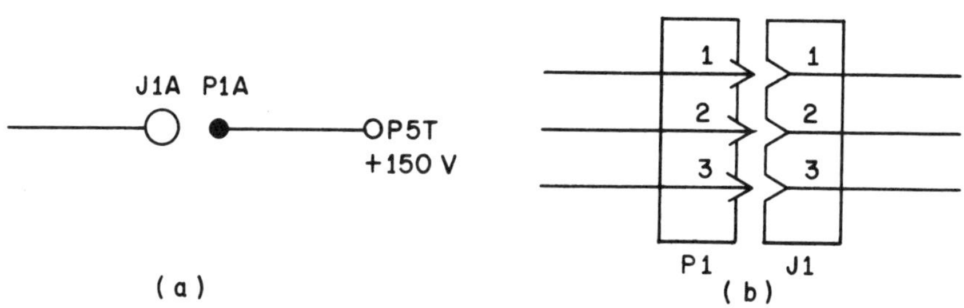

Figure 5-3 Plug and jack symbols

as J1A, for example, which would mean terminal A of jack No. 1. In the schematic, if a lead terminates in a small dot and close to it is a small circle connected to another lead, it means a plug and jack connection and should be read through as if it were a solid line, providing one is designated P and the other J, followed by their designating numbers and letters. A common jack and plug connection, for example, is J1A and P1A. By studying the schematic, it can readily be seen what resistance, or other measurements, can be made with the plug pulled, or left plugged in.

Figure 5-2b shows a small, four-prong, keyed plug and jack arrangement designated as plug 7 (P7) and jack 7 (J7).

Figure 5-2c shows a nine-prong plug and jack designated respectively as PH and JH, or plug for timing light PTL and jack for timing light JTL.

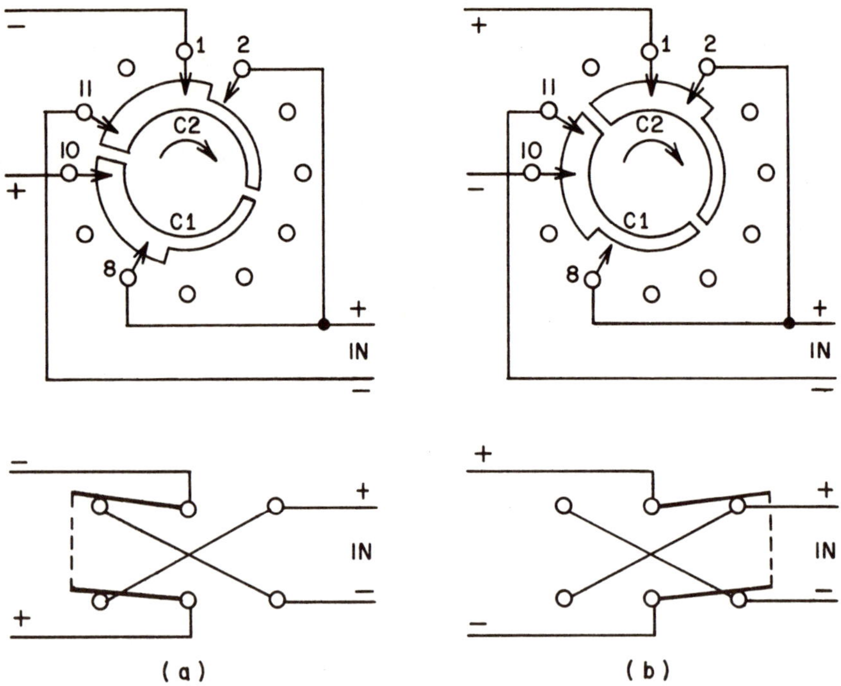

Figure 5-5 One-deck, reversing rotary switch

ple of operation of the two types of wafer switches likely to be encountered. There is an almost infinite variety of combinations of this type switch.

In tracing a circuit through these switches from the schematic only, it is necessary to visualize the connections that are made for each position of the switch, or make a separate schematic for each switch position. When tracing through the switch on a chassis, it is often simpler to turn the switch to the position desired and observe which circuits are completed, or use an ohmmeter on the lowest range. Normally, any two contacts that show a zero reading will indicate a switch connection, unless it is known that there is another closed switch in the circuit or that there is a shorted component.

Figure 5-5 shows a two-position, one-deck reversing rotary switch of the shorting type, so called because it shorts out any two or more contacts with or without having a common contact on a smaller-diameter sector.

Figure 5-5 does not have a common contact and illustrates a typical polarity reversing switch. The upper drawing shows the symbol used in schematic wiring diagrams, and the lower drawing illustrates the electrical equivalent of a DPDT cross-connected reversing switch.

In Figure 5-5b the switch is shown in the forward or straight polarity position, where the positive input IN goes through contact 2, plate C2, contact 1, and out positive. The negative input IN goes to contact 11, through plate C1, and out contact 10 to negative.

In Figure 5-5a the switch is shown in the reverse position, showing the polarity of the out connection reversed. The positive input IN now goes to contact 8, through plate C1, and out contact 10 positive (reverse of Figure 5-5b). The negative input IN goes to contact 11, through plate C2, and out contact 1 negative (reverse to Figure 5-5b).

5-6 Push Button Switch Symbols

Push button switches may be individual in that only one button operates any number of contacts and circuits, and in that they are almost always of the two-position type, "in" (ON) and "in" (OFF) to release the ON button.

Push button switches are often tied together mechanically so that any one button pushed will release a previously pushed button so that only one can be pushed at a time. There is only one OFF button for the entire switch that releases the last remaining push button so that they are all off or out.

Typical symbols of push button switches are shown in Figure 5-6. Figure 5-6a shows, in the open position, a single contact of a multiple-contact switch, and Figure 5-6b shows the same contacts in the closed position. This simplifies the schematic in that the switch contacts can be located on the schematic near the component being switched, but this leaves the problem of locating all the other contacts of the same multiple push button switch.

Figures 5-6c and d show the schematic symbol of the entire switch assembly of each push button section, which requires tracing leads to the various components being switched.

Figure 5-6c shows the switch in the OFF position, where a circuit is completed between contacts 1 and 9, and all other used contacts are open.

Figure 5-6d is the same switch as in Figure 5-6c but in the ON posi-

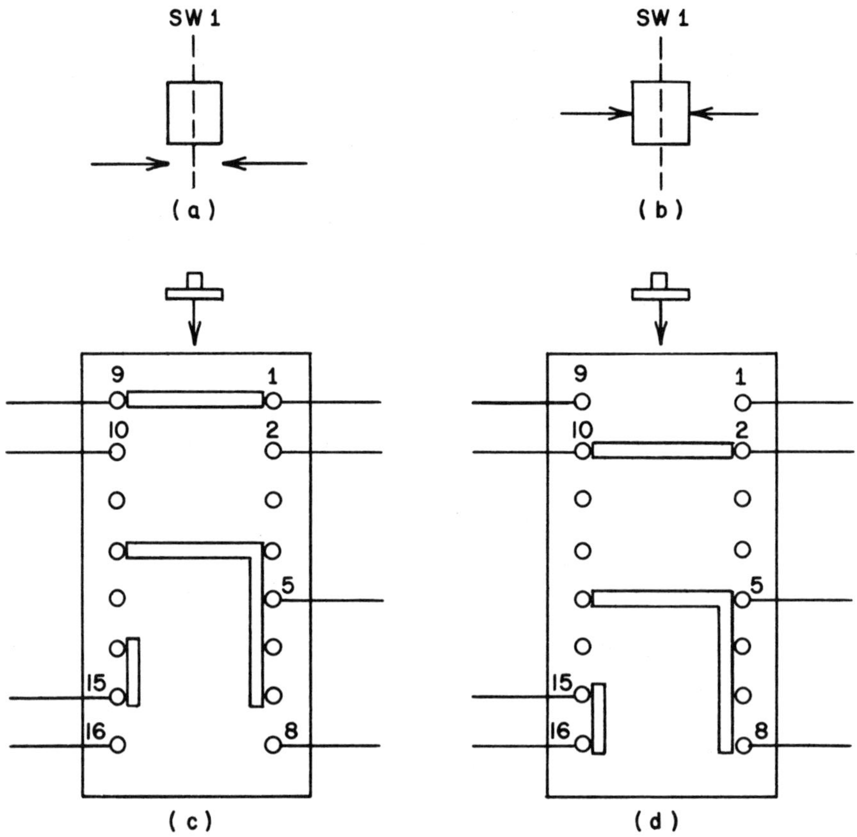

Figure 5-6 Push button switch symbols

tion, where contacts 1 and 9 are open, contacts 2 and 10 are closed, contacts 5 and 8 are closed, and contacts 15 and 16 are closed.

In tracing circuits on the schematic it is necessary either to visualize the open and closed positions or make a separate sketch showing which circuits are completed.

In tracing circuits on the chassis, it is sometimes simpler to close the switch and observe the position of the contacts if the switch is of the open type where the contacts can be seen. On an enclosed switch, an ohmmeter can be used on the lowest scale to indicate zero reading for closed circuits. A higher, or infinite, reading on the open contacts depends on the resistance of the components connected to the switch.

5-7 Tracing Circuits, General

The electronic technician's dream has always been to find it as easy to trace a circuit in the chassis as it is on the schematic wiring diagram.

At least one manufacturer, Source 15, has made this dream come true beyond the fondest hope of any technician. They have the chassis preprinted showing the schematic reference number alongside each component on the chassis. This should even simplify production assembly.

5-8 A Circuit-Tracing Example

The ability to trace circuits quickly from a chassis will prove invaluable to anyone working in the field of electronics. The following suggestions and hints should be helpful in the drawing of a schematic wiring diagram from information taken from a wired chassis or a printed circuit board.

It is usually best to start out at the input of the circuit and end up at the output, drawing from left to right. It is simple to make a circuit diagram when only one lead goes to a terminal, but when a terminal is reached that has more than one lead, show on the diagram a short stub "lead-off" from the terminal for each lead, and mark it with any color, number, or other identification shown on the lead. After following only one lead to the next terminal, and so on to other terminals, come back to each terminal that has a stub drawn on it, and complete one circuit at a time, until the circuit is completed. The use of colored pencils will also be helpful. When crossing lines, simply draw straight through, being sure to make a large dot wherever any two crossed leads have a connection. This not only saves time but also makes a neater schematic that is easier to read.

When making a layout and wiring diagram from one side of a chassis or printed circuit board PCB, to get a back-side view at the same time, do as follows. Place a carbon paper in back of the drawing paper with the carbon side facing the back of the page. The back of the page will then show the wiring and parts as they appear from the rear of the chassis or PCB. When the back-side view of an existing drawing is desired, it can be observed by holding it up to the light, and looking at the back side.

As an example of circuit tracing, let us make a schematic wiring diagram of a dc-powered timing light.

First, make a sketch of all the components shown in Figure 5-7, applying an identifying mark for each. Here, VIB is the vibrator socket, T1 the transformer, FT the flash tube, RL the relay, PB the push button, REC the rectifier stack, and C the capacitor. Show color codes or numbers appearing on all leads and assign numbers to those not marked. Mark the lead with a piece of masking tape and number it.

Start with the positive dc lead and trace to vibrator socket terminal 4, and find three other leads on this terminal. Draw out two stubs (shown in heavy lines) and proceed to trace the third lead to one side of the relay RL coil and a second lead to a resistor. Mark the resistor R1 and determine the resistance from the color code. Trace the resistor R1 to one terminal of the push button and find three other leads. Draw out three stubs as shown. Trace one stub to a connection on the metal case (ground). The relay RL armature is the only other connection that is connected to the metal case, so they are joined together. Trace out the second stub to a capacitor and mark it C1, then to a resistor and mark it R2. From resistor R2 trace to the middle connection 6 of the rectifier stack REC and find a second wire, which is traced to the red R lead of transformer T1. Now go back to the third stub on the push button PB and trace to a capacitor and mark it C2. Then trace to the back contact NC of the relay RL and find another wire, which is traced to the black B lead of the transformer T1. Trace the yellow Y lead of transformer T1 back to the front or NO contact of relay RL.

Trace the brown BR wire from the transformer T1 to vibrator VIB socket terminal 2; the other brown BR wire from the transformer T1 is traced to VIB socket terminal 3.

Now go back to the two stubs on terminal 4 of the VIB socket and trace one lead to the transformer green G lead. The other stub from terminal 4 is traced to terminal 9 of the flash tube FT socket, which should be noted protrudes through for about 3/8 in.

The second terminal of the push button switch is traced to the red lead of capacitor C. One black lead of capacitor C is traced to one of the outside terminals 5 of the rectifier stack REC, where a second lead is found which is traced to the flash tube socket FT terminal 8. The other black lead of capacitor C is traced to the other outside terminal 7 of the rectifier stack REC and a second wire is found which is traced to terminal 10 of the flash tube FT socket. The high-voltage spark plug SP

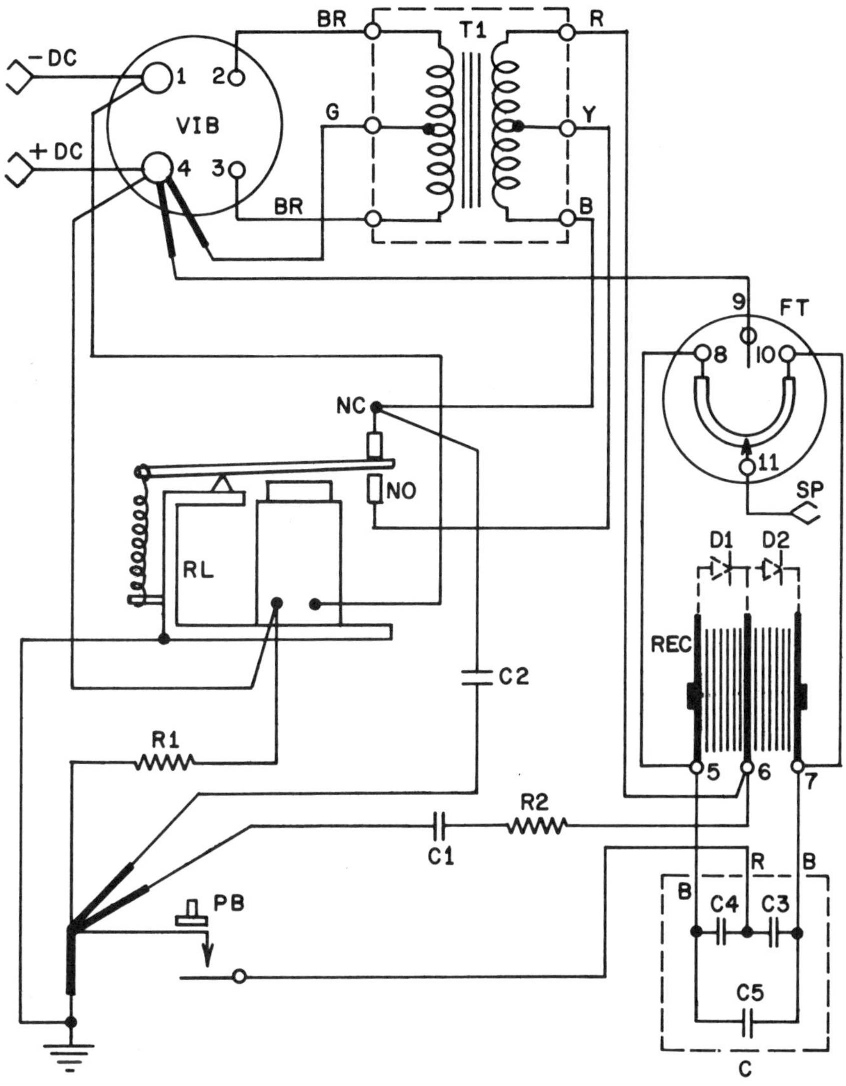

Figure 5-7 Tracing a circuit from chassis component layout

clamp and lead are traced to terminal 11 of the flash tube FT socket.

The negative dc lead is traced to the vibrator VIB socket connection, terminal 1, where a second lead is found and traced to the other coil connection of relay RL.

This completes the electrical connections. To know the internal connections and find the values of the components, it may be necessary to disconnect them, at least in part.

The internal connections of the vibrator can best be determined by referring to the manufacturer's catalog, which should show these connections for each vibrator number. Then the correct connections can be drawn. Transformer T1 can be checked with an ohmmeter. It is obvious that the low-resistance primary windings are connected to the vibrator, and that the high-resistance secondary windings are connected to the rectifier and voltage doubler. The secondary is not center-tapped in Figure 5-7; if it were, two of the leads would be the same color. Red R is the common connection, whereas the yellow Y lead is for 12-V operation and the black B lead is for 6-V operation. The windings can then be drawn in as shown.

The capacitor C can be tricky when using a capacitor checker, but with a little study the combination can be figured out. In this type of voltage doubler circuit, the two outside leads are the same color, which leaves the red lead to go to the middle connection of the two series capacitors C3 and C4. If there are only two capacitors, the capacitor value in microfarads between R and either B should be the same, and it should be one-half of this value when reading across the two B leads. However, if this reading across the two black leads is higher than one-half the value as read between R and each B lead, then a parallel capacitor is indicated, and it can be drawn in as C5.

Rectifier REC can be checked with an ohmmeter or diode tester and found to have the equivalent of two diodes D1 and D2 connected in series, as shown in dotted lines in Figure 5-7.

More complicated circuits can be traced just as easily by following this procedure. To save time, mark down any values on the components as the tracing proceeds.

6
Battery Testers

6-1 Lead-Acid Storage Battery Characteristics

The lead-acid storage battery is universally used for vehicle starting and for supplying other electrical needs of the system. The lead-acid battery is best suited for this purpose because of its low internal resistance. It can deliver maximum starting current with a minimum of voltage drop during starting.

The behavior of the storage battery while on charge is not as well understood as it should be. A thorough knowledge of its characteristics is important to adjust voltage regulators and use battery testers properly.

The battery, generator or alternator, and the regulator must be coordinated in order to ensure long battery life and good performance under a wide variety of conditions.

6-2 Variations in Battery Voltage with Change in Charging Rate and State of Charge

The curves of Figure 6-1 indicate variations in battery voltage according to the state of charge (hydrometer reading) and the charging current for 12-V batteries at 80° F. The heavy horizontal line at 14.4 V (marked REG, for regulator) is a typical voltage regulator setting, or limit, for a given size 12-V battery system. The area above this line indicates conditions which are prevented by the limiting action of the voltage regulator, and the area below the line is the "working" area of the battery. The numbers at the left of the chart indicate battery terminal voltage, and the numbers along the bottom represent the charging cur-

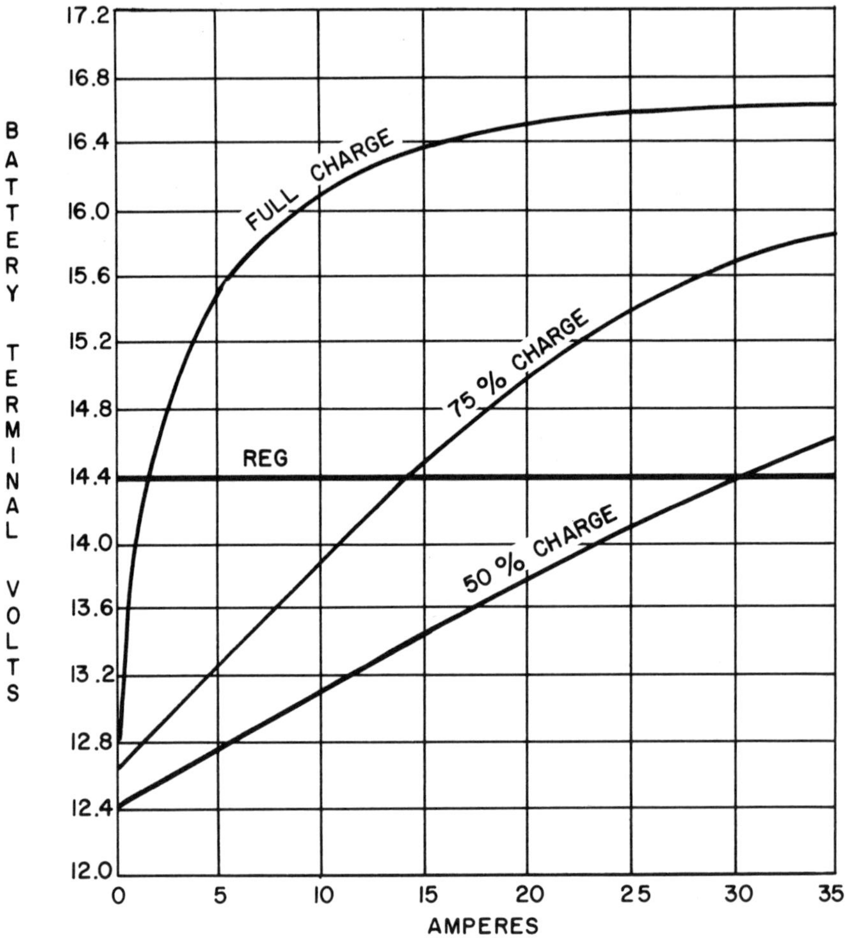

Figure 6-1 Variations in battery voltage with change in charge rate and state of charge for 12-volt battery at 80° F

rent, in amperes, from the alternator, generator, or battery charger. These curves indicate what charge rate at 80° F a battery in good condition can be expected to accept at various voltages and states of charge.

To use the chart, select a voltage and read straight across to the right until the desired one of the three "state of charge" curves is intersected. Moving straight down from the intersection, read the charge rate in amperes. The curves show that a battery with a lower percentage of full charge will accept a higher charging rate than a fully charged battery. For example, the 50% charge curve shows that a battery in this condition will accept approximately 30 amp at 14.4 V for a 12-V system and will take more at higher voltages, but this is prevented by the action of the voltage regulator (set at 14.4 V).

Even if the voltage regulator is out of adjustment (set too high), the dc generator would regulate by the current regulator to limit the current to, say, 30 amp. In the case of an alternator, which does not usually have a current regulator, the current is limited automatically by the inherent self-limiting characteristics. The current is limited in the alternator by the reactance of the windings. As the alternator is increased in speed, the frequency of the current increases proportionally. The higher the frequency the higher the impedance of the alternator windings.

As the battery becomes more nearly charged, higher and higher voltages are required to maintain the same charging rate. The voltage regulator operates to limit the voltage, so the charge rate is reduced. Referring to Figure 6-1, it can be seen that a 75% charge on the battery at 14.4 V will take a 14-amp charging current; at full charge a voltage of 14.4 will charge at only about 2 amp.

6-3 Variations in Battery Voltage with Change in Charging Rate and Temperature

A very important consideration in the charging and testing of batteries is the effect of temperature on battery characteristics.

Figure 6-2 shows the variation in battery voltage according to change in charging rate and battery temperature. Note that, with a half-charge (dotted curves) on the battery at 0° F, it will accept approximately 2 amp at 14.4 V. Under the same half-charge conditions, the battery will accept about 27 amp at 80° F, and it will accept full generator or alternator output at 120° F. This is a very serious matter in actual vehicle operation where wide variations in temperatures are encountered.

To compensate for the temperature effect on batteries, the voltage regulator must have built-in temperature compensation to permit a higher voltage cutoff at lower temperatures. At higher temperatures the cutoff voltage is reduced to prevent overcharging the battery.

Because the limits actually held by the regulator are dependent on the temperature, it is very important to check and make adjustments only at the temperature specified in the service instructions, usually

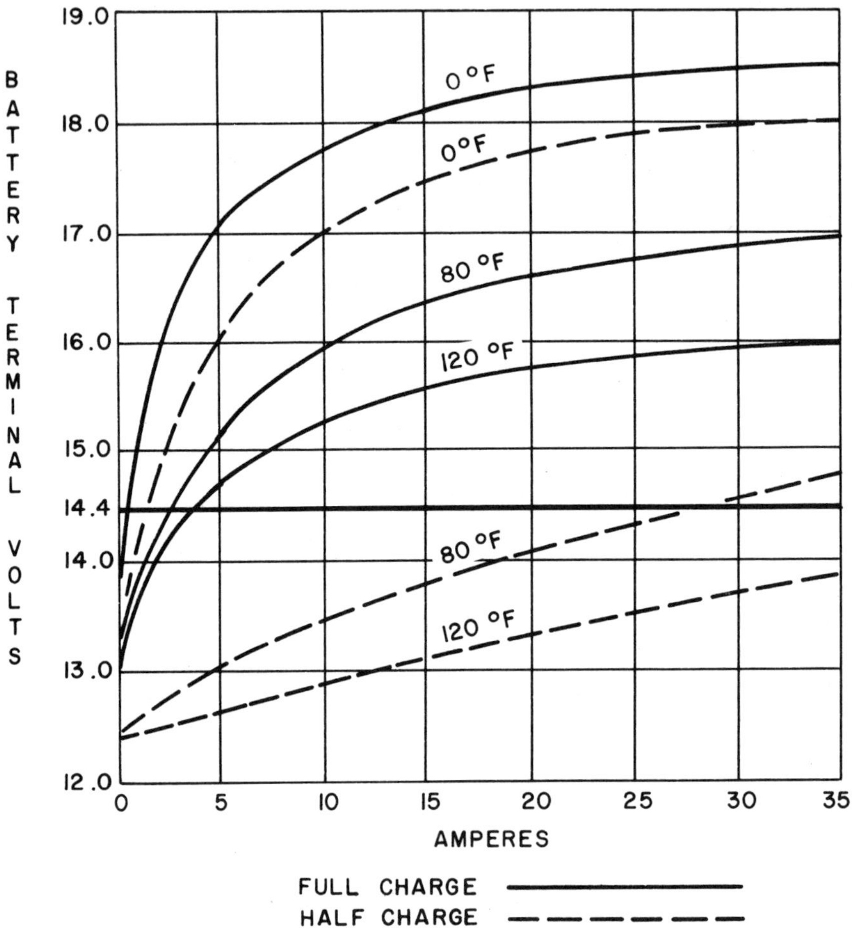

Figure 6-2 Variations in battery voltage with change in charge rate and temperature for a 12-volt battery

75–80° F ambient (air) temperature. The regulator should be operated for 10 to 15 min to normalize the temperature before making any adjustments.

Batteries should be put on a battery tester only at the temperature specified by the manufacturer (usually about 80° F) to ensure accurate results and to avoid possible damage to the battery. For open-circuit voltages and specific gravity corrections see Table 6-4.

6-4 Precision DC Calibrating Voltmeter

Anyone engaged in the calibration of battery testers and adjustment of voltage regulators requires an accurate dc voltmeter with a 0.5% guaranteed accuracy or better.

There have been no standard meters on the market that have suitable ranges and accuracy at a reasonable cost for the proper calibration and adjustment of battery service equipment.

The author has written the specifications for, and has been instrumental in having developed, the ideal dc voltmeter, designed specifically for the peculiar needs of the battery service equipment industry.

Figure 6-3 shows the scale of a three-range dc voltmeter designed for those who are concerned only with vehicle-starting batteries of 6, 12, and 24 volts.

Figure 6-4 shows the scale of a two-arc dc voltmeter having four basic ranges for batteries of 6, 12, 24 and 36 V, covering all vehicle-starting batteries, plus golf cart and other 36-volt batteries. The range selector covering these four ranges is shown in Figure 6-4a. Figure 6-4b shows two additional ranges of 8×10 (40–80 V, read on the 4–8 scale and multiplied by 10) and 16×10 (80–160 V, read on the 8–16 scale and multiplied by 10), for those who work with all battery voltages likely to be encountered or anticipated in the industrial and electric vehicle field. The detailed specifications of these meters are given in Table 6-1.

Electric meters having an accuracy of ±2% and above are manufactured with preprinted dials, which are all alike. Precision meters with ±1% accuracy, or under, require individual dials and calibration, drawn by hand. The cost to manufacture these precision, hand-drawn dial meters to specification, is about the same as an equivalent precision, hand drawn "standard range" meter (of the same accuracy) listed in meter catalogs. (By "standard range," I mean standard ranges such as 0–15 V, 0–50 V, etc.)

It is hoped that these meters will eventually be available on the meter market, but until they are, they can be made to order at a cost comparable to that of a less accurate jewel-type meter. Source 11 can furnish these meters, which are shown in Figures 6-3 and 6-4a and b and are rated in Table 6-1. They use only reliable, blank dial meter movements from such manufacturers as General Electric, Westinghouse, Weston, and others. The manufacturers equipped to make these accurately cali-

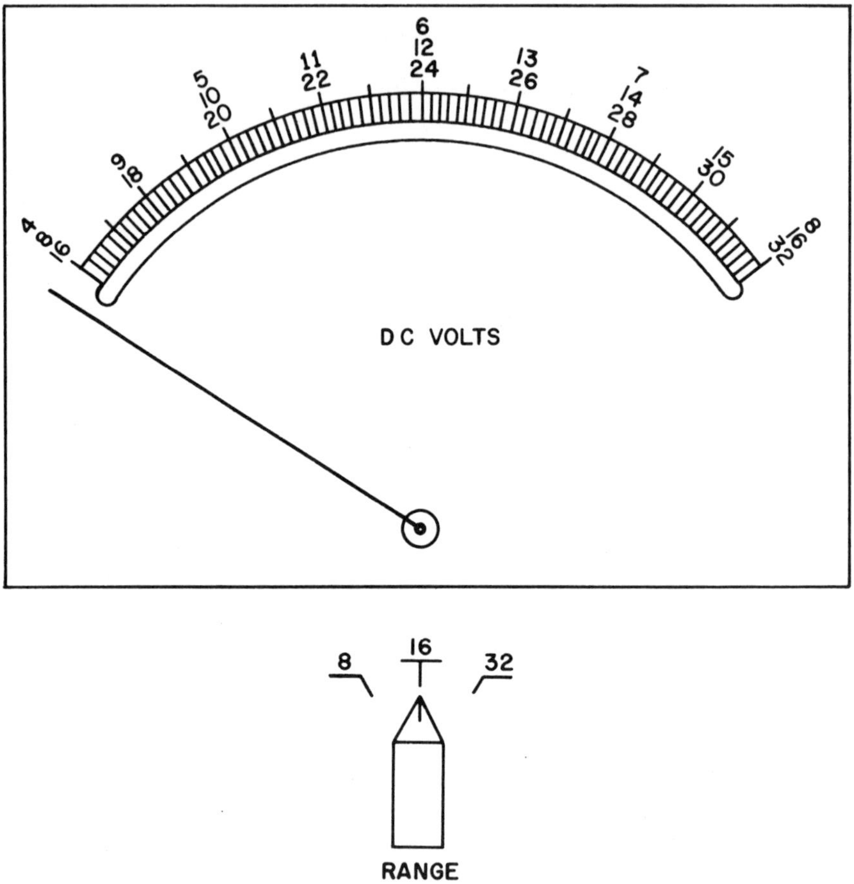

Figure 6-3 Precision dc calibrating voltmeter, one-arc scale

brated dials are familiar with the specifications and can quote a price if the figure number and preferred meter size are indicated. For example, in writing for a price, state, "Please quote price and delivery terms on the Cantonwine meter, specifications [according to Table 6-1 and Figure 6-4b], in a 6½-in. meter size."

The price is dependent on the meter size and whether it has one or two scale arcs. A meter having the additional ranges shown in Figure 6-4b (8×10 and 16×10) costs only slightly more, because it is necessary

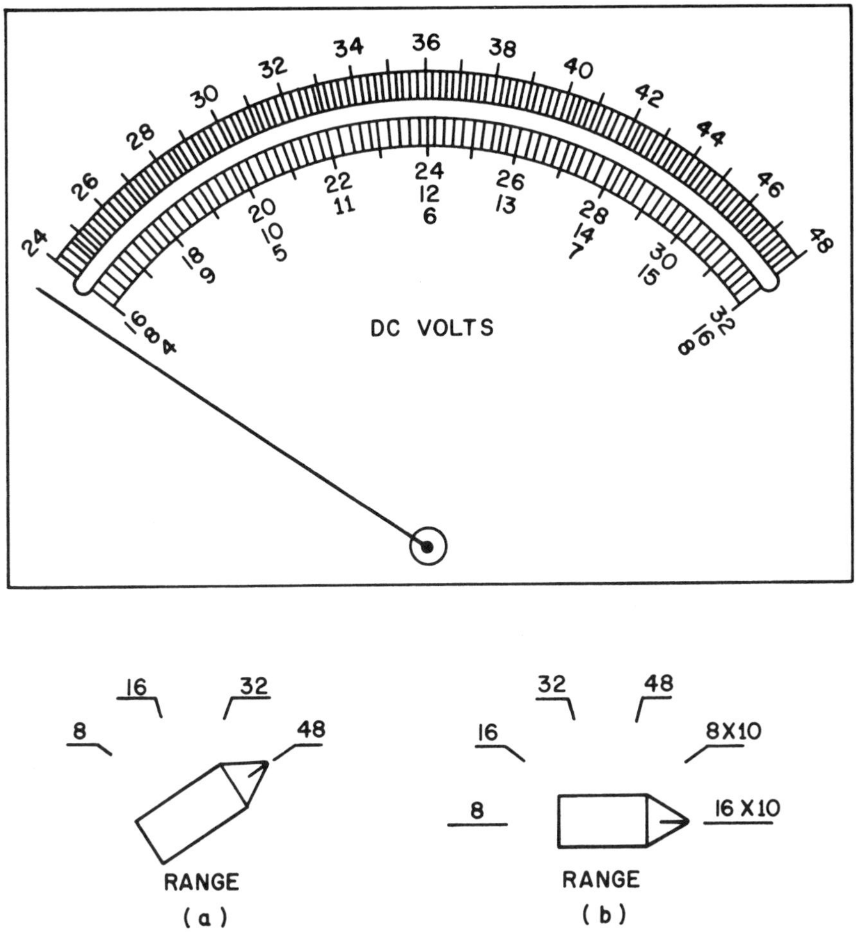

Figure 6-4 Precision dc calibrating voltmeter, two-arc scale

only to add two precision multiplier resistors and two extra positions on the selector switch.

The taut-band suspension meter is superior and costs less than the jewel type, because there are no bearings or hair spring to cause friction. The repeatability of readings is amazing, and they are more rugged and reliable over a long period of time.

Specifications for Precision DC Calibrating Voltmeter for Battery Service Equipment

			Illustration Number		
Specifications			*Fig. 6-3*	*Fig. 6-4a*	*Fig. 6-4b*
Minimum meter size (in.)			4½	5½	5½
Recommended meter size (in.)			5½–6½	6½–7½	6½–7½
Number of arcs			1	2	2
Number of ranges			3	4	6
Nominal battery volts			6	6	6
and midscale volts			12	12	12
			24	24	24
				36	36
					60
					120
Scale	*V/Div*	V@½% ±			
4–8	0.05	0.04	X	X	X
8–16	0.10	0.08	X	X	X
16–32	0.20	0.16	X	X	X
24–48	0.20	0.24		X	X
8 × 10	0.50	0.40			X
16 × 10	1.00	0.80			X
DC voltmeter			X	X	X
± ½% accuracy			X	X	X
Taut-band suspension			X	X	X
Mirror scale			X	X	X
Highly damped movement			X	X	X
Res 100–1000 ohms/volt			X	X	X

Table 6-1

6-5 Readings Below 4 V on the Precision DC Calibrating Voltmeter or Other Expanded-Range Voltmeters

The meters shown in Figures 6-3 and 6-4 are readable, as is, for voltages above 4 V. However, lower voltages from 0 to 4 V can be read with the same accuracy for calibrating cell testers, and doing "meter calibration voltage" checks as shown in Figure 6-15, column 2, for example. Unknown voltages below 4 V can also be read accurately.

Figure 6-5 shows a circuit for calibrating meters below 4 V, where a fully charged 12-V battery is connected across a potentiometer R1 (1000

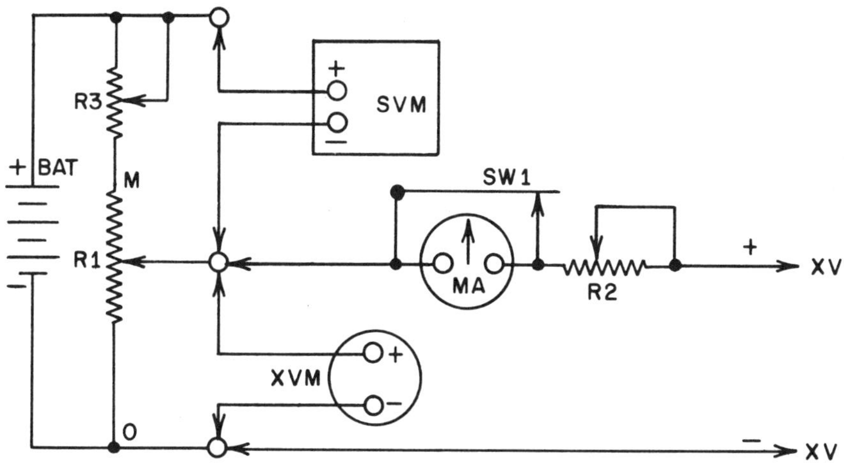

Figure 6-5 Circuit for reading voltages below 4 volts on 8- to 16-volt expanded-range meter

ohms, 10-W, wire-wound) in series with R3 (100 ohms) (optional) and standard precision voltmeter SVM, connected as shown, with the meter to be calibrated XVM connected as shown.

To calibrate meter XVM, first set the "zero" adjustment on meter XVM, unless it is an expanded range meter. Then set meter SVM on the 8–16 V range and turn R1 to zero to read full battery voltage. It is important that R1 have zero resistance at this point. Record the battery voltage, for example, 12.66 V. Determine the exact calibrating voltage to be read on the meter XVM. Subtract this value, for example, 2.05 V, from the battery voltage, 12.66, to get 10.61 V. Turn R1 toward M until meter SVM reads exactly 10.61. Meter XVM is then adjusted to read exactly 2.05 V. Some meters are provided with an adjusting screw just below the movement and accessible through a small hole covered by tape or a metal plate.

To read unknown voltages XV below 4 V, use a zero center scale milliammeter MA (0–1 ma) and rheostat R2 (10 Kohms) as a "null" meter. Switch SW1 is a N.C. "push to read" push button switch to protect the meter. Potentiometer R1 is adjusted until meter MA stands at zero center scale with switch SW1 open and rheostat R2 at zero resistance. The voltage is then read as explained above.

If unable to adjust R1 precisely, it may be necessary to include a 100-ohm rheostat R3 in series with R1 as a fine or "vernier" adjustment.

6-6 Expanded-Range DC Voltmeters

A dc voltmeter is used in a battery load tester as a separate test unit or is built into a combination battery charger and battery tester. The voltmeter leads usually go directly from the voltmeter to the battery clips by separate leads to avoid erroneous readings caused by the voltage drop in the heavy cables. A battery tester provides a means of putting a resistance load across the battery being tested and indicating the voltage under load, as well as at no load, before and after charging.

The voltmeter used to indicate the battery condition and state of charge on most late model testers is of a special design known as an "expanded-range" or "suppressed scale" dc voltmeter.

A standard voltmeter spreads the voltage from zero to full scale, over the entire scale. For a 0–15 V meter, for example, the standard, linear scale dc voltmeter is shown in Figure 6-6a. In testing 12-V batteries, voltages below 10 and above 15 (8–16 is common) are of no particular significance. In Figure 6-6a the important voltage readings occur only on the upper third of the scale. The expanded-range dc voltmeter shown in Figure 6-6b overcomes this objectionable feature by expanding the scale three times, making for readings three times more accurate. The meter in Figure 6-6b does not start registering until the voltage reaches 10 V.

This expansion may be accomplished in several ways. A bridge circuit using special resistors or lamp bulbs is in common usage. The bridge circuit may be located inside the meter case or mounted on a plate outside the meter. In the very earliest expanded-range meters, the suppres-

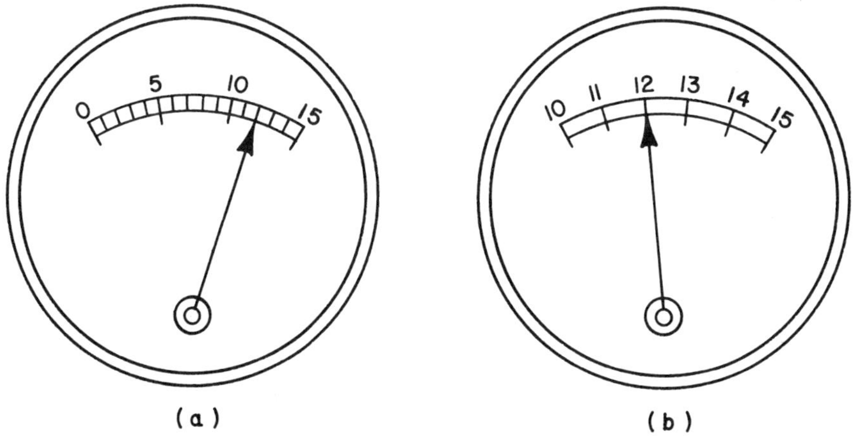

Figure 6-6 Direct current voltmeters

sion was accomplished by using the hair spring adjustment to hold the meter pointer at the low end of the scale until the voltage of, say, 10 V was reached. This system, using the jewel-type meter, had the disadvantage of banging the needle against the stop every time the voltage was suddenly removed, often damaging or bending the needle and causing inaccurate readings. These meters were, of necessity, highly damped in order to check the rebound. Now, the taut-band meter suspension system can be suppressed in this manner with dependable results.

6-7 Expanded-Range DC Voltmeter Using Zener Diodes

The development of the zener diode makes it possible to have a relatively inexpensive, accurate, and simple means of providing the expanded range on a dc voltmeter.

In Figure 6-7a, a 10-V zener diode Z is shown connected in series with the voltmeter. The meter will get no current until the voltage reaches 10 V and the zener diode starts conducting, and the meter will hold a 10-V drop across it. As the voltage is increased, the zener diode has a constant voltage drop of 10 V, within limits, regardless of the current flowing through it. At full scale of 15 V, there is a 10-V drop across the zener diode and a 5-V drop across the meter. At 10 V there is no voltage across the meter and a 10-V drop across the zener diode.

Figure 6-7b shows the same basic circuit, but for dual-voltage opera-

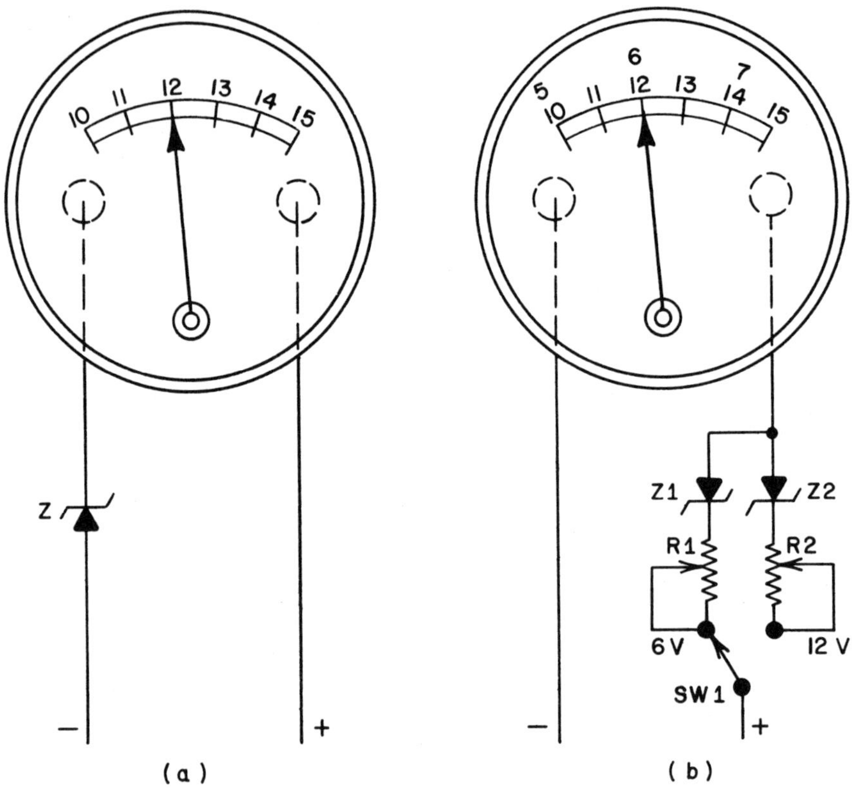

Figure 6-7 Expanded-range dc voltmeter using zener diodes

tion, such as 6 and 12 V, using two zeners. Zener diode Z1 is a 5-V or less, and Z2 is a 10-V or less. R1 is the multiplier and calibrating resistor for 6 V (5–7.5 volt range), and R2 is the multiplier and calibrating resistor for 12 V (10–15 volt range). Switch SW1 is a SPDT switch for selecting either the 6- or 12-V range. Slight value deviations of Z1 and Z2 can be compensated for by rheostats R1 and R2. Z1 and Z2 should be exactly 5 or 10 V, or less, respectively, but not over. If less, they will start conducting a little sooner.

6-8 Alternator-Regulator Battery Starter Tester

Most automotive electrical testers have incorporated in them, or as a separate unit, a dc voltmeter to read the battery voltage under various load and charging conditions.

A rather unique voltmeter is found in Source 15, Model 42-148, alternator-regulator battery starter tester, along with a new type of dc "clamp on" ammeter using the Hall effect. This ammeter is described in Sections 8-9 and 8-10.

The voltmeter section has four scales, only one of which has a suppressed or expanded range, the other three being standard linear scales.

The voltmeter section is shown schematically in Figure 6-8, and the expanded range circuit only is shown in Figure 6-9.

The two voltage connections to the unit are the large battery clips BAT positive and Bat negative. One-half of each clip has a heavy conductor to 500-amp carbon pile load-control rheostat R20. The other half of each clip carries a smaller conductor for voltage circuits only. Another pair of voltage clips VL is provided for measuring volts only in the "volts test selector" position, which is position No. 3 of switch SW1. Switches SW2 and SW3 are ganged together, with position No. 1 for 2 V, No. 2 for 20 V, No. 3 for 40 V, and No. 4 for the expanded range of 13 to 15.6 V. When switches SW2 and SW3 are in position No. 4, switches SW4 and SW5 close only on the expanded range. The actual tester uses a push button arrangement, but it is simplified here to make the circuit easier to follow.

Meter M1 is protected against over voltage and reverse voltage by the two diodes CR1 and CR2 operating in the barrier region. Diode CR2 is in the correct polarity to conduct on a positive voltage to the positive of meter M1, but a diode (even in correct polarity) does not start to conduct until the voltage reaches the barrier region of 0.2 to 0.5 V. The voltage required to give full-scale reading across meter M1 is somewhat less than this, so unless the voltage across meter M1 exceeds this value, diode CR2 will not conduct. Similarly, for reverse polarity, diode CR1 is connected to conduct and protect the meter if the voltage exceeds a safe value in case the polarity is reversed. It also protects the meter on the expanded range, where it is connected at reverse polarity.

The voltage ranges of 2, 20, and 40 are in normal circuitry, with only one calibrating potentiometer R12 for all three ranges, this should be done before calibrating the expanded-range scale.

The fourth expanded range, however, uses a bridge circuit as shown in Figure 6-9, and a zener reference diode CR3 to prevent the meter from reading until the voltage reaches 13 V. It then reads on up to 15.6 V full scale with 2.6 V spread over the entire scale for more accurate

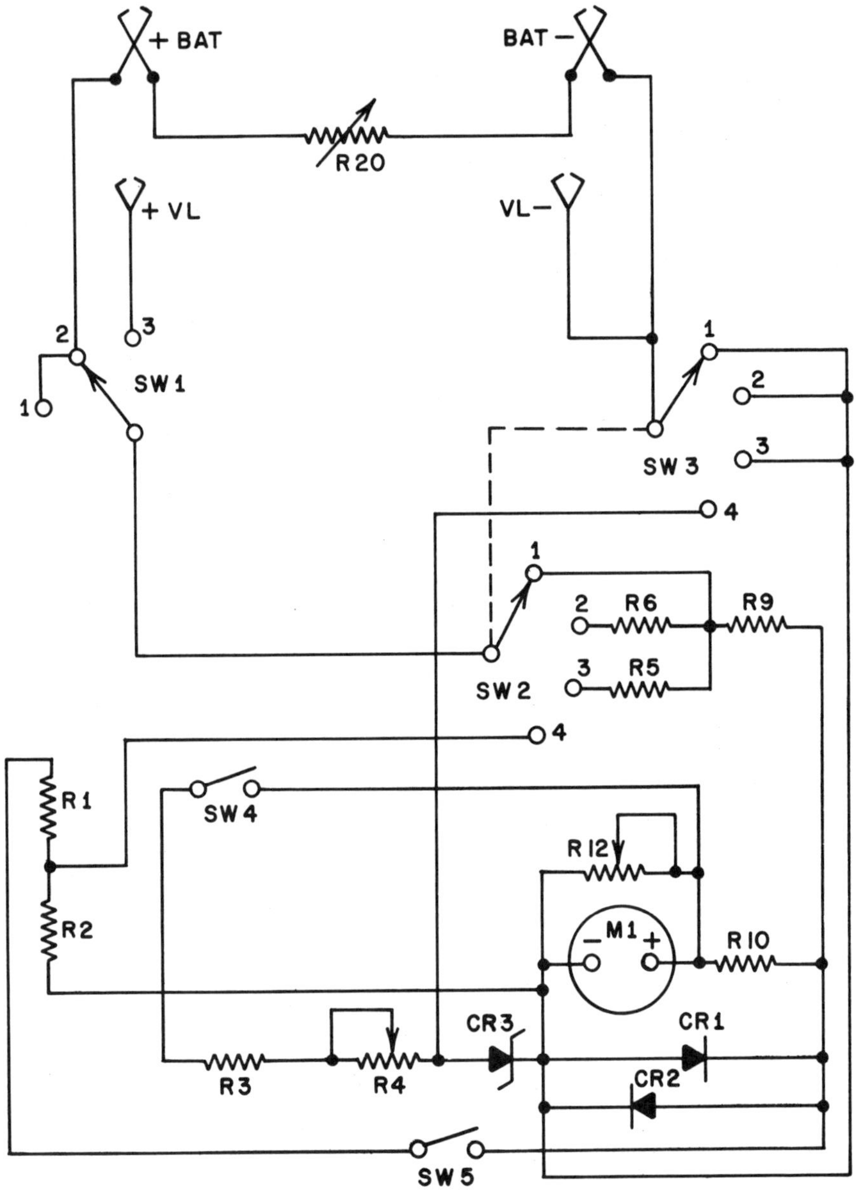

Figure 6-8 Alternator-regulator battery starter tester, voltage section

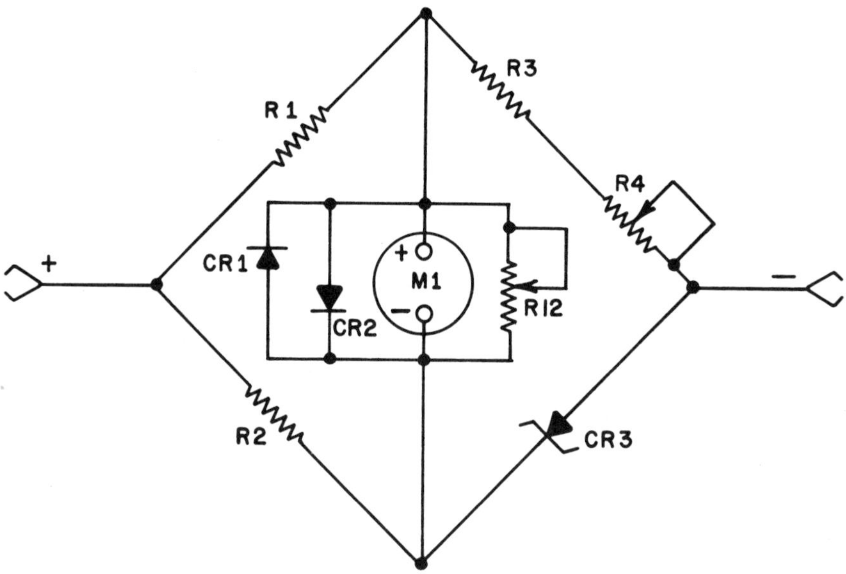

Figure 6-9 Expanded-range voltmeter section of Figure 6-8

reading in adjusting voltage regulators. The expanded range is calibrated by applying exactly 13 V to the volts test leads and adjusting R4 until the meter reads exactly 13 V, rechecking also at 15 V.

Zener diode CR3 has a zener voltage of 13 V, or slightly less than 13 V. The difference is balanced out by R4 so that when CR3 starts to conduct through the meter, through R1, and back to positive, the meter will read 13 V.

6-9 Battery Load Tester, Source 10

The battery load tester shown in Figure 6-10 provides a reading of battery voltages under no load and under charge, and at a specified amperage load on the battery. It can test 6- or 12-V batteries. The scale of the dc voltmeter VM is usually calibrated in volts, and it always indicates battery condition and state of charge by color. It usually has a colored scale for calibrating and adjusting voltage regulators on vehicles.

To operate this tester, first select the battery voltage under test by setting the test switch, consisting of SW2 and SW3 ganged together as a DPDT switch, to either 6 V or 12 V. This selects the correct meter range

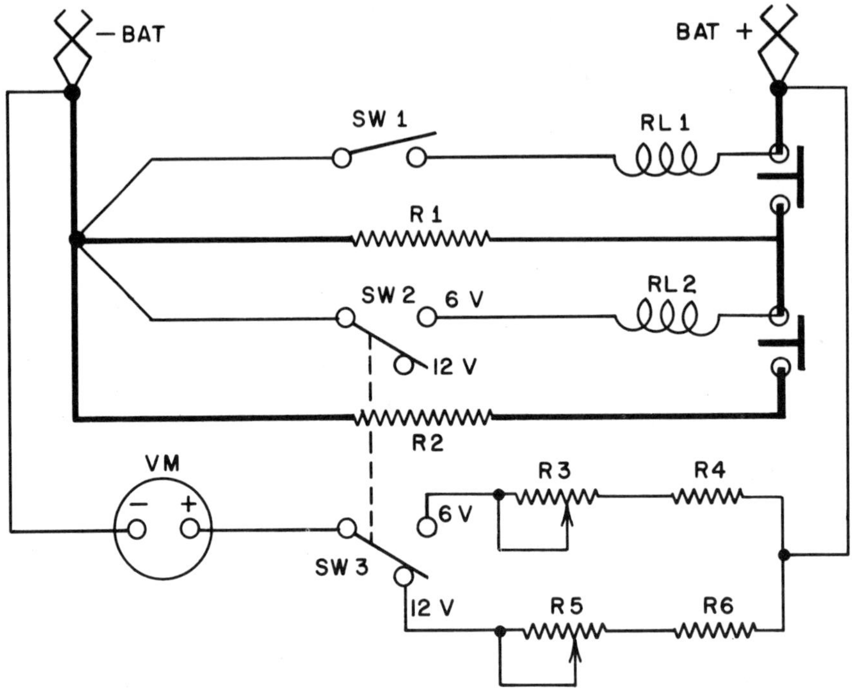

Figure 6-10 Battery load tester, Source 10

on the test meter and also connects the correct resistance load. When SW1 is set to test, this energizes solenoid RL1, which closes the contacts and places load resistor R1 across the battery. If the test switches SW2 and SW3 are in the 12-V position, then solenoid RL2 is not energized, and only resistance R1 is across the battery for the 12-V test. If test switches SW2 and SW3 are in the 6-V position, then solenoid RL2 closes the contacts and places load resistance R2 in parallel with load resistor R1 and across the battery. A 6-V battery requires a higher load test current than does a 12-V battery.

It should be noted that small wires from the test meter circuit go directly to the battery test clips to give accurate voltage readings at the battery terminals. This avoids error due to voltage drop in the cables. Cables, after long usage, may have many broken strands and offer a higher resistance than when new. These small wires are usually built into the jacket of the heavier cable.

Rheostat R3 (25 ohms) is a calibrating resistor, and R4 (39 ohms) is a multiplier resistor for 6 V. Rheostat R5 (40 ohms) is a calibrating resistor, and R6 (100 ohms) is a multiplier for 12 V.

6-10 Combination Charger and 421 Battery Tester, Source 15

The battery tester shown in Figure 6-11 is a type "421" tester and is considered a more accurate and reliable test instrument than any discussed so far. It has a built-in 14-amp battery charger and is designed for 12-V batteries only. This tester has been specifically designed to perform the 421 battery test. The 421 test came about as the result of testing and actually tearing down several thousand batteries of all sizes and brands, in all conditions and states of charge.

The 421 test is based upon an analysis of differential open-circuit terminal voltages. The terminal voltage is measured after the battery has been conditioned with a given discharge of 50 amp for a specific period of 15 sec, and again after the battery has been conditioned with a given charge of 14 amp, also applied for a specific period of 45 sec. The difference between these two readings is then compared with a standard colored meter scale to determine battery condition. The standard is based on voltage differences characteristic of good batteries and of unsatisfactory batteries. Testers capable of performing this 421 test can be recognized by the "421 Test" seal appearing on the instrument panel.

Referring to Figure 6-11, the following prescribed procedure is printed on the front panel of the tester. Explanations have been introduced here, where appropiate, to further explain the functioning of each component at all stages of the test.

"421" Test Procedure

1. Check battery for visual defects, and add water if needed. CAUTION—Be sure: all lights and accessories must be off.
2. Plug power lead into a 115-V outlet.
3. Connect the red clamp to the positive battery post, the black clamp to the negative post. Twist clamps to make good connections.

The SELECTOR SWITCH, SW3A and SW3B, selects 421 test (T), turns OFF (O), and connects the tester as a regular battery charger, charge (C), at 14 amp, 12 V. The selector switch is a DPDT toggle switch having a middle OFF position.

4. To test, move selector switch SW3 to TEST position. Rotate

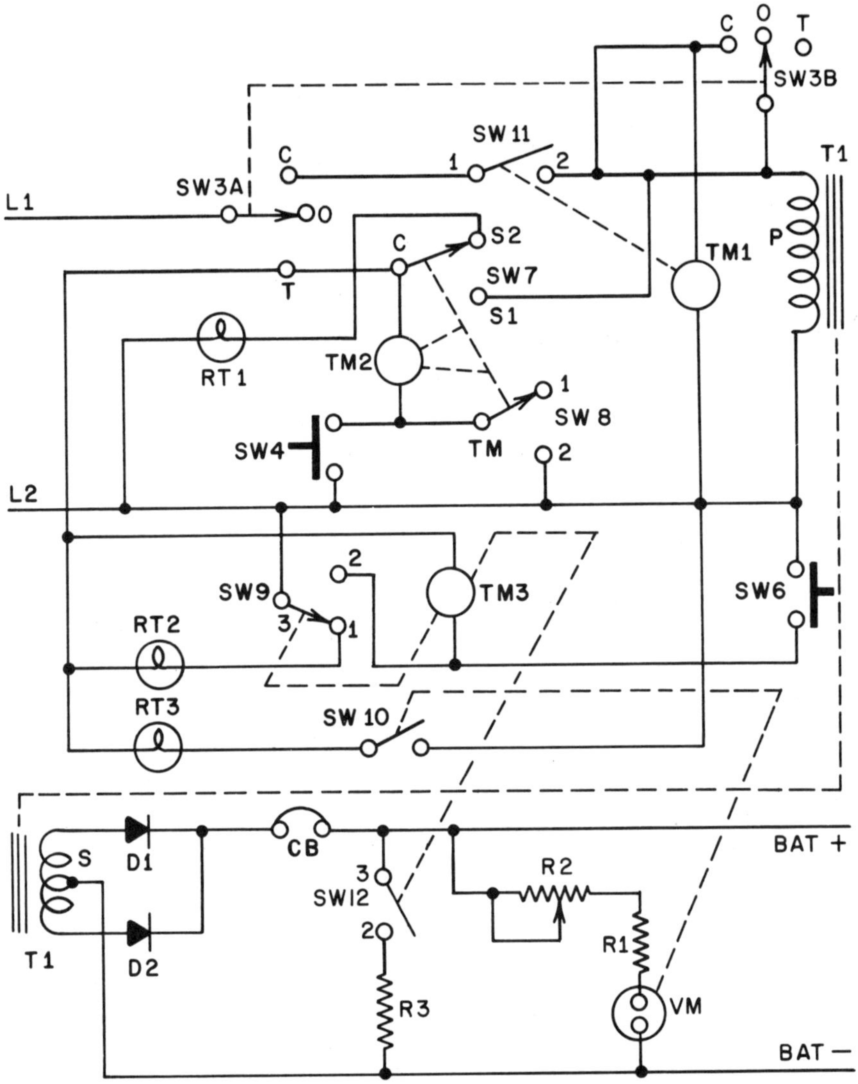

Figure 6-11 Combination charger and 421 battery tester, Source 15

TEST INDICATOR on the meter (a small knob on the front of the meter) to UP position. All three test lights, SET, RESTART, and CONDITION should be lit. The test indicator (SW10) on the meter must be turned to the up position so that the restart light RT3 goes on; otherwise, the restart light could not give OFF indication in step 7. Switch SW10 is a microswitch mounted on the meter and is closed, turning on restart light RT3 when the test indicator is set to intersect the SET LINE and meter pointer at over 11 V on the meter scale. Switch SW10 is open, and the restart light RT3 does not light when the test indicator is set to intersect the set line and meter pointer below 11 Volts on the meter scale.

5. Depress SET BUTTON SW6, and hold it down until SET LIGHT RT2 goes out. The set button closes the circuit connecting the load timer motor TM3 across the 115-V ac line. After a few seconds, switch SW9, which is cam-operated and driven by load timer motor TM3, moves movable contact 3 from contact 1 to contact 2, which turns off set light RT2 and connects contact 3 to contact 2, which shunts set button SW6 and keep TM3 running, even though set button SW6 is released. At the same time, load timer switch SW12 moves contact 3 from contact 1 to contact 2, placing the 50-amp load resistor R3 across the battery for 15 sec. Press and hold the set button until the set light goes out. If the set button is not held down until the set light goes out, the timer and cycle will stop. After the set light goes out, the 50-amp load is applied for 15 sec, after which it is turned off, but the set light stays off for 5 sec more while the battery voltage stabilizes.

If the timer stays on and does not turn off the set light within a few seconds, immediately disconnect the battery, and check the cam on the load timer. Load timers have been known to "hang up" on the cam front and stall the timer motor, leaving the load resistor connected across the battery. Usually a small dab of grease on the steep part of the cam will cure this. If it does not, dress off the cam to remove any burrs. The recommended grease is Molykote G lubricant, available from any machinery supply house. Run the tester several times through the load test cycle with the battery disconnected, otherwise as in step 5, to be sure the timer is operating without fail. Better yet, disconnect the leads of the timer motor TM3 and apply 88 V through a Variac, potentiometer, or other adjustable ac source. If it operates for several minutes and starts on 88 V, it can be considered to be dependable on 115 V ac. The torque at 88 V is only 58.5% of the torque at 115 V. The torque varies, not proportionally, but as the square of the voltage.

The large contacts are 2 and 3 of load timer switch SW12, and a load timer switch SW9 is a small microswitch; each is driven by load timer motor TM3 and by two separate cams. After 20 sec, timer TM3 returns the contacts of SW9 and SW12 to the original position, as shown in Figure 6-11.

If the load timer should be left "hung up" with the load resistor R3 still connected, do not try to charge the battery or conduct step 7, since the extra load on the rectifiers would overload them or kick out the 30-amp circuit breaker CB.

6. When SET LIGHT comes on again, immediately rotate TEST INDICATOR until it crosses the meter pointer at the SET LINE.

7. If RESTART LIGHT goes off, proceed to step 9. The restart light will not go off if the setting of the test indicator in step 6 was over 11 V.

8. If restart light stays on, repeat steps 5, 6, and 7. If it stays on after repeating these steps three times, proceed to step 9.

9. Depress CONDITION BUTTON and hold it down until CONDITION light goes off. The condition button turns on condition timer TM2, which starts a 14-amp, 45-sec charge cycle. Charge starts when condition light RT1 goes off, and stops when the condition light turns on. When condition button SW4 has started the condition timer motor TM2, after a few seconds, the timer-driven cam switch SW7 moves contact C away from contact S2 and to contact S1. This disconnects condition light RT1 and it goes out. Contacts C and S1 apply 115 V ac to the primary of transformer T1 and puts a 14-amp charge on the battery through silicon diodes D1 and D2 in a conventional, center-tapped, full-wave charger circuit. At the same instant, the condition timer motor operates switch SW8 by moving contact TM from contact 1 to contact 2, which shunts the condition button switch SW4. Motor TM2 will continue to run, through the cycle, even when condition button switch SW4 is released. SW4 can be released as soon as, but not before, condition light RT1 goes out. The condition timer motor continues to run for 45 sec, and shuts off to stop the charging cycle by moving contact C away from S1 half-way between S1 and S2, but it continues to run for an additional 15 sec to stabilize the voltage. Then SW8 turns movable contact TM from contact 2 to contact 1, stopping the motor TM2. At the same time, movable contact C of SW7 moves to contact S2, turning on the condition light RT1.

10. When CONDITION light turns on again, in about 1 minute, immediately note where the meter pointer crosses TEST INDICATOR line. A reading in the left portion of the red area indicates low voltage after charge, possibly caused by a short in the battery.

A reading in the right part of the red area indicates high voltage after charge, caused by sulfation or high battery resistance. A reading in the green area indicates a good battery and, if recharging is necessary, the approximate time in hours required to charge.

11. If the reading is in the red, replace the battery. If in the green, the battery is OK; recharge for the number of hours shown on the upper scale.

To Charge Batteries

1. With the SELECTOR SWITCH in OFF position, connect cable clamps to battery at correct polarity, and plug in ac. Move selector switch to TEST position; all three lights should glow before switching to CHARGE position. Move the SELECTOR SWITCH to CHARGE position.

2. Set CHARGE TIMER TM1 to the required number of hours. In Figure 6-11, the 12-hr timer switch SW11 is a cam-operated switch driven by the charge timer motor TM1. When the charge timer dial is turned clockwise, it closes contacts 1 and 2. This connects the charge timer motor and transformer T1 primary across the 115 V ac line, and the motor continues to run the number of hours set on the dial. This provides a 14-amp (at start) 12-V charge to the battery.

The test voltmeter VM can be calibrated to read the exact voltage across the cable clamps, with the selector switch at the OFF position, by adjusting R2, which is located inside the case and has a screw driver slot in a stub shaft. A precision, 0.5% voltmeter should be used to calibrate the meter VM, if available. However, fairly accurate results can be obtained by using a fully charged battery in good condition that has been allowed to stabilize, after charge, for 24 hours, with no additional charge or discharge during the 24-hr period. At 77° F and 1.260 specific gravity, the battery should have a voltage of 2.11 V per cell, or 12.66 V for a 12-V battery. If the meter VM on the tester is fairly close (within 0.1 V), it is better to leave it as it is, unless a precision calibrating meter is used. To calibrate, adjust rheostat R2 to read exactly 12.66 V on meter VM.

6-11 Simple Hand-Held Battery Tester, Source 10

Figure 6-12 shows a small, hand-held voltmeter used to test battery voltage under starting load, as well as for setting voltage regulators on vehicles. This tester is unique in that it uses no switch to read 6 or 12 V; it is necessary merely to reverse the connections. There is a meter prod MP protruding from the case of the meter and a test prod TP at the end of a flexible cable. To read on the 6-V scale, connect the meter prod MP to the battery positive post, and the test prod TP to the battery negative post. On 6 V the current flows through the meter prod MP and silicon diode D1 (but it is blocked by diodes D3 and D2), through the positive meter VM terminal, through the 6-volt multiplier-potentiometer R1, through diode D4 to the test prod TP, and back to the battery negative post. To read on the 12-V scale, reverse the prod

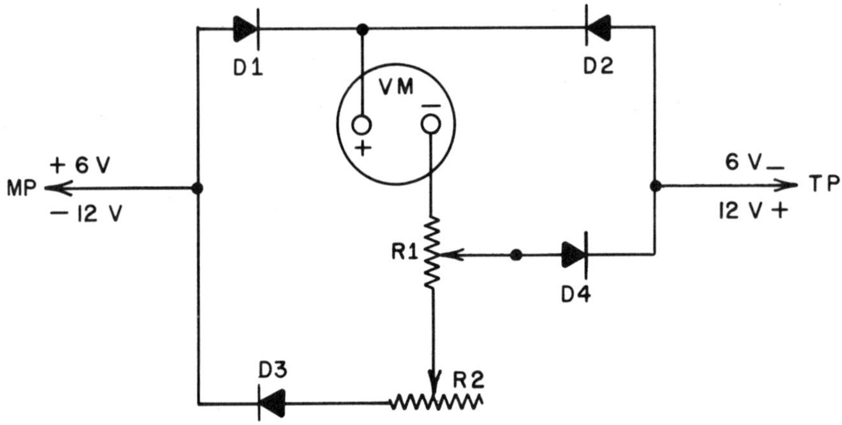

Figure 6-12 Simple hand-held battery tester, Source 10

connections used on 6 V. The test prod TP is connected to the 12-V positive post, and the meter prod MP is connected to the negative battery post. On 12 V, the current flows through the test prod TP, through diode D2 (D4 and D1 are blocking), through the positive meter terminal, through R1, R2, and D3, and back to the negative post through meter prod MP.

Diodes D1, D2, D3, and D4 are all alike—silicon diodes rated at 500 ma and 50 PIV or higher.

Potentiometer R1 is the 6-V calibrating resistor, rated at 50 ohms. It should be adjusted first, in case there is a slight reverse leakage in diode D4. Rheostat R2 is the 12-V calibrating resistor, and is 100 ohms.

6-12 Battery Condition Tester, Source 10

Figure 6-13 shows a battery condition tester with a timed load of 250 amp at 6 V, and 180 amp at 12 V, having an expanded scale meter and an automatic 6/12-V selector relay RL3.

The 6/12-Volt relay RL3 has one set of two contacts—1 and 2, and one set of three contacts—3, 4 and 5. With the coil of RL3 deenergized, or on 6 V, the contacts do not move but remain in the 6-V position as

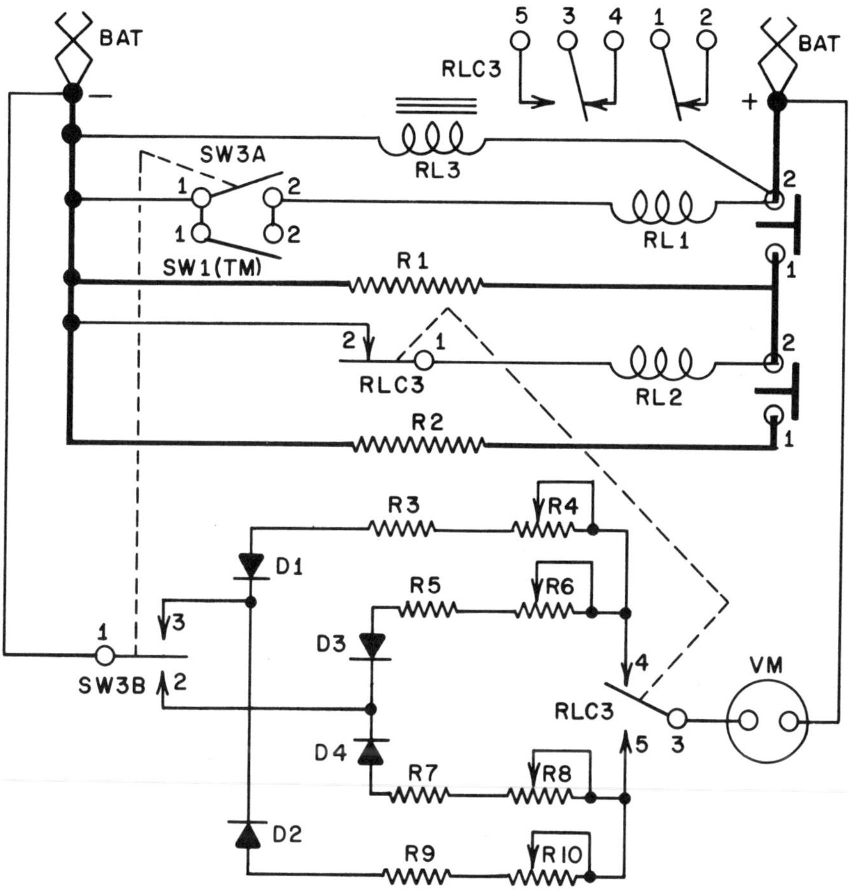

Figure 6-13 Battery condition tester, Source 10

shown. However, 12 (9.5–10) Volts will move the armature, move contact 1 away from contact 2, and move contact 3 away from contact 4 to contact 5, thereby setting the load and meter on the 12-V range. Solenoid RL1 closes on the BATTERY CONDITION, or load test, for both 6 and 12 V, either by the timer TM contacts (SW1 contacts 1 and 2) or by the battery condition switch SW3A contacts 1 and 2, or both. This puts the 180-amp discharge resistor R1 across the 12-V battery. For 6 V, switch RLC3 has contacts 1 and 2 connected, which energizes solenoid RL2 and places discharge resistor R2 across R1 in parallel and across the battery for the higher 250-amp load on 6 V.

The STATE OF CHARGE and BATTERY CONDITION switch (SW3A and SW3B) is a DPDT switch. When in the state of charge position, contacts 1 and 3 of SW3B are closed, placing the voltmeter VM across the battery for STATE OF CHARGE and OPEN CIRCUIT (O.C.) readings. The BATTERY CONDITION position closes contacts 1 and 2 of SW3A, energizing the discharge solenoids, and it also closes contacts 1 and 2 of SW3B, energizing the test meter VM through a lower-resistance multiplier network so that the condition voltage will come up higher on the scale. This is an expanded-range meter. The four silicon diodes D1, D2, D3, and D4 are all alike and are used to simplify the mechanical switching of the voltage multiplier circuits of the voltmeter.

The switching is automatic. For example, on the 6-Volt O.C. and state of charge position, the current flows from the positive cable clamp through the meter, through switch RLC3 contacts 3 and 4, through R4 (25 ohms), R3 (33 ohms), and diode D1, through contacts 3 and 1 of SW3B, and back to negative (D2 blocks). On the 6-V battery condition position, the current flows from the positive cable clamp through the meter, through RLC3 contacts 3 and 4, through R6 (25 ohms), R5 (18 ohms), and D3, through SW3B contacts 1 and 2, and back to the negative cable clamp.

On the 12-V state of charge or O.C. position, the current flows from the positive battery clamp, through the meter, through RLC3 contacts 3 and 5, through R10 (40 ohms), R9 (100 ohms), and D2, through switch SW3B contacts 3 and 1, and back to the negative battery clamp. On the 12-V battery condition position, the current flows from the positive battery clamp, through the meter, through RLC3 contacts 3 and 5, through R8 (40 ohms), R7 (75 ohms), and D4, through SW3B contacts 2 and 1, and back to the negative battery clamp.

The test procedure and interpretations of the readings are printed on the front of the panel.

Solenoids RL1 and RL2 are 6-V intermittent duty units, but they will withstand 12 V for the few seconds they are energized.

6-13 Battery Overall Tester, Source 10

Figure 6-14 shows the schematic of a battery overall tester, Source 10, Model 260N.

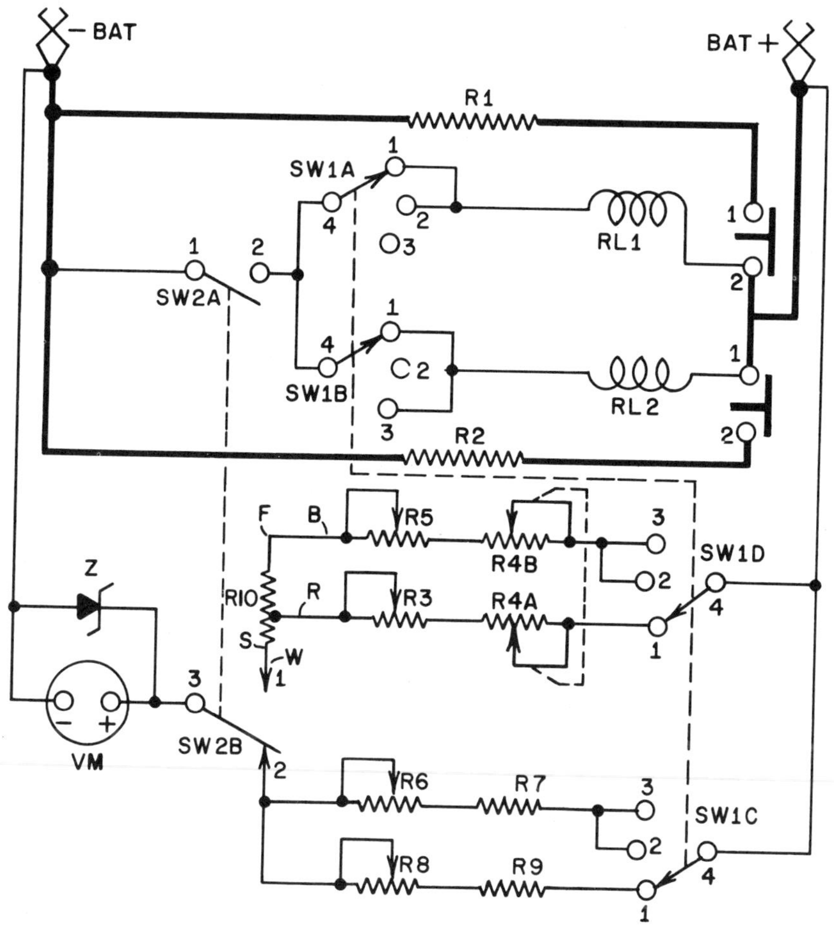

Figure 6-14 Battery overall tester, Source 10, Model 260N

The prescribed test procedure is found printed on the front panel of the tester, and here is how it works.

Test #1—Battery Charge

Set CAR BATTERY switches SW1A, B, C, and D, to the car type (6-V or 12-V compact, or 12-V standard). When contact 4 of each of the switches—SW1A, B, C, and D—is set on contact 1, the circuit is set up for 6-V batteries; when set on contact 2, the circuit is set up for 12-V compact car batteries; and when set on contact 3, the circuit is set up for 12-V higher-capacity standard car batteries.

Connect the red clamp to the positive battery post and the black clamp to the negative battery post.

Depress load switches SW2A and SW2B for 15 sec to remove the surface charge. If switches SW1A and SW1B are in position 1 for 6 V, solenoid RL1 and RL2 are energized, and discharge resistors R1 and R2 are connected across the battery in parallel at 165-amp load. With switches SW1A and SW1B in position 2, only solenoid RL1 is energized, and load resistor R1 is connected across the battery, placing a load of 150 amp on 12-V compact car batteries. With the switches SW1A and SW1B in position 3, only solenoid RL2 is energized, placing discharge resistor R2 across the battery at 180 amp for standard size 12-V car batteries. At the same time switches SW1C and SW1D connect the test meter through series resistors across the battery.

In position 1 (6 V), SW1D connects the positive clamp through contacts 4 and 1 of SW1D, through R4A (32 ohms), R3 (8 ohms), R10 (see Table 6-2) red to white, through contacts 1 and 3 of SW2B, through the test meter, and back to the negative clamp. A zener diode Z is placed across the test meter to protect against reverse polarity and voltages above 16 V. R10 is a temperature-compensating probe, consisting of a sealed spool of nickel wire whose resistance changes with the temperature of the electrolyte in the battery. This probe changes the calibration of the load scale, depending on its temperature, and must be placed in any cell during the load test only. It is not in the circuit during the open circuit (OC) test. Refer to Table 6-2 for temperature-resistance values.

On positions 2 and 3, switch SW1D connects the positive clamp to contact 4 with 2 and 3, through R4B (90 ohms), R5 (25 ohms), R10 black to white, through SW2B contacts 1 and 3, test meter, and back to the negative clamp.

Temperature-Compensating Resistor Rio, Figure 6-14, Source 10

Temperature (° F)	*12-V Compensating Resistor (ohms, white-to-black)*	*6-V Compensating Resistor (ohms, white-to-red)*
62	53.53	23.1
64	53.82	23.1
66	54.11	23.3
68	54.50	23.4
70	54.79	23.5
72	55.08	23.7
74	55.37	23.85
75	55.515	23.9
76	55.66	23.9
78	55.95	24.00
80	56.24	24.20
82	56.53	24.35
84	56.82	24.40
86	57.11	24.60
88	57.40	24.65
90	57.69	24.80
92	57.98	24.90
94	58.27	25.00
96	58.56	25.20
98	58.85	25.30
100	59.14	25.40

Table 6-2

After depressing load switch for 15 sec to remove surface charge, remove vent caps and check water level. This will allow time for the battery voltage to settle.

Read center BATTERY CONDITION scale. For a red reading—recharge battery; for a green reading—charge is OK.

Test #2—Battery Condition

NOTE: Perform this test only if test #1 indicates 25% or more charge in battery. If less than 25% charge, recharge battery, then perform #2 test.

The temperature-compensating probe R10 must be inserted into any cell of the battery to compensate for temperature of electrolyte.

Connect red clamp to positive post of battery and black clamp to negative post.

Set car battery test switches SW1A, B, C, and D as in test #1.

Set PERCENT CHARGE switch (indicator) to the reading (25% or more) obtained in test #1.

Depress LOAD SWITCH and read lower center BATTERY CONDITION scale while depressing the load switch.

Set PERCENT CHARGE switch (indicator) to AFTER CHARGE arrow if battery has been fully charged. If the reading is red, replace battery; if green, battery is OK.

Test #3—Charging System, Upper Right Scale

For all batteries found in SAFE condition and fully charged, set battery switch SW1 as in tests #1 and #2. Connect the red clamp to positive and black clamp to negative battery posts. Turn all electrical accessories off. Warm up engine, and operate at fast idle for 5 min. Then, with car still at fast idle, read upper right-hand scale.

If meter reads red high—the regulator needs adjustment to prevent over-charging and boiling, or damage to the battery.

If it reads green OK—the alternator, or generator, and regulator are OK.

If it reads red low—either the fan belt is slipping, or the alternator or generator or regulator is faulty.

Check the fan belt first.

Check generators by disconnecting the field (F) lead at the regulator and touching the disconnected wire, first to ground, then to the terminal on the regulator marked GEN or ARM.

Check alternators by disconnecting the field (F) lead at the regulator and touching the disconnected wire to the IGN or SW terminal on the regulator.

If the generator and alternator tests result in a green OK or red high reading, replace the regulator. If not, replace the generator or alternator.

To check or calibrate the meter, refer to Figure 6-15b for the correct voltage, with tolerances, at check points shown in Figure 6-15a.

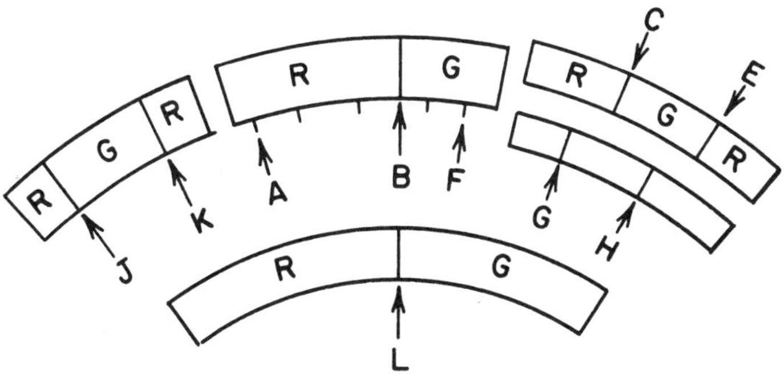

(a)

POINT	METER CALIBRATION VOLTS	INSTRUMENT CALIBRATION	
		6 VOLT SCALE	12 VOLT SCALE
A	1.95 ±.02	5.85 ± .06	11.70 ±.12
B	2.05 ± .02	6.15 ± .06	12.3 ±.12
C	2.3 ±.05	6.9 ± .15	13.8 ± .30
E	2.52	7.55 ± .30	15.1 ± .60
F	2.11	6.33 ± .06	12.66 ± .12
G	2.25	-------	13.5 ±.30
H	2.47	-------	14.82 ± .60
J	1.48	4.44 ± .6	8.88 ±1.2
K	1.83	5.49 ±.15	10.98 ±.30
L*	1.74/1.60	5.22 ±.06	9.60 ±.12

* AT 75 DEGREES F., AND WITH COMPENSATOR SET AT 75 %; SEE TABLE 6-3 FOR OTHER TEMPERATURES

(b)

Figure 6-15 Battery overall tester, Figure 6-14 dial and calibration chart, Source 10, Model 260N

Test #4—Starter

Battery must be in SAFE condition and fully charged.

Set BATTERY switch for correct battery.

Connect red clamp to positive and black clamp to negative battery posts. Turn all electrical accessories off.

Crank engine with remote starter button, leaving ignition key off. Engine may be cranked with the key or car starter button, but the center wire should be removed from the ignition coil to prevent the engine from starting.

If meter reads red low—check for bad bearings or defective armature in starting motor.

If it reads green—starter is OK.

If it reads red high—check for worn brushes, loose connections, or bad cables.

Test #5—Short or Leakage in Car's Electrical System

Turn off all accessories in car.

Remove ground cable from battery. Set BATTERY switch for correct battery voltage.

For cars with a negative ground, connect black tester clamp to the negative battery post, and connect the red clamp to the ground cable or to the motor block.

For cars with positive ground, reverse the connections above. That is, connect the red tester clamp to positive battery post and the black clamp to ground, or motor block.

Any reading at all indicates a drain on the battery. Touch the ground cable to its battery post to wind the electric clock, if any. Then if the meter continues to read, there is a short, or a fault in the car's electrical system.

When checking systems for short or leakage, leave the meter connected to indicate when fault is located.

In tests #3, #4, and #5, the test meter is connected and used on the open circuit OC connection, and meter multiplier resistors R6 (40 ohms), R7 (100 ohms), R8 (25 ohms), and R9 (39 ohms) are in the circuit. When the load switch is depressed, the voltage drops, and to bring the pointer on scale, lower resistance multipliers are used, R4A (32 ohms), R4B (90 ohms), R3 (8 ohms), R5(25 ohms), and R10 (only used on load test, resistance values in Table 6-2 for various temperatures).

Figure 6-15a shows the expanded-range meter scale, and Figure 6-15b shows the calibration voltages for the test points indicated. Table 6-3 shows the calibration for the "L" scale for various temperatures, and Table 6-2 shows the resistance values of R10 for various temperatures.

Calibration, Model 260N, Source 10, Load Circuit Only

	Volts DC for "L" Scale Reference	
Temperature (°F)	*12 Volts (0.0138 V/°F)*	*6 Volts (0.0057 V/°F)*
62	9.42	5.15
64	9.45	5.16
66	9.48	5.17
68	9.50	5.18
70	9.53	5.19
72	9.56	5.20
74	9.59	5.21
75	9.60	5.22
76	9.61	5.22
78	9.64	5.23
80	9.67	5.24
82	9.70	5.25
84	9.72	5.26
86	9.75	5.27
88	9.78	5.28
90	9.81	5.29
92	9.83	5.30
94	9.86	5.31
96	9.89	5.32
98	9.92	5.33
100	9.94	5.34

Table 6-3

All of these figures and tables apply to the schematic of Figure 6-14.

In Figure 6-15, the two columns to the right give the overall instrument calibration for 6 and 12 V. However, the column on the left shows the calibration for the meter only, which should be performed first. These meters are usually adjustable with a small screwdriver through a small hole located on the bottom just back of the panel. The hole is usually covered with a metal plate or tape.

In calibrating the instrument, it is better to disconnect the load resistors and solenoid coils so that the dc supply will not be overloaded or drop off. Disconnect the heavy cable that goes to the common load resistor and the small wires that go to the solenoid coils.

6-14 Battery Load and Cell Tester, Source 10

Figure 6-16 shows a battery tester with two prods, PR1 and PR2, for reading individual cell voltages on all but hard-top batteries.

The prods are used to measure the voltage of a single cell at a time, or on the car starter load test, or on an open circuit, or charging test.

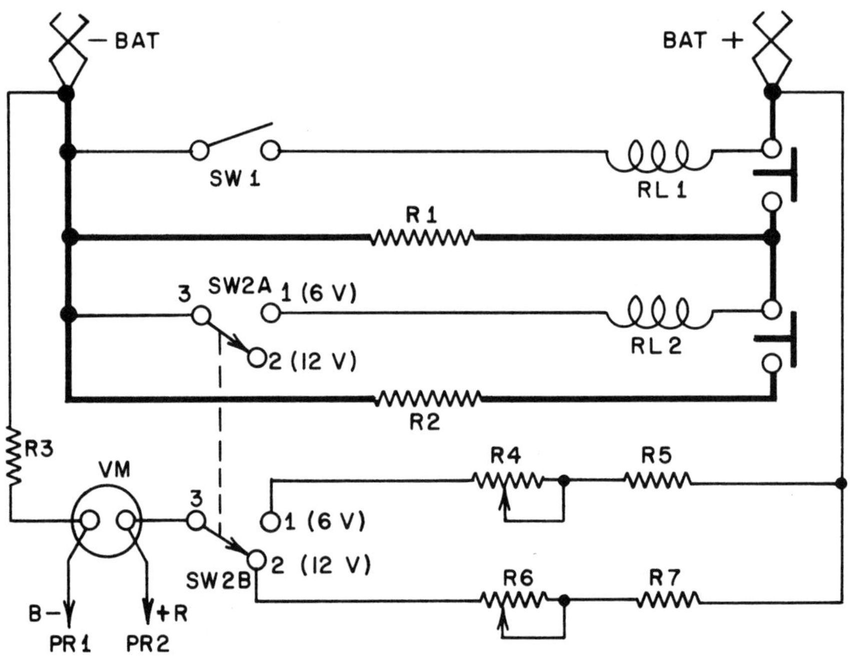

Figure 6-16 Battery load and cell tester, Source 10

The voltage of all cells should read the same. A low or high reading indicates a bad cell. Nominal open-circuit (unloaded) voltage should be about 2 to 2.11 V per cell. DO NOT ATTEMPT TO USE THE PRODS WHILE THE LARGE CLAMPS ARE CONNECTED TO THE BATTERY TERMINALS. The bottom scale indicates the cell voltage from 1.4 to 2.6 V. There is no calibration for the meter alone, but this is not important, because on the cell readings the main concern is the comparison of the voltages between cells, and not necessarily the exact voltage. However, a calibration is provided for the entire instrument in the 6- and 12-V ranges.

The prescribed procedure for use is as follows.

Set battery selector switches SW2A and SW2B to the voltage of the battery, 6 or 12 V.

1. Raise timer switch SW1 handle, which runs for 15 sec and shuts off. When SW1 closes, it energizes solenoid RL1 and places discharge resistor R1 across the battery for either 6 or 12 V. If switches SW2A and SW2B are in the 6-V position, solenoid RL2 is also energized, placing both R1 and R2 across the battery in parallel. The voltmeter VM then reads the voltage across the battery. When SW2 is in the 6-V position, the multiplier resistors R4 (25 ohms) and R5 (9.1 ohms) are in the meter circuit. When SW2 is in the 12-V position, resistors R6 (40 ohms) and R7 (75 ohms) are in the meter circuit.
2. Read the upper center scale after the timer shuts off.

After Charging Battery—

1. Raise timer switch SW1.
2. After the timer shuts off, raise timer switch again.
3. Read left scale while timer is on.

Generator-Regulator Test

1. Turn off all lights and accessories.
2. Read battery voltage in upper right scale.

The meter can be calibrated at the line between LOW and SAFE under the generator-regulator test, at 6.75 V on the 6-V position and 13.5 V on the 12-V position. At the line between SAFE and HIGH the voltage should be 7.1 and 14.2, respectively. Rheostat R4 (25 ohms) is the calibrating rheostat for 6 V, and R6 (40 ohms) is the calibrating rheostat for 12 V.

Resistor R3 (30 ohms) serves to prevent a direct accidental short, in case the clips are left on the battery during the cell test.

6-15 Battery Condition Tester, Source 5

The battery tester shown in Figure 6-17 has a voltage-sensitive relay RL3 for automatic battery voltage selector, which holds the contacts N.C. as shown on 6 V, but 12-V (9.5–10 V) batteries will operate relay RL3 to open the N.C. contacts 2-3 and 4-5 close, N.O. contacts 1-2. On 6 V, solenoid RL1 closes, placing load resistor R1 across the battery when discharge switch SW1 is pushed, and meter VM reads the battery condition. Meter multiplier resistor R4 is short circuited by N.C. contacts 4 and 5 on 6 V, but is in the circuit for 12 V.

On 12 V, relay RL3 opens contacts 4-5 and 2-3 and closes contacts 1

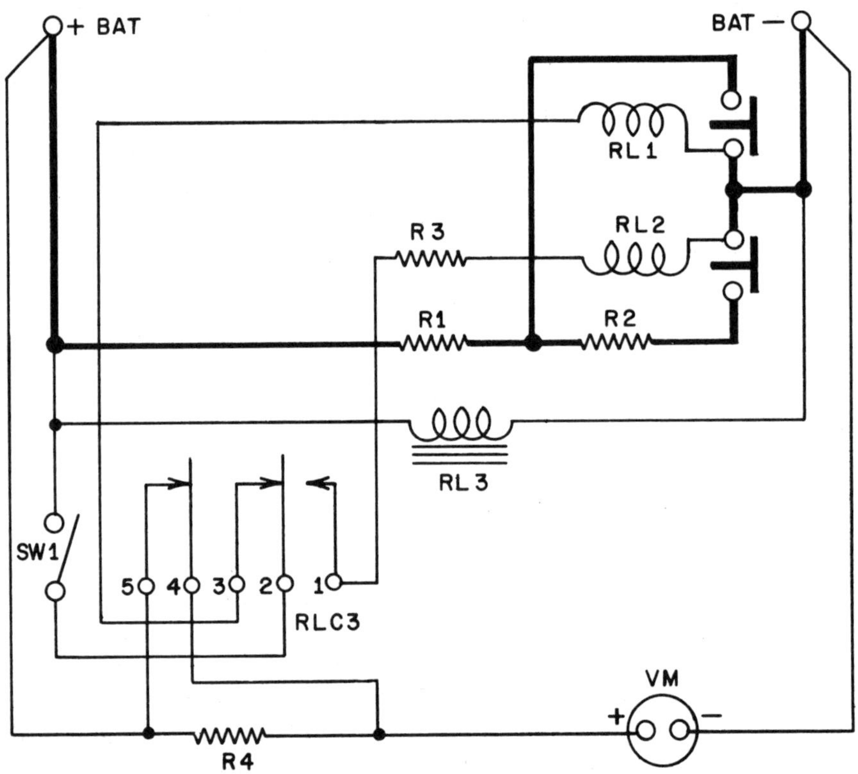

Figure 6-17 Battery condition tester, Source 5, Models T2 and T3

and 2, so that when SW1 is pushed, the 12-V solenoid RL2 closes and places load resistors R1 and R2 in series across the battery. With the discharge switch SW1 open, the voltmeter VM reads the voltage of the battery under no-load conditions.

Meter calibration is shown in Figure 6-18 for Source 5, Models T2 and T3, and in Figure 6-19 for late Model T3 tester. There is no calibrating rheostat for the voltmeter, but it can be adjusted by adjusting

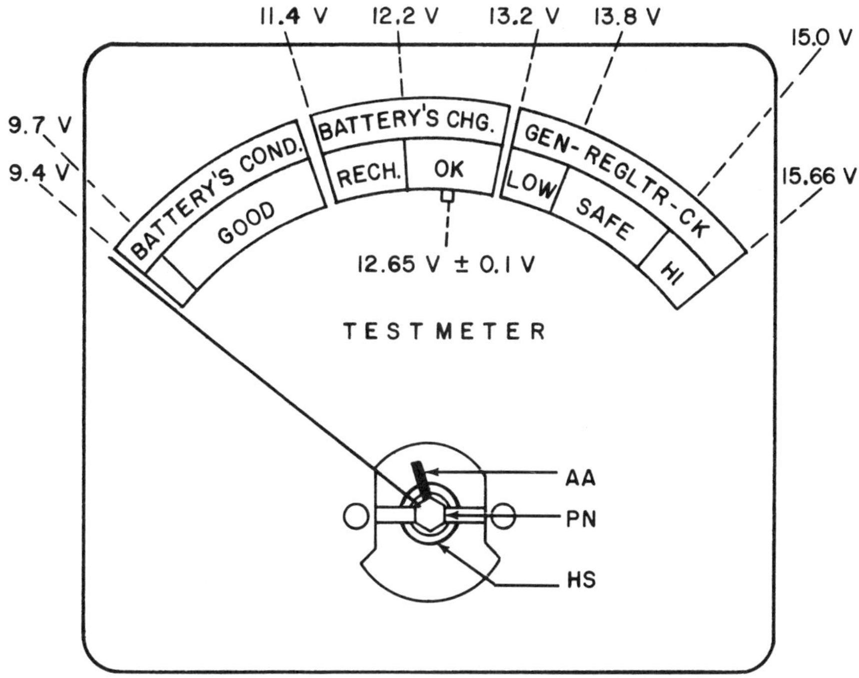

Figure 6-18 Meter calibration, battery condition tester, Source 5, Models T2 and T3

the screw head protruding from the face of the meter cover. If there is no adjusting screw, remove the plastic cover and adjust the hairspring HS tension by turning adjusting arm AA, which extends from the pivot point, interconnected to the hairspring. It should not be necessary to loosen the pivot nut PN to make this adjustment.

A 0.5% calibrating dc voltmeter is recommended to make these adjustments at the voltages shown, but in the absence of this, the following procedure is recommended by Source 5 for their instruments.

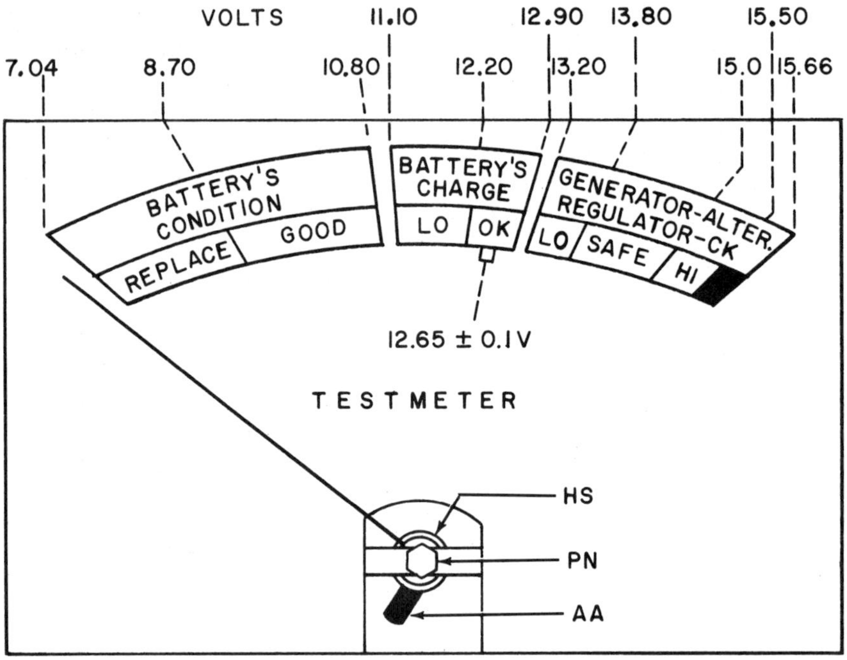

Figure 6-19 Meter calibration, battery condition tester, Source 5, late Model T3

Connect a good fully charged 12-V battery with a specific gravity of 1.250 or more (see Table 6-4), at room temperature (77° F), that has been standing idle without charge or discharge for at least 24 hr but not more than 48 hr. The meter should read, or should be adjusted to read, in the middle of OK under BATTERY'S CHARGE. This should be 12.65 V plus or minus 0.1 V. More precisely, the voltage should be 12.2 V, when using a calibrating voltmeter, at the line between RECHARGE (Figure 6-18) or LO (Figure 6-19), and OK under BATTERY'S CHARGE. Other calibrating points are shown with the correct voltage.

These are expanded-range dc voltmeters; therefore, do not attempt to set the pointer at the far left of the scale as in a zeroed-type meter, except at the voltages indicated.

6-16 Meterless Battery Cell Comparison Tester, Source 5, Model T5

Meterless battery testers are more portable, less fragile, and less expensive than those using calibrated meters. These operate by reading a

Full-Charge Voltages of Batteries at 77°F After 24 Hours *

Nominal Battery Voltage	*Number of Cells*	*Voltage at 2.11 V/Cell*	*Nominal Battery Voltage*	*Number of Cells*	*Voltage at 2.11 V/Cell*
6	3	6.33	24	12	25.32
12	6	12.66	30	15	31.65
18	9	18.99	36	18	37.98

* At 1.260 specific gravity. Allow 0.01 volts per cell for every 0.010 specific gravity change above or below 1.260.

Table 6-4

dial-calibrated potentiometer, and use SCR, or zener diodes and transistors, and usually are temperature-compensated by a thermistor. A typical Source 5, Model T5, meterless tester is shown in Figure 6-20. Complete instructions for its use are printed on the panel. The calibration can be checked and recalibrated by adjustments provided. Number 1 on the dial represents 2.15 V and No. 21 represents 1.95 V, with each numbered division in between representing 0.01 ($1/100$) Volt. Number 11 on the dial is 2.05 V and is also the dividing line between CHARGE OK and NEEDS RECHARGING. The prods are placed in the electrolyte of two adjacent cells, one at a time, and the readings compared. Thus cell readings can now be taken on hard-top batteries without access to the between-cell straps.

The dial is turned to No. 1 and the button pressed. Then the dial is slowly turned CW until the SCR1 fires and lights the pilot light PL. The voltage of the cell is then read on the dial.

Resistor R1 is a temperature-compensating thermistor. The battery BAT is a 1.5-V size AA dry cell, internal to the tester, and should be replaced when the light becomes dim.

To replace the battery or to make internal repairs or adjustments, the tester case must be opened. The case is made in two parts, which are locked together at the four bottom corners with nylon pressure studs. These are removed by inserting a screwdriver between the two parts of the stud head and prying upward.

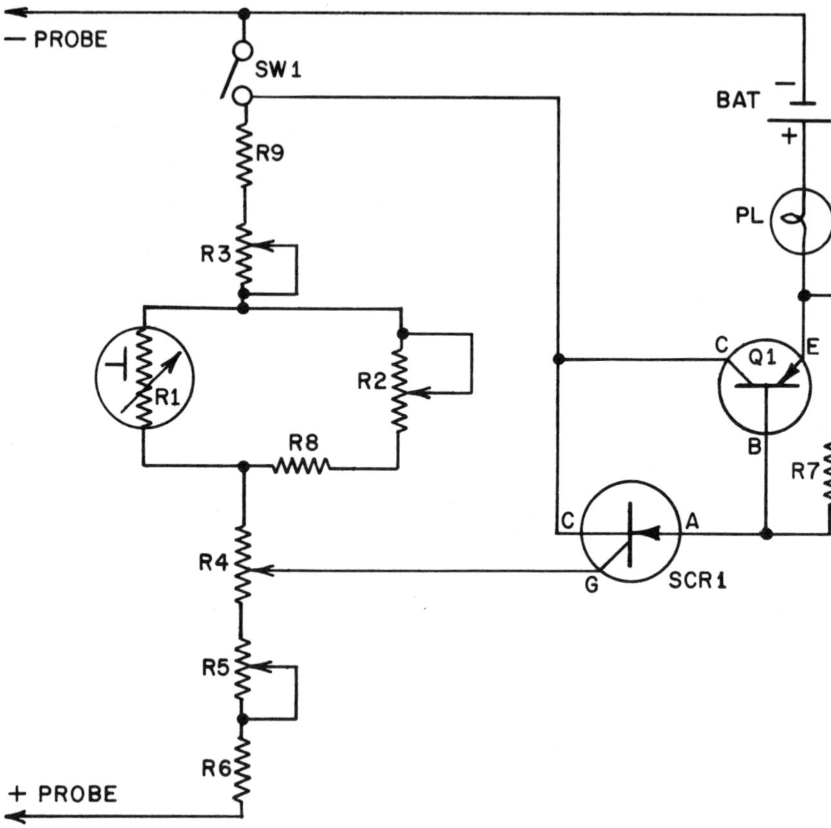

Figure 6-20 Meterless battery cell comparison tester, Source 5, Model T5

6-17 421 Battery Tester, Source 5

All 421 battery testers perform the same functions and cycles, but the various manufacturers have different ways of accomplishing the same thing.

Source 5, Model T4, battery tester is a 421 test unit and is shown schematically in Figure 6-21. The complete operating instructions are on the panel, and the dial of timer TI is clearly marked as to the cycle. The pilot light PL burns at each of the three READING points. If it lights any other time, or when first hooked up, it indicates reverse polarity. Relay coil RL1 is connected in series with diode D3 so that with correct polarity relay RL1 does not operate because diode D3 blocks. The relay RL1 contacts, RLC1, stay in the N.C. position, which connects the ac power to the timer TI contact 2. On reverse polarity, relay RL1, with diode D3 now conducting, opens the N.C. contacts and closes the N.O. contacts. This disconnects the timer and power to the charger and connects pilot light PL across the ac line.

Solenoid RL2 is the load test contactor that connects load resistor R2; CB1 is a circuit breaker in the charging line; D1 and D2 are the charging diodes through transformer T1; the test meter VM is across the battery. To check the calibration of the meter, the number scale of 0–80 corresponds with voltages of 8–16 V; each division above zero represents one tenth of one volt (0.10). For example, 40 on the scale would be 12.0 V, 41 on the scale would be 12.1, and so on. There is no calibrating rheostat for the meter, but it can be adjusted as shown in Figures 6-18 and 6-19.

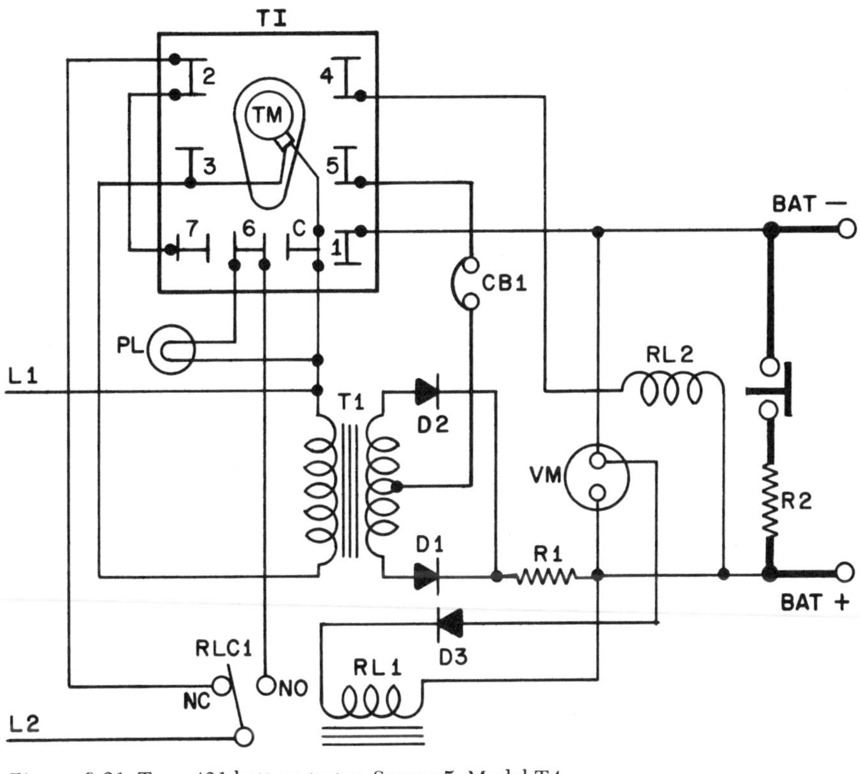

Figure 6-21 Type 421 battery tester, Source 5, Model T4

6-18 Calibrating a DC Voltmeter or Test Meter

In the absence of an accurate (0.5% or better) calibrating standard voltmeter, a dc voltmeter can be calibrated to be reliably accurate at one point by using a lead-acid battery known to be in good condition. The battery should be fully charged, with a specific gravity of 1.260 at room temperature (77° F). It should have been standing idle without charge or discharge for at least 24 hr, but not more than 48 hr. The voltage should be 2.11 V per cell. Table 6-4 shows the voltages for common nominal battery voltages.

From the examples of battery test meter calibration given above, it can be seen that any battery tester can be calibrated, or checked, at one point even if its scale is not calibrated in volts. At the colored part of the dial, or at the point on the scale indicating STATE OF CHARGE, such as 100% charge, OK, and full charge, the meter should read correctly at the voltages shown in Table 6-4. At temperatures above 77° F the voltages may be lower, and below 77° F the voltages may be higher. This is because the specific gravity changes with the temperature. Therefore, no temperature correction is necessary unless it has changed the specific gravity from 1.260. Allow 0.01 V per cell for every 0.010 specific gravity change above or below 1.260. For example, if the specific gravity is 1.270, add 0.01 V per cell; if 1.250, subtract 0.01 V per cell. The specific gravity also depends on the level of the electrolyte in the cells. If the level is low, it is more concentrated and will read higher than it will when the electrolyte is at its normal level. An accurate hydrometer is required to make these corrections: read directly; do not correct to 80° by thermometer.

It is easy to remember how temperature affects the hydrometer reading. The lower the temperature, the more the electrolyte "shrinks," thereby making it more concentrated, resulting in a higher specific gravity. Similarly, at higher temperatures, the electrolyte expands and becomes less concentrated, resulting in a lower specific gravity.

6-19 Battery Hydrometers

The two hydrometers used in automotive service are the battery electrolyte hydrometer and the radiator antifreeze hydrometer. They both work on the same principle but are calibrated for different values.

The hydrometer works on the principle than an object immersed in a liquid will displace a volume and weight of the liquid in direct proportion to the volume and weight of the object. For example, 1 cubic foot of water weighs 62.4 lb. Any object having a volume of 1 cubic foot will weigh 62.4 lb less when completely immersed in water. A float calibrated in specific gravity will sink deeper into the less dense (lighter) liquids and rise higher in liquids of greater density (heavier). Early users of the hydrometer principle in this country were the pioneers. The recipe for the correct amount of salt to add to water to make a brine to "corn" beef stated, "Place an egg in the water and add salt, stirring constantly, until the egg rises to the surface of the brine."

Battery hydrometer floats are calibrated to read specific gravity (water has a specific gravity of 1.000) at only one temperature, usually 80° F. If used in an electrolyte of any other temperature a correction must be applied to bring the reading to what it would be at 80° F. This procedure is for use in determining "state of charge" by the hydrometer method. If the state of charge is determined by the "open-circuit" voltage method, then the temperature is important only as it affects the hydrometer reading. If the hydrometer reading is 1.260 (as actually read without any temperature correction) regardless of the temperature, then the values shown in Table 6-4 will be correct. Correction factors are applied to the open-circuit voltage depending on the specific gravity, and not on the temperature primarily. Allow 0.01 V per cell for every 0.010 specific gravity change above or below 1.260.

Battery hydrometers are usually equipped with a built-in thermometer showing the calibration corrections to make. A simple hydrometer can be used if an accurate thermometer is placed in the same cell. It is the temperature of the electrolyte that determines the amount of correction, not the surrounding air temperature.

The correction for temperature is 0.004 points of specific gravity for every 10° F deviation from the standard base of 80° F. For example, to find the specific gravity of an electrolyte at 80° F when it reads 1.250 at 20° F—

$$80-20=60$$
$$60/10=6$$
$$6\times 0.004=0.024$$
$$1.250-0.024=1.226$$

For electrolytes warmer than 80° F, ADD the specific gravity correction to the reading to get the corresponding value at 80° F.

7
Vacuum and Pressure Gauges

7-1 Vacuum and Pressure Tester Construction

Vacuum and pressure gauges have long been used in automotive service. For most uses, the gauges should be accurate, and should be checked periodically against a standard gauge of known accuracy.

The arrangement shown in Figure 7-1 will not only calibrate vacuum and pressure gauges but will also serve many other useful purposes around the automotive service shop. This unit was designed by the author for his own use.

Anytime that a known vacuum or pressure is needed, the unit shown in Figure 7-1 will provide just the right amount of vacuum or pressure required to perform the following operations, and many more.

—Check the vacuum advance diaphragm for degree advance for various vacuums; also test for leaks.

—Check the pressure-operated stoplight switch for proper functioning at a given pressure.

—Use the vacuum to suck out fluids and sediment from carburetor bowls, etc., catching the fluid in glass receptacle REC. Evacuate air conditioning systems.

—Use as a portable air blow-out hose with valve V6; or with an air chuck in its place, inflate tires, operate pneumatic jacks and the like.

Other vacuum- and pressure-operated devices can be tested. Some are listed here.

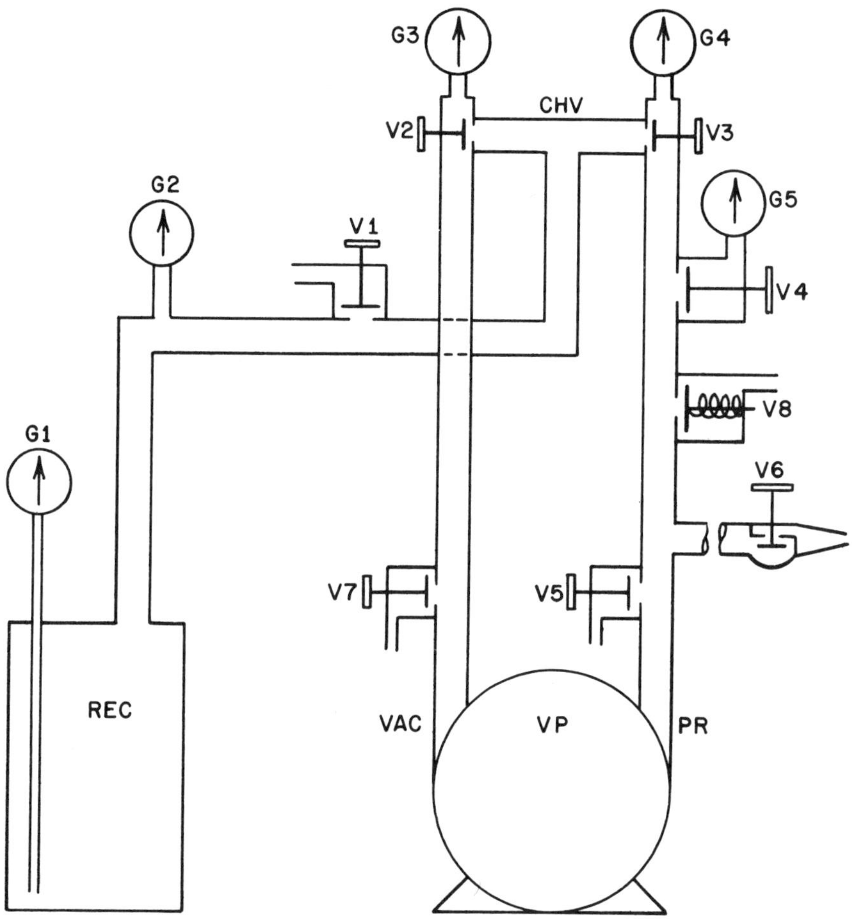

Figure 7-1 Vacuum and pressure tester construction

1. Park brake release
2. Dash pot
3. Automatic transmission modulator valve
4. Door locks
5. Trunk locks
6. Radio antenna
7. Heater and air conditioner controls
8. Headlight doors

9. Windshield wipers
10. Power brakes

In Figure 7-1, VP is a motor-driven vacuum and pressure pump. The rotary vane type, such as used in refrigeration compressors, is best since it is steady, without pulsations. A piston type should have vacuum and pressure chambers and a check valve on the vacuum line, if one is not provided internally. PR is the pressure outlet, and VAC is the vacuum inlet to the compressor. The unit CHV can be a standard refrigeration charging valve assembly or equivalent, with a vacuum and low-pressure gauge G3 (which can be omitted), and a high-pressure gauge G4. (If the accuracy is not known, substitute a standard gauge 0–300 lb.) Gauge G5 is the high-pressure gauge under test and gauge G1 is the vacuum–low-pressure gauge under test. Gauge G2 is a standard compound pressure and vacuum gauge of known accuracy having a 0–30-in. vacuum scale and a 0–15 lb pressure scale.

Valve V5 is a pressure bleeder valve for adjusting the pressure. It is closed during all full-pressure applications and open when on vacuum.

Valve V7 is closed during all vacuum applications and open during pressure uses.

Valve V4 is closed at all times except when checking a gauge in position G5 when valve V3 is closed. Valve V5 is used as a bleeder valve to adjust the pressure.

Valve V3 is closed except when calibrating gauge G1 on the 0–15 lb scale. Before opening valve V3 for this test, open bleeder valve V5 until the pressure on gauge G4 is below 10 lb.

Valve V2 is opened for vacuum when valve V7 is closed.

Valve V1 is a vacuum or low-pressure bleeder valve, used when calibrating unknown gauge G1 against the standard gauge G2.

The receptacle REC can be a glass quart or gallon jar having a screw-on top with an airtight gasket. The vacuum inlet can be ¼-in. copper tubing for a slip-on rubber hose connection for gauge G1 and other vacuum applications.

A float valve in the receptacle as an extra precaution will prevent fluids from being sucked into the pump. Otherwise, the fluids should be emptied frequently.

A fine screen should be placed over all vacuum inlets such as in the receptacle tube that goes to the pump, and over the inlets of valves V1 and V7. An alternative would be to place a filter in the pump suction line.

For full pressure, open valve V7 and close valves V4, V5, and V3. If a pressure relief valve V8 is not installed, it will be necessary to watch the gauge and bleed off by opening valve V5.

For full vacuum, open valves V2 and V5 and close valves V1, V3, and V7.

The vacuum test connection is at the connection of gauge G1.

The pressure test connection is at the connection of gauge G5.

7-2 Types of Gauges

The most common gauges used in automotive testing are pressure gauges for measuring cylinder compression having a scale of zero to 250 or 300 pounds per square inch. Compound gauges are vacuum-pressure gauges for measuring fuel pump vacuum and pressure, and also for measuring intake manifold vacuum, having a double scale from zero to 10 or 15 pounds per square inch, and from zero to 30 in. of mercury vacuum. Gauges are also used for servicing air conditioners.

For cylinder compression testing, the gauge should have a built-in check valve, or a check valve on the plug end of a connecting hose. The purpose of the check valve is to hold the pressure after each piston stroke until it reaches a maximum. The gauge is used in a spark plug hole while the engine is cranked by the starter. The gauge should register an equal pressure rise for each compression stroke, and should reach its highest reading for each cylinder in the same number of compression strokes.

Available from a few manufacturers, including Source 19, is a hand-held compression gauge that has a built-in recording chart as well as a dial indicating pressure. A chart can be made for each cylinder and kept with the car's service record.

The common gauges have a coiled tube of thin metal that tends to straighten out when under pressure and coil up when a vacuum is applied. This coil is connected to the gauge pointer by a linkage and gear. There is usually no calibrating adjustment except a zero adjustment in some gauges.

Standard instruments are available for accurately checking gauges. For checking and calibrating pressure gauges a dead-weight system is used. For example, a 100-lb weight on a free-floating piston with an area of one square inch would exert a pressure of exactly 100 pounds per square inch. For calibrating and checking vacuum gauges, a mercury column is used. The mercury column is a U-shaped glass tube partly filled with mercury and with one open end. The vacuum is placed on the other end of the tube and the difference between the two mercury levels, in inches, indicates the inches of vacuum. Accurate gauges are also available and are called "standard" gauges; they should be handled with care and used only for calibration purposes.

7-3 Altitude Correction for Vacuum Gauges

The vacuum gauge measures the difference between the pressure inside the intake manifold and the atmospheric pressure outside. Consequently, since atmospheric pressure decreases with altitude, normal vacuum readings will decrease similarly. Generally speaking, the vacuum will read 1 in. lower for each 1000 ft above sea level. Table 7-1 shows the approximate readings in inches of vacuum for various altitudes which will be obtained on a normal engine.

Altitude Correction for Vacuum Gauges

Altitude	*Inches Vacuum*
Sea level to 1000 ft	18–22
1000 to 2000 ft	17–21
2000 to 3000 ft	16–20
3000 to 4000 ft	15–19
4000 to 5000 ft	14–18
5000 to 6000 ft	13–17

Table 7-1

7-4 Auto Air Conditioner Service Unit

All refrigeration and air conditioning servicemen are familiar with the standard charging valve assembly consisting of two gauges, two valves, and three connecting hoses. Other pieces of equipment required for service are separate units, such as evacuating pump, charging cylinder and scales, thermometers, and tachometer.

A complete self-contained auto air conditioner service unit is available from at least one manufacturer, Source 15, Model 40-171. This unit uses solenoid valves operated by push buttons, interlocked to prevent errors in correct valve operation. It also is more rapid than manually operated valves.

Figure 7-2 shows the "plumbing" hook-up of the system. Figure 7-3 shows the printed circuit board PCB of the power supply and program selector. Figure 7-4 shows the meter scale selector PCB.

In Figure 7-2, V1 to V8 are solenoid-operated valves, as shown schematically in Figure 7-3. On check valves CHV1 to CHV3 an arrow indicates the direction of flow. Also shown are systems high-pressure gauge

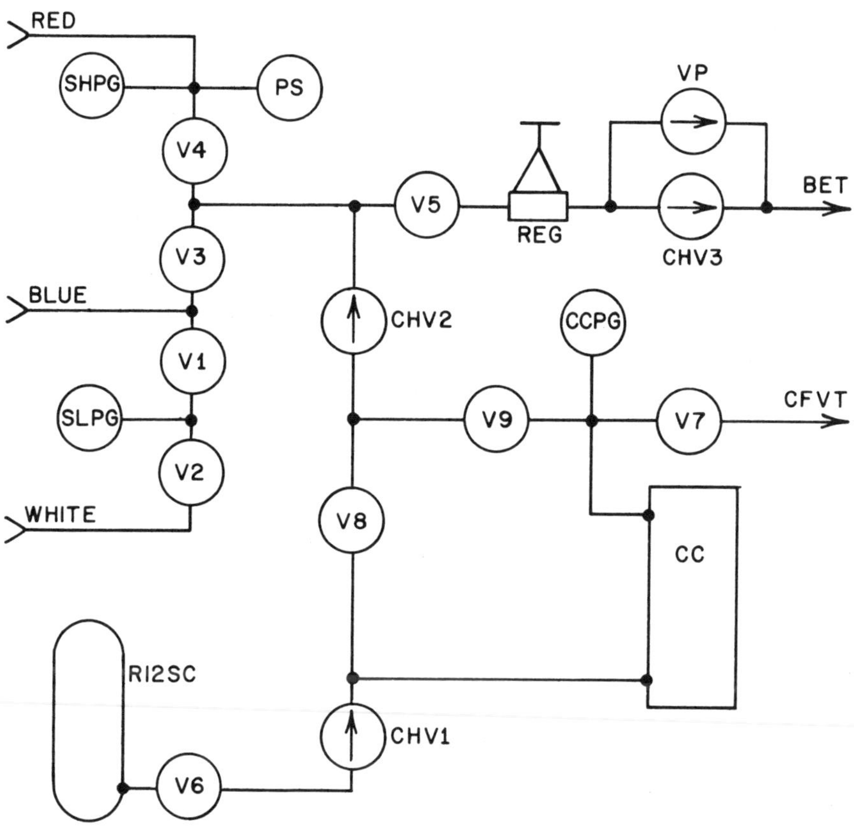

Figure 7-2 Block diagram of "plumbing," auto air conditioner tester, Source 15, Model 40-171

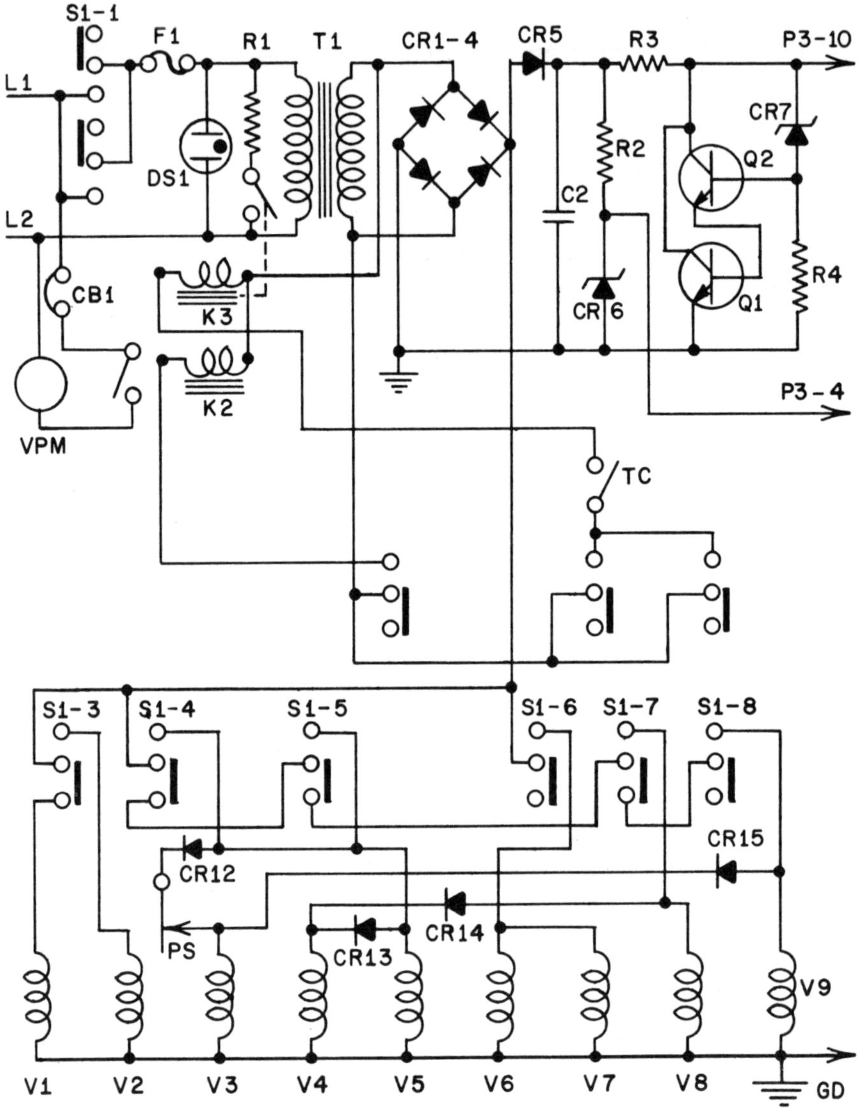

Figure 7-3 Power supply and program printed circuit board for auto air conditioner tester, Source 15, Model 40-171

SHPG, systems low-pressure gauge SLPG, charging cylinder pressure gauge CCPG, refrigerant R12 supply cylinder R12SC, charging cylinder CC located in test unit, pressure switch PS, pressure regulator REG, vacuum pump VP, systems high-pressure hose connection RED, systems

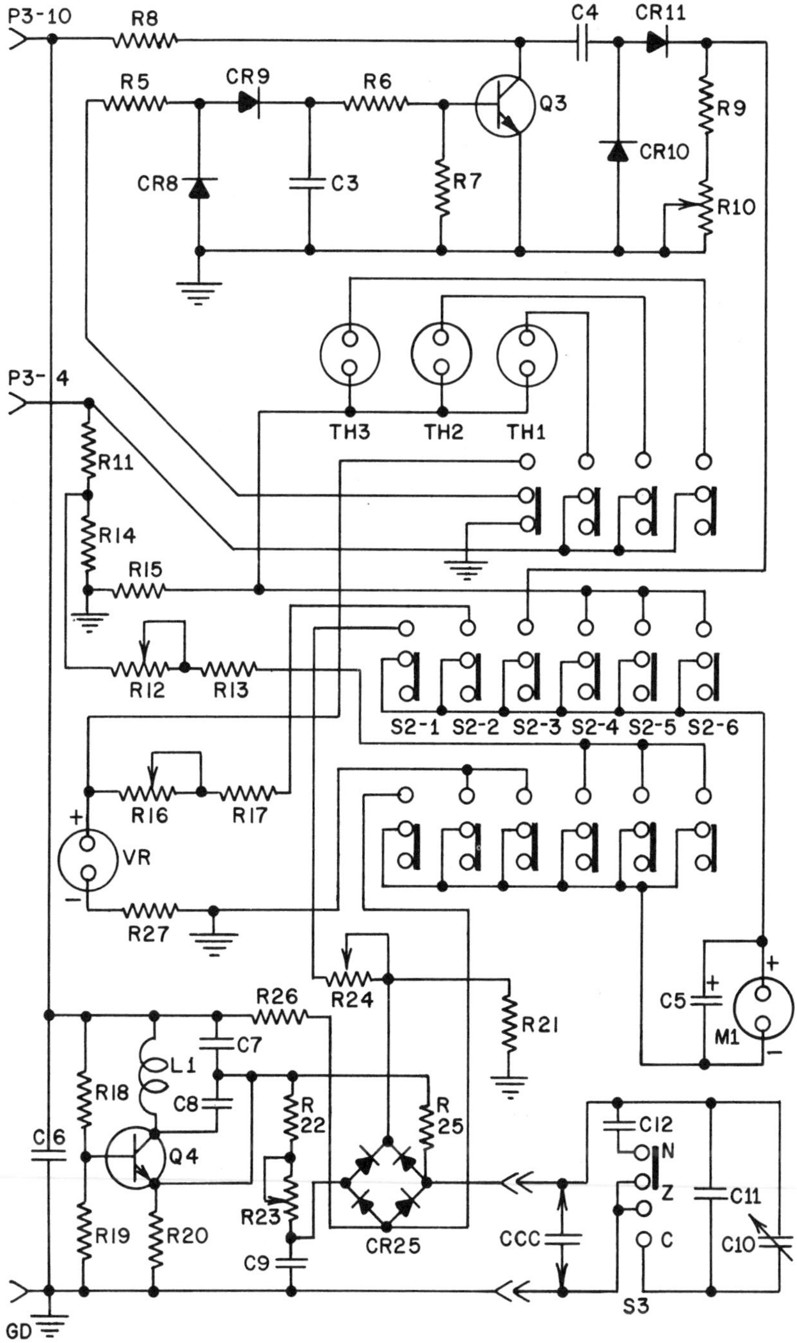

Figure 7-4 Meter scale selector printed circuit board for auto air conditioner tester, Source 15, Model 40-171

low-pressure hose connection BLUE, auxilliary low-pressure hose connection WHITE for use on low side of units having an evaporator pressure regulator (EPR), bleed and evacuator tube BET, and cylinder fill vent tube CFVT.

In Figure 7-3 the PROGRAM SELECTOR switches S1-1 to S1-8 perform the following functions.

Program Selector OFF *Switch S1-1.* This switch position removes ac power and closes all solenoid valves. A gauge reading is provided on the SYSTEMS HIGH (red hose) line in all PROGRAM SELECTOR positions. This switch does not have to be on or out to read dc voltages.

Program Selector SYSTEMS TEST *Switch S1-2.* This switch position has no electrical contacts, but it mechanically releases the OFF button to turn the unit on. Solenoid valve V1 is opened to provide a gauge reading on the SYSTEM LOW (blue hose line). The SCALE SELECTOR switch functions may be selected at any time and are not programmed by the PROGRAM SELECTOR switch.

Program Selector EPR TEST *Switch S1-3.* This is a spring-loaded momentary switch position. When the button is pressed, solenoid valve V1 is closed and V2 opened to provide a gauge reading on the AUX. LOW (white hose) line.

Program Selector SYSTEMS BLEED *Switch S1-4.* This switch position opens valves V4 and V5, allowing SYSTEM HIGH (red hose) to bleed through the regulator REG, check valve CHV3, and out the bleed tube BET located at the bottom of the unit. When the pressure on the system high (red hose) reaches 30–40 lb, the pressure switch PS contacts close to open valve V3. This allows the SYSTEM LOW side to bleed slowly so compressor oil is not lost.

Program Selector SYSTEM EVACUATE *Switch S1-5.* This button actuates the same valves as for the SYSTEM BLEED position, plus the vacuum pump relay K2. When the vacuum pump starts, check valve CHV3 closes, allowing system evacuation through the regulator and valves V5, V4, and V3.

Program Selector CYLINDER FILL *Switch S1-6.* This is a spring-loaded momentary switch position. It opens valves V6 and V7, allowing liquid refrigerant R12 from the supply cylinder to flow through check valve CHV1 into the charging cylinder CC. Valve V7 allows gas to escape from the top of charging cylinder CC as liquid enters at the bottom. The quantity of refrigerant in the charging cylinder is indicated on the

POUNDS REFRIGERANT scale of the multiscaled meter M1 when the REFRIG. SCALE SELECTOR button is depressed.

Program Selector LIQUID CHARGE *Switch S1-7.* This button opens valves V4 and V8, allowing liquid refrigerant from the bottom of the charging cylinder CC to enter V8, check valve CHV2, V4 and into the system high side. The quantity of refrigerant placed into the system is monitored on the POUNDS REFRIGERANT scale of the multiscale meter.

This switch position also energizes a thermostatically controlled relay K3 that energizes the charging cylinder heater R1 circuit. The heater raises the refrigerant temperature and pressure to shorten the LIQUID CHARGE time.

Program Selector VAPOR CHARGE *Switch S1-8.* This button opens valves V3 and V9, allowing refrigerant gas from the top of charging cylinder CC to enter V9, check valve CHV2, V3 and into the system low side. The quantity of refrigerant placed into the system is monitored on the POUNDS REFRIGERANT scale of the multiscale meter. This switch position also energizes the charging cylinder CC heater R1 to shorten the VAPOR CHARGE time.

Meter Scale Selector Switches S2-1 to S2-6. In Figure 7-4 the meter scale selector switches S2-1 to S2-6 perform the functions of selecting the four scales and circuitry for the multiscaled meter plus three different temperature readings from three probes, as follows.

REFRIG. (S2-1). This switch position actuates circuitry for measuring the liquid refrigerant in the charging cylinder CC. This circuit consists of 33-Kc Colpitts oscillator working into a rectifier circuit CR25 (A2M4B) and an RC bridge circuit consisting of C9 (470 mmfd), R22 (2.2 Kohms), R23 (2 Kohms, zero pot.), R25 (4.7 Kohms), and a variable charging cylinder CC capacitance represented as CCC. A change of liquid refrigerant level in the charging cylinder causes a corresponding change in capacitance, thus unbalancing the bridge circuit and changing the meter reading. For calibration, a factory-adjusted variable capacitor C10 (16-150 pF), and capacitor C11 (160 pF) are substituted into the circuit to represent 3 lb of refrigerant.

DC VOLTS (S2-2). This button couples dc volts at voltmeter leads through voltage receptacle VR, through VOLTS calibration pot R16 (50 Kohms), multiplier resistor R17 (180 Kohms 5%) and the multiscaled meter M1. The voltmeter scale is divided into 0.4-V increments from 0 to 20 V dc.

TEST RPM (S2-3). This button couples the primary signal at the voltmeter leads to the tachometer circuit input at resistor R5 (3 Kohms). The tachometer is calibrated by the TACH calibration pot R10 (3 Kohms) so that the center of TEST RPM blocks represent 1500 rpm for four-, six-, and eight-cylinder engines.

TEMP. PROBE 1 (S2-4). TEMP. PROBE 2 (S2-5). TEMP. PROBE 3 (S2-6). These three buttons select circuitry for measuring the temperature at probes plugged into inputs TH1, TH2, and TH3. The disc thermistor probe sensors have approximately 1000 ohms resistance at room temperature (77°F), and have a negative temperature coefficient of resistance. The circuit is a resistance bridge and is calibrated by TEMP calibration pot R12 (10 Kohms).

Check-Out Procedure

The unit shown in Figures 7-2, 7-3, and 7-4 can be checked out for proper operation without extra equipment other than a stopwatch or a timepiece with a sweep second hand and a cylinder of R12 refrigerant gas as follows.

1. Connect a tank of liquid refrigerant R12 to the supply input fitting.
2. Connect the power cord into a 120-Vac 60-Hz, three-prong U-grounded outlet.
3. Press SYSTEM TEST (S1-2) and REFRIG. (S2-1) buttons. For check-out procedure, the charging cylinder should contain 1 to 2 lb of refrigerant, as indicated on POUNDS REFRIGERANT meter. If it does not, press CYLINDER FILL (S1-6) button and watch the pounds refrigerant meter until 1 to 2 lb are indicated. Pressing the cylinder fill (S1-6) button checks operation of solenoid valve V7 by audible discharge of refrigerant vapor at bottom of stand, and V6 by upward movement of POUNDS REFRIGERANT meter.
4. Close valve on R12 supply tank. Remove charging hose between it and the input manifold. Connect same hose between SYSTEM LOW and SYSTEM HIGH fittings on the rear of the unit.
5. Depress SYSTEM EVACUATE S1-5 button. Vacuum pump should run and SYSTEM LOW PRESSURE gauge should drop to at least 26 on VACUUM scale within 20 sec by stopwatch.
6. Note VACUUM scale reading in step 5, then press SYSTEM TEST(S1-2) button. After 2 min there should be no change in the reading. This procedure checks vacuum pump operation and the unit valves and fittings for leakage.
7. Depress LIQUID CHARGE (S1-7) button. Pressure readings will be in-

dicated on SYSTEM HIGH PRESSURE and SYSTEM LOW PRESSURE gauges, checking V8 and V4 operation. Heater operation is noted by a slow increase in SYSTEM HIGH PRESSURE and CHARGING CYLINDER PRESSURE gauges. After 2 to 3 min the charging cylinder pressure gauge should level off at 110 to 125 pounds per square inch. If the system high pressure gauge rises to 200 psi or higher, it indicates a defective thermostat which must be replaced before proceeding.

8. Remove white hose from AUX. LOW fitting at rear of unit. While watching the auxiliary low fitting, press ERP TEST (S1-3) button and note momentary discharge of liquid refrigerant R12. *Caution:* Stand clear of fitting. Hold button until SYSTEM LOW PRESSURE gauge drops to zero. This checks operation of solenoid valves V2 and V1.

9. While watching sweep second hand of a clock or stopwatch, press the SYSTEM BLEED (S1-4) button. Refrigerant should be expelled from bleed hose at the bottom of the unit. The SYSTEM HIGH PRESSURE gauge and then the SYSTEM LOW PRESSURE gauge should drop to about 40 lb. Fluctuation of the system low pressure gauge in the 30- to 40-lb range is normal. At this time the pressure switch should close, energizing V3, and both gauges should drop to zero.

> NOTE: It should take 20 to 30 sec for gauges to reach zero. This checks the operation of the regulator. If it takes less than 20 sec for gauges to reach zero and the regulator has not been tampered with, replace the regulator. If it takes longer than 30 sec to reach zero, regulator may require adjustment, or V4 and/or V5 inoperable (not getting electric power, defective coil or mechanical problem with valve mechanical operation).

10. Press SYSTEM TEST (S1-2) button. Remove jumper hose between SYSTEM LOW and SYSTEM HIGH fittings on the rear of the unit. *Caution:* Be sure pressure reads zero on both gauges before disconnecting hose.

11. Reconnect jumper to input fitting attached to charge cylinder. (Straight end of hose fitting connects to tester input fitting, and angle end of hose fitting goes to the R12 supply tank).

12. With one finger on the LIQUID CHARGE (S1-7) button and one finger on the SYSTEM TEST (S1-2) button, press the liquid charge (S1-7) button until liquid refrigerant is seen coming out of SYSTEM HIGH fitting at the rear of the unit. Quickly press the system test (S1-2) button to shut off refrigerant. This checks operation of V8 and V4.

13. Hold a strip of paper over SYSTEM LOW fitting and press VAPOR CHARGE (S1-8) button. Paper should move outward, indicating release of refrigerant R12. Quickly press SYSTEM TEST (S1-2) button to shut off refrigerant. This checks correct operation of V9 and V3.

14. Press LIQUID CHARGE (S1-7) button until CHARGING CYLINDER PRESSURE gauge and POUNDS REFRIGERANT meter reaches zero. This removes refrigerant from the unit for service operations, or when shipping unit back to customer.

Component Values for Figures 7-3 and 7-4

Component values for Figures 7-3 and 7-4 not previously mentioned are as follows.

Resistors (in ohms): R2 (1000), R3 (270), R4 (1000), R6 (2200), R7 (15,000), R8 (6800), R9 (1000), R11 (4000, 1%), R13 (15,000), R14 and R15 (750, 1%), R18 (12,000), R19 (8200), R20 (1000), R21 (120,000), R24 (5000 calibrate), R26 (220,000), R27 (120).

Capacitors (in microfarads): C2 (200, 50 V), C3 (0.15), C4 (0.068), C5 (100, 6 V), C6 (0.22, 200 V), C7 and C8 (0.022), C9 (470 pF), C12 (68 pF).

Diodes: CR1-CR4 (30S1), CR5 (M100G), CR6 (1N4372A zener), CR7 (1ZM20T5 zener), CR8-CR15 (M100G), CR16-CR24 (M100G, not shown but connected across solenoid valve coils V1-V9 at reverse polarity to discharge coils when circuit is broken).

Transistors: Q1 and Q2 (2N3414), Q4 (2N3566).

Choke: L1 (25 mH).

Each of the switches S1-1 to S1-8 (Figure 7-3) and switches S2-1 to S2-6 (Figure 7-4) in vertical alignment (one above the other) are operated by the one designated push button. For example, push button S2-3 (Figure 7-4) is a 3PDT switch.

8
Generator–Alternator–Regulator Testers

8-1 General Operating Principles

To test generators and alternators for output, a test stand is required, consisting primarily of a constant speed electric motor (having provision for speed adjustment) to drive the generator or alternator, a clamp to hold the generator or alternator, load batteries, load resistors, carbon pile rheostat, ammeter, field rheostat, voltmeter, and tachometer. Some may have a torque arm to test the output of starters with a scale to indicate the foot-pounds of torque.

The adjustable speed motor drive, up to 3 hp, should be reversible. The speed control may consist of a three- or four-step pulley on the motor connected by V-belt to the pulley on the generator or alternator. The more elaborate speed controls use a two-speed motor with a pair of adjustable V-pulleys that give a wide range of constant speeds. It is connected to the generator or alternator by means of a flexible coupling with a variety of adapters to drive the generator or alternator from the pulley end, with or without the pulley mounted.

Those testers that have provision for testing the torque of starting motors may consist of a simple torque arm clamped to the starter drive and a spring scale to read the foot-pounds torque. The more elaborate starter torque testers may have a back-geared motor, turning at about 1 or 2 rpm, mounted on the torque arm so that it will permit the starter shaft to turn slowly so that the torque variations can be read as the starter armature commutates over each bar. The pound scale may be an electrical strain gauge having an electric meter calibrated in pounds or foot-pounds. The meter may be indicating or recording or both.

8-2 Typical Motor-Driven Generator-Alternator-Regulator Tester

The electrical circuit of a typical Source 1, Model E241, "electrical laboratory," generator-alternator-regulator tester is shown in Figure 8-1.

The tachometer section consists of a rotating magnet M which is mounted on the motor-driven drive shaft and it induces a voltage in tachometer coil TC, which is indicated on the tachometer T calibrated in rpm. This circuit is described in detail in Chapter 12, Dwell Tachometers.

The ammeter A is a millivoltmeter connected across the shunts SH. It has a zero center scale so that the polarity of the connections can be ignored. Also, if ammeter A is plugged into a circuit that may be on charge or discharge, the meter will automatically reverse to the other side of zero. Shunt SH is in the form of a round, tapered plug P1 with a heavy lead terminating in a clip T3, with two small leads tapped off shunt SH. These two small leads are connected to the ammeter through plug P14 and jack J14.

Shunts are available to give a full-scale meter reading on 10, 50, 100, 500, and 1000 amp. Each shunt assembly is complete with plug P1, shunt SH, clip T3 (lug terminal on 500- and 1000-amp shunts) and plug P14. Plug P1 is plugged into any one of the jacks J1, J2, J3, J4, or J5.

Jacks J1 to J5 are connected as shown, with jack J2 connected to the negative battery BAT and jack J4 connected to the positive battery BAT through battery selector switch SW3 to select 6 V, position 1, 12 V, position 2, and 24 V, position 3. Position 4 is a spare to connect additional batteries to obtain any desired voltage, such as 32, 36, or 48 V. The battery negative is permanently connected to fixed load FL resistor R12. The battery positive is permanently connected to variable load VL resistor R13, which is a heavy duty carbon pile rheostat.

There is a connecting link on the ground GD jack J3 that can be swung to connect either the positive or the negative battery to ground. Ground is connected to the frame of the machine to make a connection to the clamping device that holds generator, alternator, and regulator. There is a connecting link on the jack J1 that can be connected to the battery negative to place the variable load R13 across the battery. A connecting link on jack J5 can be connected to the battery positive to place the fixed-load resistor R12 across the battery.

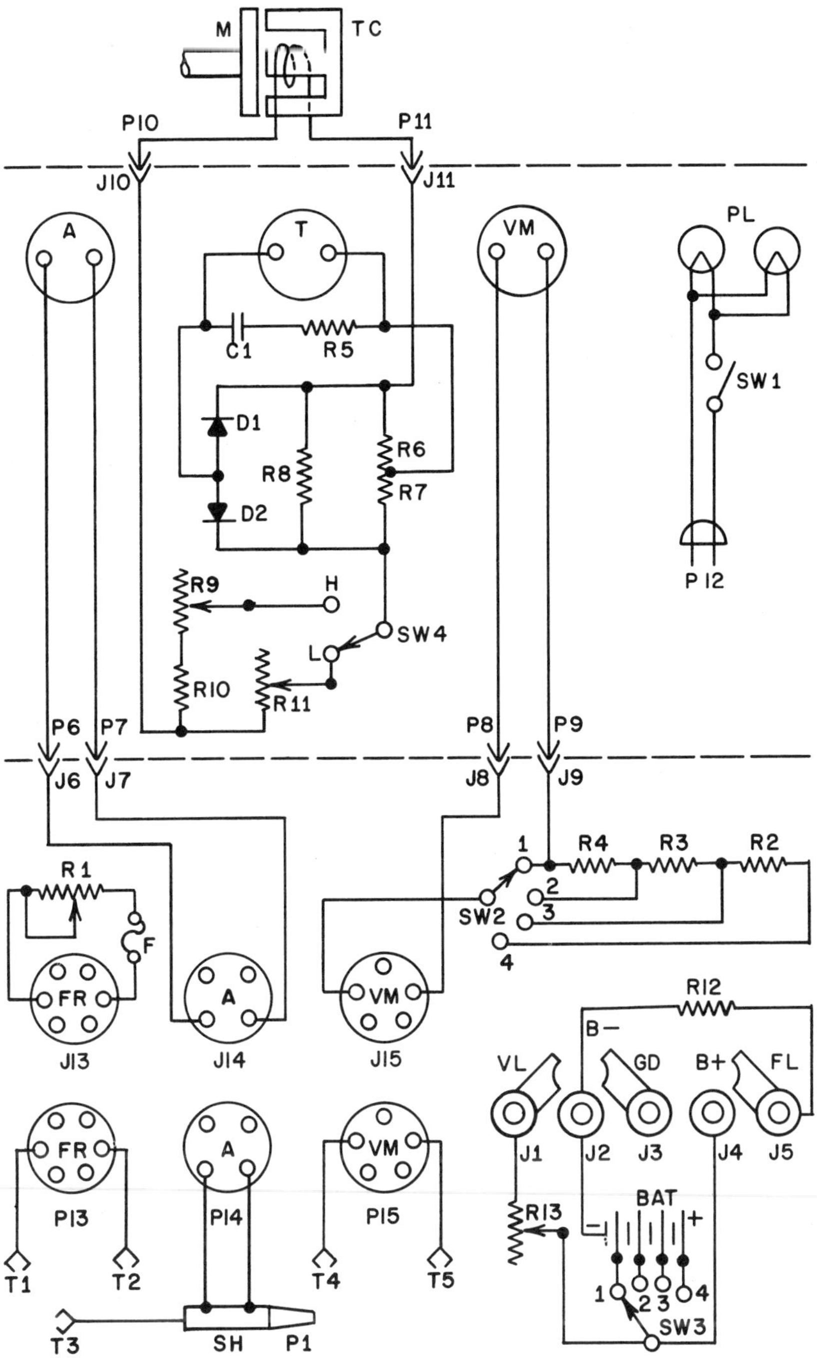

Figure 8-1 Generator-alternator-regulator tester, Source 1, Model E241

After selecting the correct shunt for the current involved, plug P1 can be inserted into any one of the jacks J1 to J5 and read the current. (Plug P14 must be plugged into jack J14.)

Resistors R12 and R13 can also be connected in series with a starting motor or other load. When the carbon pile rheostat R13 is placed in series with the generator, or alternator output and battery, it permits increasing or decreasing the voltage of the generator or alternator above the battery voltage for adjusting the voltage regulator. When R13 is connected across the battery, the battery voltage can be reduced when desired by adjusting the voltage regulator.

A 30-ohm field rheostat FR (R1) terminates at jack J13 through a 5-amp fuse F. When the field rheostat is used, plug P13, having two leads terminating in clips T1 and T2, must be plugged into jack J13. The field rheostat is connected by clips T1 and T2 in series with the generator or alternator field and battery to adjust the current output.

The dc voltmeter VM section has a jack J15, and when reading voltage it is necessary to plug in plug P15, consisting of two small leads terminating in clips T4 and T5. Switch SW2 selects the desired meter range. Position 1 is direct reading to the voltmeter calibrated 0–10 V, position 2 inserts series resistor R4 (1000 ohms) to read 0–20 V, position 3 inserts an additional series resistor R3 (3000 ohms) to read 0–50 V, and position 4 inserts an additional resistor R2 (5000 ohms) to read 0–100 V.

There are two panel lights PL that are controlled by switch SW1 when plug P12 is plugged into 115 V ac.

Voltage regulators can also be tested and adjusted on this machine. The regulator is clamped to a grounded plate that is adjustable to be in the same angle as when mounted on the vehicle, and connections are made to the generator or alternator. Speeds, voltages, and currents can be read and adjusted with complete flexibility under all conditions.

Normally, the batteries will not be kept in an equal state of charge from testing. The batteries should, therefore, be serviced and tested with a hydrometer at least once a week. They should be placed on an "equalize" charge for several hours at a rate of about 5 to 10 amp to bring up the weakest cells to full charge without undue gassing of the other cells. The charger may be built into the machine, or an auxilliary charger can be used.

8-3 Carbon Pile Rheostats

Carbon pile rheostats are universally used in battery load testers and testers for generators, alternators, regulators, and starters.

Carbon pile rheostats are so called because they consist of a pile of carbon discs that are compressed together by a knob-operated screw clamp. As the pressure is increased, the resistance decreases, and as the pressure is eased off, the resistance increases.

Carbon pile rheostats are used for heavy currents at relatively low resistance values and are economical in cost and space for the same capacity. Normally they are used for intermittent duty, but they can be

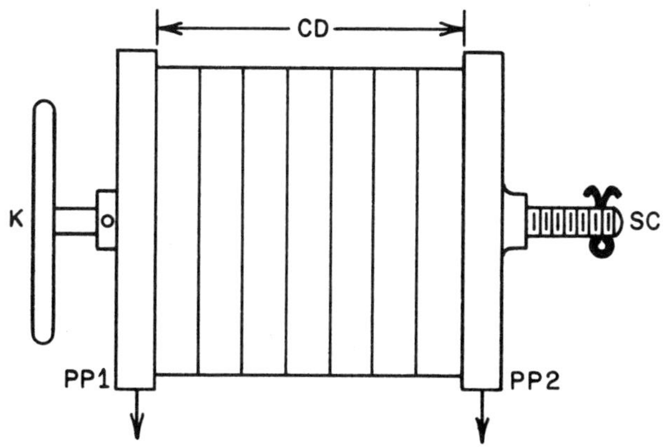

Figure 8-2 Carbon pile rheostat construction

designed for continuous duty by providing for the dissipation of heat through fins and forced air cooling. They also have the advantage of giving a stepless, or smooth, change in the resistance over the entire range.

Figure 8-2 shows the usual construction for a single pile. The carbon discs CD, or washers, have a hole in the center for screw SC, which is turned by knob K. The pressure plates PP1 and PP2 are the electrical connections to the rheostat. As knob K is turned, it rotates screw SC in the threaded pressure plate PP2, thereby compressing the pile of carbon discs CD to change the resistance.

The larger-capacity rheostats may have two or more of these piles side by side, with one knob K and screw SC operating the pressure plates in unison. The electrical connections may be connected so that the separate piles are in series for a higher resistance, or in parallel to carry a higher current at lower resistance.

These carbon pile rheostats are easily serviced in the field. They can be disassembled for resurfacing or replacing any of the parts, which are available from most sources. The carbon discs and pressure plates can be resurfaced, unless they are very badly pitted or burned. Use a perfectly flat surface, such as a surface plate or a piece of *plate* glass, and fine sandpaper. With a rotary motion, sand each surface flat and clean, and wipe off before reassembling. Any parts that will not respond to this treatment should be replaced.

The carbon pile rheostat can be damaged if the screw is turned tighter than necessary, it will never reach zero ohms no matter how tight. If the maximum wattage capacity is exceeded, the unit may be damaged. If a high voltage is applied with little pressure on the plates, the carbon discs will arc badly and cause poor contact, or the discs may break when compressed again.

Frequently the carbon discs have alternate plates or discs of different diameters to secure greater radiating surface for cooling. These should be replaced in the same arrangement.

Carbon Pile Rheostat Design Data

The following design data can be used to design special rheostats, or the figures can be used to estimate the rating of an existing one.

Carbon plates, either circular or square, varying in thickness from 1/8 to 3/8 in., are satisfactory. Since practically all of the resistance is in the contacts between successive plates, the plate surface must be ground very flat and smooth. The thicker the plates, the stronger they are; also, they will have a greater heat-radiating surface because most of the heat is generated at the point of plate-to-plate contact.

The current-carrying capacity for such rheostats varies from 5 to 12 amp per square inch, with an average of 7 amp per square inch.

For temperature rises ranging from 40 to 85°C, the radiating coefficient is from 0.002 to 0.005 W per square inch of radiating surface per degree Centigrade rise.

For pressures ranging from 7.5 to 180 pounds per square inch, the variation in resistance per square inch per contact, based on average results, is shown in Table 8-1, where the pressure is in pounds per square inch, and the resistance is in ohms per square inch *per contact.*

Resistance of Carbon Pile Rheostats for Various Pressures

Pressure	*Resistance*	*Pressure*	*Resistance*
7.5	0.023	50	0.0047
10.0	0.0185	60	0.0041
12.5	0.0145	70	0.0036
15.0	0.0125	80	0.0032
20.0	0.0100	90	0.0029
25.0	0.0082	100	0.0027
30.0	0.0071	120	0.0024
35	0.0063	140	0.0021
40	0.0057	160	0.0018
45	0.0052	180	0.0015

Table 8-1

8-4 Meterless Alternator-Regulator Tester

Figure 8-3 shows a typical Source 9 alternator system with a plug-in harness jack J1A (alternator) that normally is plugged into the voltage regulator plug P1R (regulator), as shown in Figure 8-4. Figure 8-5 shows a special plug-in type tester, manufactured by Source 1, using only indicator bulbs and no meters, for quickly testing these plug-in type Source 9 systems.

In Figure 8-3 the battery BAT is the vehicle battery, IS is the ignition switch, AT is the accessory terminal, CIL is the charge indicator light shunted by resistor R1 (15 ohms), which goes to IA, the vehicle ignition connection to jack J1A.

The ALT in dotted lines represents the alternator, which has a three-phase stator STC, rotating field coil FC, and six diodes D1 to D6. The alternator output A goes to terminal AA of jack J1A, the neutral N of the star-connected stator coil STC goes to terminal SA of jack J1A, and the field F terminal goes to terminal FA of jack J1A.

In Figure 8-4 the voltage regulator plug P1R has the same spacing and terminal designations to fit the alternator jack J1A. FR is the field relay that shorts out the charge indicator lamp CIL (Figure 8-3) when the voltage across the field relay coil FR is sufficient to close the con-

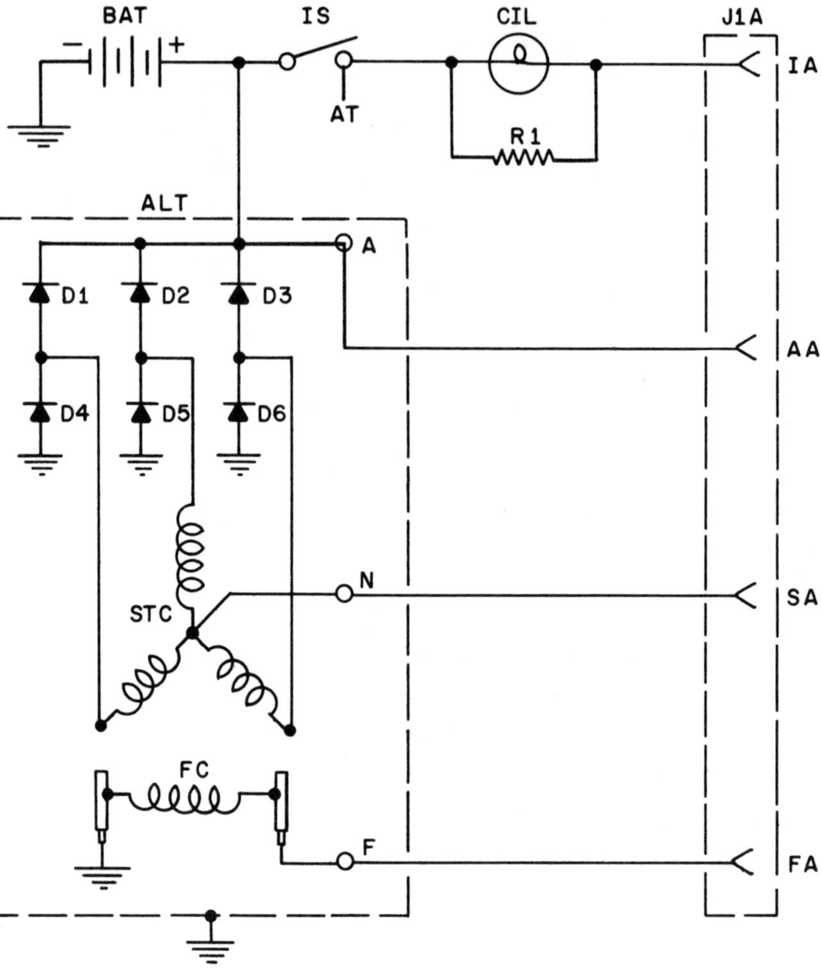

Figure 8-3 Alternator system circuit, Source 9

tacts. The field relay derives its voltage from the neutral terminal of the alternator through terminal S (SR, SA, and N) and is one-half the battery voltage. This connection amounts to the unused center tap of the secondary of a single-phase bridge rectifier circuit.

The voltage regulator relay VR has the upper contacts and the common contact normally closed to give full voltage across the field terminal F. When the voltage gets high enough, these contacts open, in-

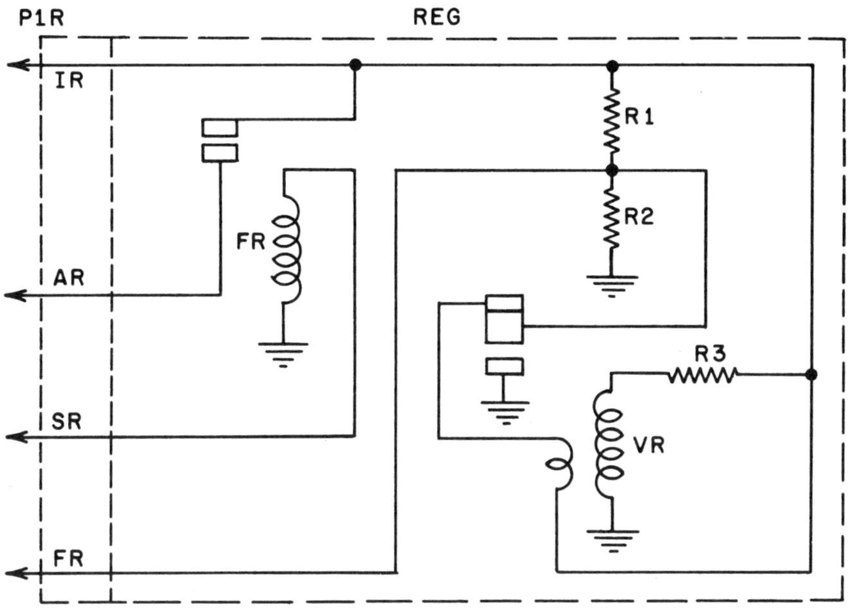

Figure 8-4 Alternator regulator circuit, Source 9

serting a resistor R1 (10.5 ohms) in series with the field. If the voltage becomes too high, the movable contact makes connection with the bottom contact, which is grounded, thus removing all voltage from the field. Other values are, resistors R2 (50 ohms) and R3 (16 ohms). At the voltage regulation point, the points will vibrate rapidly. On the tester this may show up as a flickering of the red REG bulb.

Figure 8-5 shows the plug-in tester typical of Source 1, Model RE20-22.

To use the tester, the jack is pulled out of the regulator and the tester is inserted between the alternator and regulator. Plug P1T is plugged into jack J1A, and the plug on the regulator P1R fits into jack J1T of the tester.

The following operating instructions and test procedure, with results, are printed on the back of the tester case.

Operating Instructions

1. Connect tester to voltage regulator and harness.
2. Connect clip to regulator mounting screw (ground).

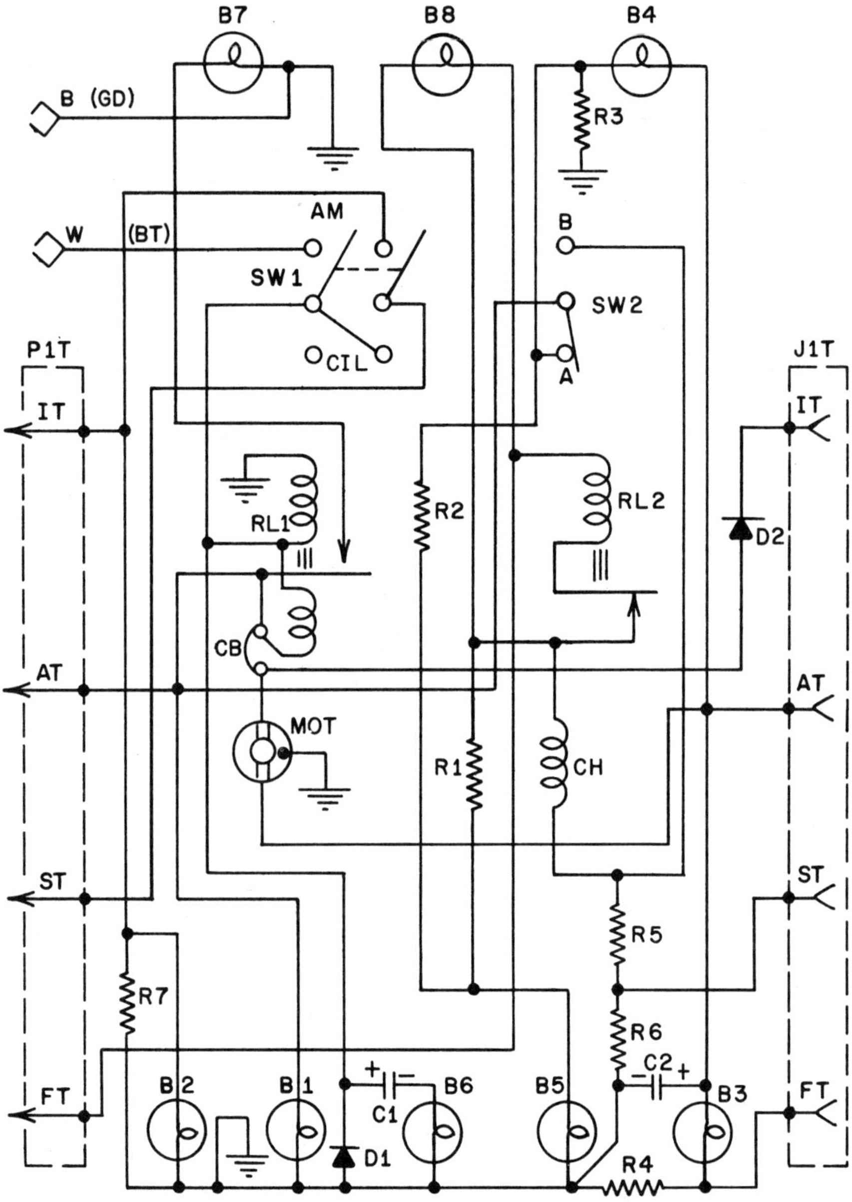

Figure 8-5 Meterless alternator-regulator tester, Source 1, Model RE20-22

3. In all indicating lamp systems (idiot lights) and all 1963 systems (Source 9), both ammeter and indicating lamp, set circuit selector switch to "indicator lamp" position. On 1964 and later model ammeter systems, set the circuit selector switch to AMMETER position, and connect the stator test lead (bullet connector) to the stator wire bullet connector on the regulator end of alternator harness (white with black tracer).

> NOTE: On models not using stator wire, connect stator test lead with adapter (included with tester) to stator terminal on alternator.

4. Start engine and adjust speed to 1000–1500 rpm (fast idle) for all tests. Charge indicator lamp should be on. In the following Test A and Test B, bulbs B1 to B8 have been added to indicate the reference numbers as shown in Figure 8-5.

Test A

Bulb	*Color*	*Indicator*	*Test Indication*
B1	Green	BAT	On—Circuit OK
			Off—Open circuit
B2	Green	IGN	On—Normal, circuit OK
			Off or dim—Open circuit
			On bright—Shorted circuit
B3	Green	REG	Off—Regulator defective
B4	Red	REG	On—Regulator defective

Test B

Bulb	*Color*	*Indicator*	*Test Indication*
B5	Green	FIELD	On—Circuit OK
			Off—Open circuit
B6	Green	STATOR	On—Alternator operating
			Off—Defective circuit or alternator or loose belt
B3	Green	REG	Off—Defective regulator
B7	Red	DIODE	On—Open or shorted diode
B8	Red	SHORT	On—Field shorted or grounded
B3	Red	REG	On—Defective regulator

In Figure 8-5, tester plug PIT and tester jack JIT have the same spacing and nomenclature as in Figure 8-3 and 8-4.

The labeling and color of bulbs B1 and B8 are shown above under Tests A and B. Bulbs B2 and B6 are No. 47 lamps, and all the others are No. 1815 lamps.

Switch SW1 is a DPDT slide switch to select for systems using ammeters, position AM, or for systems using charge indicator lamps position CIL.

There are two clips on leads from the tester. A black lead B (GD) is the ground lead, and it is connected to the regulator-mounting screw. The other clip W (BT) has a bullet connection splicer BC and is referred to in the Operating Instructions above.

Switch SW2 is a SPDT slide switch with a normally closed position A for Test A and a momentary contact position B for Test B.

Relay RL1 (200 ohms, 100 ohms each opposing coil) operates to light bulb B7 (Red–DIODE) when the voltage across the alternator neutral N and terminal ST is higher or lower than normal, as indicated by an open or shorted diode. This is a differential relay that closes on voltages below 5.5 V or above 6.5 V to light bulb B7 (with a 12-V system).

Bulb B6 (Green-STATOR) is also in circuit with relay RL1 coil, having a diode D1 and capacitor C1 (25 mfd) connected as shown. Bulb B6 will light only when there is a component (rectified ac) coming from the alternator to indicate stator condition.

A 12-V dc motor MOT is connected, through a circuit breaker CB, between the alternator A terminal and regulator A terminal. This motor has a very low resistance when not running, and it allows current to pass and operate the relays in the voltage regulator. However, if there is a short or a low resistance in series, the motor runs, drawing low current. This motor is not used to drive anything in the tester.

Relay RL2 has a very low-resistance coil of only a few turns of heavy wire, and it is connected to act as a vibrator on Test B to indicate a short or ground in the alternator field circuit, lighting lamp B8. This relay sparks violently on a low-resistance short; switch SW2 should, therefore, be released from position B immediately if the SHORT lamp lights to avoid burning the contacts.

Other component values are, resistors (in ohms), R1 (350), R2 (247), R3 (1.0), R4 (10), R5 (18), R6 (22), R7 (15); capacitors, C1 and C2 (25 mfd 25 V); CH is an RF choke and resistor, wrapped around resistor R1.

Diode D2 is for blocking purposes, allowing positive direct current to enter the voltage regulator at terminal I, but not allowing dc to flow from other sources in the voltage regulator back through terminal I to enter the tester circuits at this point.

8-5 Shop Checker and Calibrator for Meterless Alternator-Regulator Tester

To service, adjust, and calibrate these meterless alternator-regulator testers shown in Figure 8-5, it is not always practical to have on hand an automobile equipped with the plug-in alternator and regulator system.

Figure 8-6 shows schematically the author's design of an alternator-regulator simulator that can be built from readily available material. This is connected and used in the same way as the alternator system shown in Figure 8-3.

The charge indicator light CIL consists of a No. 53 bulb B1 and a shunting resistor R5 (15 ohms).

Resistors R2 and R3 (50 ohms, 5 W) hold the average dc volts across the battery to one-half of that value which is applied to terminal S. Potentiometer R1 (100 ohms, 10 W) is connected as shown, with a dc voltmeter VM to vary the voltage at terminal S for calibrating and checking the differential relay RL1 (Figure 8-5). The red DIODE light on the tester should not light when this voltage at S is one-half the battery voltage, or for 0.5 V either side of the midvoltage. For example, if the voltage across the battery is 14 V, the red DIODE light should not light at any voltage between 6.5 and 7.5 V, but should light at voltages below or above these limits. The differential relay RL1 has two coils wound in reverse direction, so that when the voltage across each coil is the same, there is no magnetic pull to the armature, but when the voltage across one is below 6.5 V and across the other is above 7.5 V, the armature should move and close the contacts to light and red DIODE bulb.

To simulate an off balance condition, such as an open or shorted diode in the alternator, switch SW3 places zero voltage across one coil of relay RL1 and full battery voltage across the other coil to insure closing the contacts to light the red DIODE bulb.

The four switches SW1 to SW4 are labeled to show the handle positions as they appear on the actual tester, but the contact position is in the opposite direction to the handle position. For example, switch SW1 is shown in the (ON) position, SW2 (ON), SW3 (OK), and SW4 (OK).

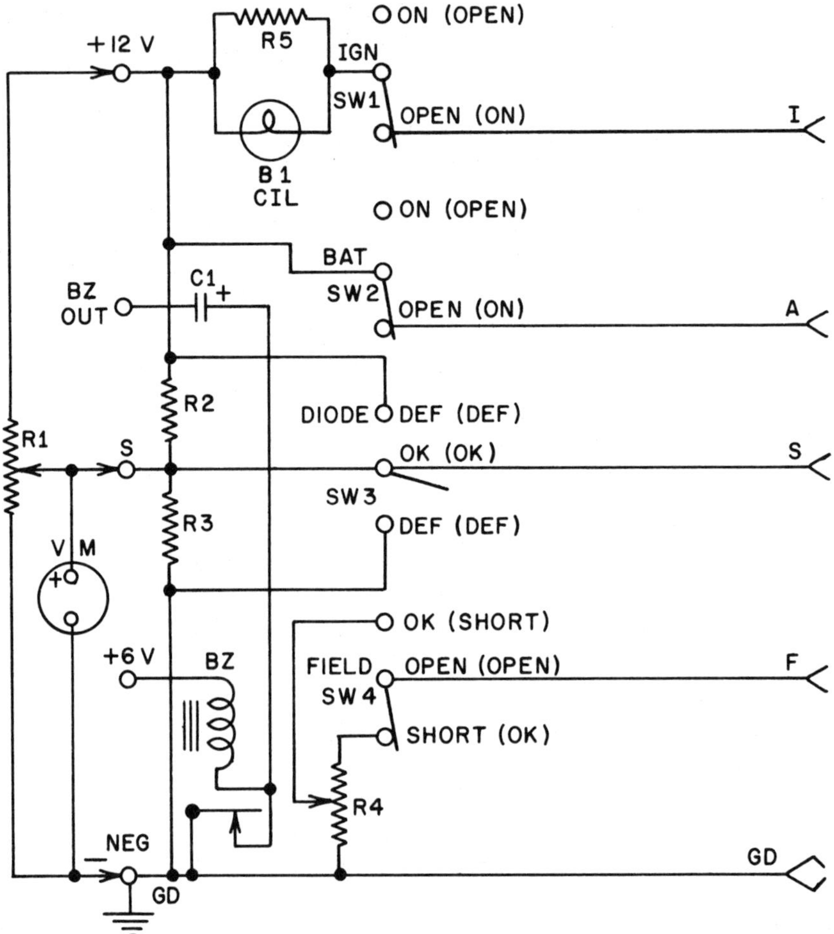

Figure 8-6 Alternator simulator and calibrator for meterless alternator-regulator tester

Switch SW1 (SPST) is the ignition IGN test switch to simulate a normal ON position and an OPEN circuit position.

Switch SW2 (SPST) is the battery BAT test switch to simulate a normal ON position and an OPEN circuit position.

Switch SW3 (SPDT with OFF position) is the DIODE test switch to simulate a normal OK middle position and two defective DEF positions. One defective position would indicate a shorted diode; the other would indicate an open diode.

Switch SW4 (SPDT with an OFF position) is the FIELD test switch to simulate a normal OK position, an OPEN position, and a SHORT (shorted or grounded) field position. Resistor R4 is a 4- ohm, 3-amp resistor that can be made from a coil of nichrome wire mounted on an asbestos board, and it simulates the alternator field, which does not draw enough current to operate relay RL2 (Figure 8-5) to light the red SHORT lamp. To simulate a shorted field, a clip is attached to R4 at a point that will leave about 2 or 3 ohms when switch SW4 is moved to the SHORT position. This draws enough current to operate relay (buzzer) RL2 and light the red SHORT lamp. Remember, the switches make contact to the opposite position of the handle label shown in Figure 8-6. The switch contact positions are shown in parentheses( ).

To simulate the stator test, it is necessary to apply a pulsating voltage to terminal S of relatively high frequency as is normally obtained from the N terminal of the alternator. The buzzer BZ is a 6-Volt, high-frequency vibrator that feeds pulses to terminal BZ-OUT through capacitor C1 (25 mfd, 25 V). This is fed directly to terminal S when the circuit selector switch of the tester is at the INDICATOR LAMP position.

With the circuit selector switch of the tester at the AMMETER position, terminal BZ is connected to the bullet connector of the tester (or to the white-with-black tracer wire with a clip-adapter) instead of directly to S. This checks out the continuity of this circuit. In either of these connections the green stator light should light.

The positive 6 V should be supplied from the midterminal of the 12-Volt battery. On hard-top batteries, this 6-V tap can be obtained by drilling or cutting through the insulation and installing a screw in the lead strap between cells 3 and 4.

To complete the simulator, a standard Source 9 voltage regulator known to be in good condition and adjustment, is used. The cover is removed so that defects, such as opening and closing contacts, can be simulated.

A battery charger should be connected across the 12-V battery in order to raise the voltage to simulate the alternator when it is running and charging. The voltage must be built up sufficiently to operate the voltage regulator relay. If it is not, it will show up as a defective regulator on the tester. If the voltage regulator is operating just at the regulating point, it will cause the red REG light to flicker.

A common ground circuit must exist between all units, battery negative, simulator, tester, and regulator.

8-6 Construction and Use of Armature growler

The armatures and field coils of automotive starters and generators can be tested for shots on a growler.

A growler consists of a laminated iron core C, Figure 8-7a and b, and a winding GW, which is connected to the 115-V ac power line. It should never be plugged in or turned on without armature A or flux bar FB in position, and even then it should not be used for more than a few minutes at a time, since it will overheat.

Figure 8-7a shows a growler with an armature A having a winding coil AW in position. As the flux path FP is generated by the growler coil GW, it goes through the armature, and as the flux reverses 60 times a second it generates a voltage in the armature coils. A hacksaw blade B placed on the armature slots as shown will vibrate only if the armature coil winding AW is shorted, or if there is short between two adjacent commutator bars. If the armature is shorted, there will be at least two slots which indicates vibration of the hacksaw blade. If only one coil is shorted, these two slots indicate exactly which one.

8-7 Testing Field Coils on a Growler

Figure 8-7b shows how an armature growler can be used for testing field coils for shorted turns. Prepare a flux bar FB to lay on the growler. It should make good contact with the growler and cover as

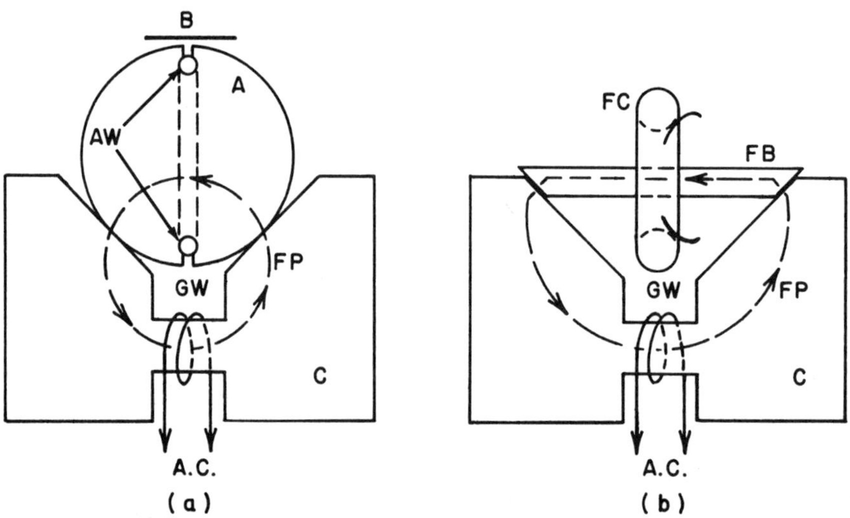

Figure 8-7 Armature growler for testing armatures and field coils

much area as possible. To test a field coil for shorts, insert the flux bar FB through the field coil and place it on the growler. The field coil should have enough clearance so that it does not touch the growler or flux bar. When the field coil is held lightly between two fingers, it will vibrate only if it is shorted. Be sure the two leads are not touching. Connecting the two leads together will cause it to vibrate since the whole coil is shorted.

8-8 Growler and Grounds Tester

Open circuits and grounds are checked on growlers equipped with a continuity tester using either a meter or a lamp bulb. One type is shown in Figure 8-8, where switch SW1 is a SPDT with an OFF position, GW is the growler coil, T1 is a 1:1 ratio isolation transformer with a

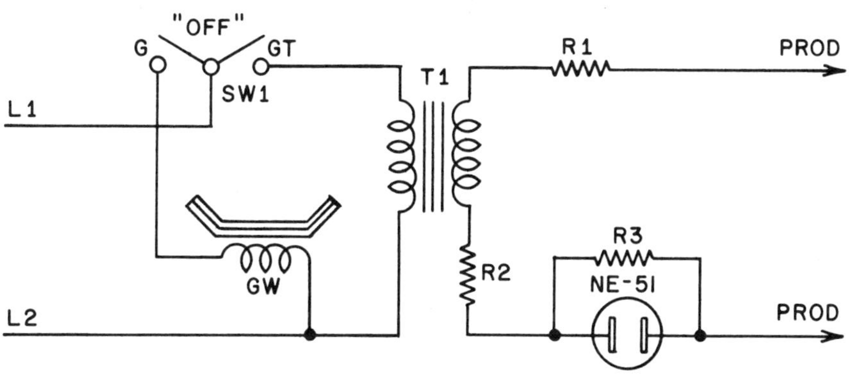

Figure 8-8 Armature growler and grounds tester

120-V secondary, and resistors R1, R2 (10 Kohms), and R3 (15 Kohms) are current-limiting resistors. NE-51 is a small bayonet base neon lamp that lights in case of a ground or high leakage across the prods. Switch SW1 connects only the growler on position G and only the ground test on position GT, and it is OFF at midposition.

To test for grounds or high leakage, connect one Prod to the commutator of the armature and the other to the armature shaft or laminations. The lamp NE-51 should not glow if the insulation is within safe limits.

8-9 Ammeter Types

Direct current ammeters are used as a service tool to measure engine starter current, generator or alternator charging current, and load currents.

A standard ammeter must be placed in series with the circuit to measure amperes, which requires opening the circuit and reconnecting it to include the meter.

Special induction ammeters have been used that read the current when held next to the cable. The cable is fitted into a U-shaped trough attached to the back of the meter. A small magnet inside the meter is attached to the meter pointer and is held at zero by a hairspring. As current flows through the external cable, the magnetic flux created tends to move the magnet at right angles to the cable against the hairspring. The greater the current, the greater the movement. The dial is calibrated in amperes. This type is neither very accurate nor dependable, because any magnetic material, such as an engine block or a steel frame, tends to deflect the magnet and give an erroneous reading. They are used for dashboard ampere indicators, however, because they are calibrated to read consistently even in the presence of adjacent magnetic materials that remain in a fixed position.

8-10 Clamp-Around DC Ammeter Using Hall Effect

Clamp-around ammeters for reading alternating current have been in use for a long time, but not until recently has there been available a clamp-around ammeter for reading direct current. The clamp-around ac ammeters work on the transformer principle and are not affected by adjacent magnetic material, but they cannot measure direct current.

Now it is possible to have a clamp-around dc ammeter that is accurate and not influenced by adjacent magnetic materials or fields, because of a relatively recent development in the field of electronics, called the Hall effect.

At least one manufacturer, Source 15, uses this in a combination alternator-regulator battery starter tester, Model 42-148. The ammeter section is described here and the rest of the unit is described in Section 6-8.

The principle of operation of this ammeter is shown in Figure 8-9a and b. The heart of the Hall effect ammeter is a linear semiconductor device called a Hall generator (HG1). It is essentially a solid state device that provides a voltage output Vh at terminals R3 and Y2 proportional to the current IC that passes through terminals G1 and W4 at right angles, and to the magnetic field F that passes through the Hall generator HG1 at right angles to the plane of HG1. In this particular application, the current IC is maintained at a constant value; therefore the voltage output Vh is directly proportional to the current through the current-carrying conductor L. The current through conductor L produces a magnetic flux F in the iron core C in proportion to the current flowing through it. Core C is made in the form of jaws that open and close around conductor L. The voltage Vh is connected to a voltmeter calibrated in amperes.

The three-dimensional vector diagram of Figure 13-lb is the same as

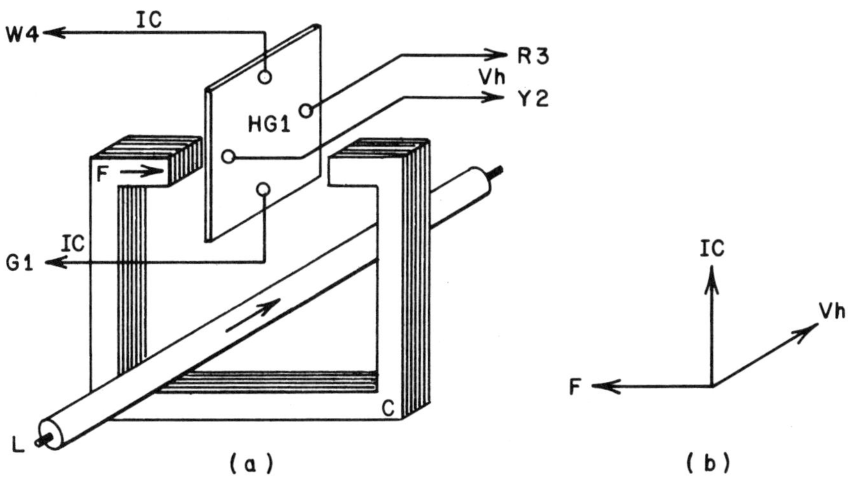

Figure 8-9 Hall effect ammeter generator

the old familiar right- and left-hand rule for motors and generators, except that there is no motion of the conductor through a flux path. Instead, the electrons produce a voltage Vh and are developed in the Hall generator HG1 at terminals R3 and Y2.

This is only one of the many possible applications for the Hall effect ammeter. Anyone interested in learning more about this device and its characteristics can obtain a Hall effect kit from an electronic supply house, such as Source 2. This kit contains the Hall effect generator, transistors, magnets, and other essential components with a book of instructions.

8-11 Typical Starter Tester Using Hall Effect Ammeter

The Hall effect ammeter section of Source 15's Model 42-148 tester is shown schematically in Figure 8-10a. The ampere pickup PU clamp-around assembly and cable plug are shown in Figure 8-10b clamped around current-carrying cable CA.

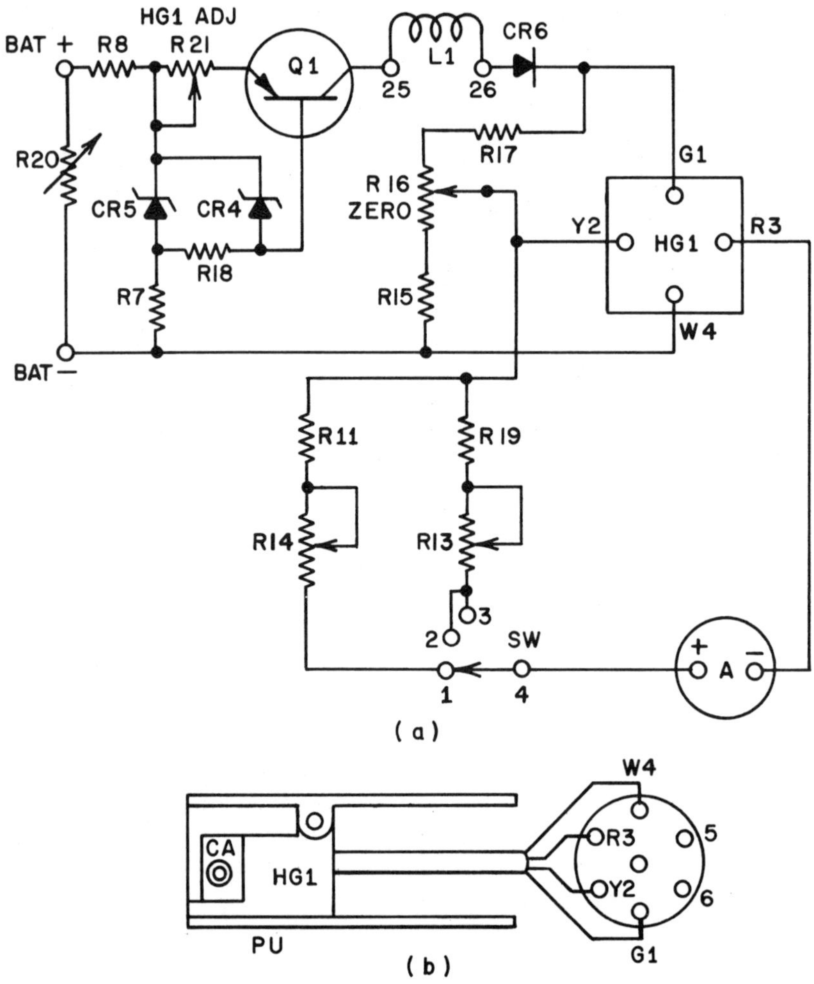

Figure 8-10 Hall effect ammeter circuit section of tester, Source 15, Model 42-148

The constant current for the Hall effect generator HG1 is regulated by transistor Q1 and associated circuit components R7, R8, R18, R21, CR4, and CR5. Power for operation is supplied from the 12-V battery positive and negative. Current for the Hall generator is set at 112 ma through calibrating coil L1 by adjusting R21 (HG1 ADJ) potentiometer. Zener diode CR4 holds Q1 base 3-V negative with respect to the positive battery clip. A decrease in battery voltage will tend to decrease the current through R21 and R8—effectively changing the bias on Q1, which, in turn, changes resistance to maintain constant collector current. The net result is that for every change of current through R21 and R8 there is an equal and opposite change in the transistor Q1 junction resistance. R20 is a 500-amp carbon pile rheostat for battery load testing and for calibrating the ammeter. The scale of ammeter A is a zero center dial with a 0–100 amp and a 0–500 amp range on each side of zero, so it is not necessary to observe polarity. The clamp pickup PU unit may be clamped around either cable or reversed.

To Calibrate the Ammeter

The electrical components, such as resistors, diodes, transistor, and calibrating potentiometers, are mounted on a printed circuit board accessible through the top cover. If service is required on any component, the manufacturer recommends that the entire printed circuit board be replaced. Mounted on the inside of the cover is the ampere calibration coil L1 which has a dc resistance of 11 ohms, and is plugged into the circuit board at terminals 25 and 26.

The actual tester uses several push button switches for the various functions, but for the ammeter section, switch SW is shown in simplified form. On position 1 of SW, ammeter A is connected for 0–500 amp through resistors R11 and R14. On positions 2 and 3 ammeter A is connected for 0–100 amp through resistors R13 and R19.

With a fully charged 12-V battery connected to the battery positive and negative, and with load rheostat R20 turned counterclockwise to remove any load, proceed to calibrate ammeter A as follows:

1. With the amperes pickup PU not clamped to any cable, adjust R16 (ZERO) until ammeter A is in the center of the scale at zero.

2. Remove the amperes calibration coil L1 plug at the printed circuit board. Connect a standard milliammeter to the exposed pins, terminal

25 positive and 26 negative. Turn R21 (HG1 ADJ) potentiometer for a reading of 112 ma. Remove the milliammeter and reconnect the coil plug.

3. Clamp the amperes pickup PU around the amperes calibration coil L1.

4. With switch SW in position 2 or 3, adjust R13 for a reading of 50 on the 0–100 amp scale.

5. Connect a standard ammeter in series with one of the battery cables. Clamp the amperes pickup PU around one of the battery cables. Turn the load control R20 for a reading of 50 amp on the standard ammeter. The tester ammeter A should read 50 amp on the 0–100 amp scale.

6. Press the BAT STARTER 500-amp test selector button (switch SW in position 1). Turn the load control R20 for a reading of 200 amp on the standard ammeter. Adjust R14 for a reading of 200 on the 0–500 amp scale of ammeter A.

7. Turn load control R20 for a reading of 50 amp on the standard ammeter. The tester ammeter A should read 50 amp, plus or minus (±) one division on the 0–500 amp scale.

This calibration is for one side of the meter. To double check the reading on the other side of zero, reverse the clamp and repeat the calibrating operations at each calibration point.

The unit is now calibrated, and is used by clamping the ampere pickup PU around any conductor for reading dc amperes. The battery cables must be connected to the vehicle battery and appropriate push buttons depressed. The load resistor R20 should always be in a CCW position except for battery load testing and calibrating.

8-12 Voltage Comparison Method of Testing Current Drawn by Starter Motor

A simple procedure for measuring the amperes drawn by the starting motor without opening the circuit to install an ammeter is to use the voltage drop comparison method as follows:

1. Connect a dc voltmeter across the battery terminals.

2. Connect an ammeter of suitable range in series with a heavy-duty carbon pile rheostat. One terminal of the meter is connected to one ter-

minal of the rheostat and the other is connected to a short cable ending with a heavy-duty battery clamp. The other terminal of the carbon pile rheostat is also connected to a short cable and heavy-duty battery clamp.

3. Read the battery voltage before turning on the starter, for example, 12 V.

4. Turn on the starting motor and, while cranking, note the reading of the voltmeter, for example, 9 V.

5. Turn off the starting motor and connect the two battery clips from the ammeter and rheostat (should be in CCW position) across the battery. Adjust the carbon pile rheostat until the voltmeter reads the same as when the starter motor was cranking, which was 9 V.

6. Quickly read the ammeter. The reading should be the same number of amperes the starter was drawing to pull the battery voltage from 12 V to 9 V. Repeat step 4. The voltage should still read close to 9 V if the battery is in good condition.

Any battery tester equipped with a carbon pile rheostat, ammeter, and voltmeter can be used to make this test.

9
Exhaust Gas Analyzers

9-1 General Principles

Combustion of the fuel-air mixture in an internal-combustion engine must be complete to attain maximum performance and efficiency as well as to prevent unnecessary air pollution. The completeness of combustion is determined by an analysis of the contents of the exhaust gases.

Oxygen is essential to any combustion process. Air is the most economical source of oxygen, but it contains only 20.9% oxygen by volume. The remaining part of air is composed of nitrogen, 78.1% and other inert gases, 1%.

Carbon and hydrogen are the combustibles, or inflammable elements, found in common fuels. A compound of these elements is called a hydrocarbon. With complete combustion, these hydrocarbons combine with oxygen in the air to form carbon dioxide (CO_2) and water (H_2O). This accounts for the water and steam expelled from the exhaust pipe, which causes muffler corrosion.

Incomplete combustion results in the formation of carbon monoxide (CO). It is caused by the lack of sufficient oxygen, insufficient time, poor mixing of air and fuel, and too low a temperature to burn the CO to CO_2. In extreme cases, free, unburned carbon in the form of black soot remains.

For every type of fuel there is a known percentage of carbon dioxide in the exhaust gases when complete combustion occurs. Any percentage less than that indicates incomplete combustion, or excess air, which dilutes the percentage.

If pure carbon is burned completely and without excess air, the oxygen of the air will be converted into carbon dioxide, and because the oxygen constitutes 20.9% of the air, the gases will contain 20.9% CO_2.

The hydrocarbons used for fuel in internal-combustion engines contain hydrogen, which combines with the oxygen of the air to form water; thus, the carbon dioxide content of the exhaust gases will be lower than if the fuel were pure carbon. The carbon dioxide content will, therefore, be less than the ideal 20.9%. The average CO_2 content in gasoline, butane, and propane is about 9%; that of natural gas, 12%; and Diesel oil, 15.7%.

Exhaust gas analyzers measure the percentage of CO_2 and/or CO in the exhaust gases. The amount of CO_2 and CO indicates the completeness of combustion, whereas the amount of oxygen indicates the amount of excess air. It is necessary only to measure the amount of CO_2, because if the CO_2 is at its maximum value, the other elements will be correct for optimum combustion.

It is very important to know that an excess of oxygen or air will show up as a reduced percentage of CO_2 due to dilution.

A low CO_2 reading can mean either a lack of oxygen (incomplete combustion—too rich a mixture) or an excess of oxygen (too lean a mixture). For this reason, some of the modern, electrically operated exhaust analyzers will reverse their readings after reaching the peak CO_2 percentage, and precautions must be observed in making adjustments. Others that have a bridge circuit that is "balanced to zero" at an air-fuel ratio (AFR) reading of 13 will then read up for a lean mixture but will reverse at an AFR of about 14.7, and will read down for a rich mixture.

The modern exhaust gas analyzer is known by many names, such as combustion analyzer, carburetor analyzer, CO_2 analyzer, and CO analyzer. They all measure the amount of oxygen by measuring the CO_2 or CO content of the engine exhaust gases.

The analyzer consists of an electrical cell which changes its electrical characteristics when exposed to exhaust gases. This cell is in circuit with a sensitive meter with a dial calibrated in air-fuel ratio. This indicates the volume of air in cubic feet being mixed with one cubic foot of fuel in gaseous form. The dial may be calibrated for various fuels, such as gasoline, propane, butane, and diesel fuel, because each type of fuel requires a different air-fuel ratio and may have a different CO_2 content.

The air-fuel ratio for gasoline, butane, and propane is shown in Table 9-1.

Exhaust analyzers should be recalibrated from time to time to check their accuracy.

To calibrate an exhaust gas analyzer that measures the percentage of CO_2 a standard gas with a known percentage of CO_2 must be used. Such a gas is available from Source 1 for example, that contains a mixture of 6.5% CO_2 plus or minus 0.25% and the rest nitrogen. Each cylinder of compressed gas is marked by the supplier with the specific per-

Air-Fuel Ratios for Gasoline, Butane, and Propane Gas

Air-Fuel Ratio (Meter Reading)	*Gasoline*	*Butane*	*Propane*
11.0	11.0	11.1	11.1
11.5	11.5	11.6	11.7
12.0	12.0	12.2	12.3
12.5	12.5	12.8	12.9
13.0	13.0	13.3	13.4
13.5	13.5	13.9	14.1
14.0	14.0	14.4	14.6
14.5	14.5	15.0	15.3

Table 9-1

centage of CO_2 within the above tolerance range (6.25, 6.5, 6.75, etc.). It is rechecked and marked by Source 1 with the proper air-fuel ratio calibration for the specific mixture, as applied to their instruments.

These cylinders of calibrating gas, at the present time, are available only in a size B tank, which contains enough gas for approximately 150 calibrations. The cost of these tanks including regulator and deposit is a deterrent to all but the largest users. The author has suggested that calibrating gas be made available in small individual CO_2 ampules such as used in charging soda dispensers, inflating life rafts, and loading BB guns. These could be supplied in the standard 8500-mg. or the giant

12,000-mg. size. A pressure regulator would not be required. A simple piercing device could discharge the contents of the ampule into a balloon of sufficient volume to make at least one calibration. This would be less costly and more portable and would appeal to the occasional user.

9-2 Thermal Conductivity Method of Gas Analysis

The Wheatstone bridge principle is used in some form or other in most electrically operated gas analyzers.

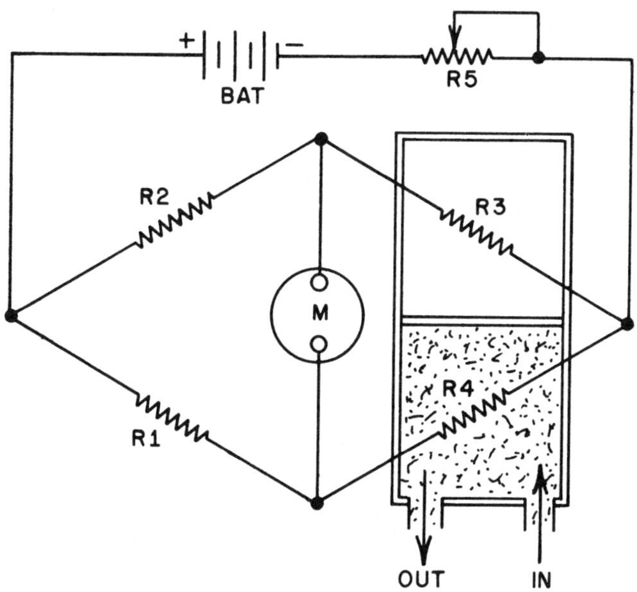

Figure 9-1 Thermal conductivity principle of gas analyzers

Figure 9-1 shows the circuit used in the thermal conductivity method of determining air-fuel ratio. This method is based on the principle that different compositions of gases have different values of thermal conductivity. This circuit shows a means of measuring the resistance of a heated conductor exposed to air as compared with the resistance of a similarly heated conductor exposed to the exhaust gas of an engine. The resistance is measured by means of a Wheatstone bridge circuit which responds to any small changes in resistance caused by thermal conductivity of the gas.

In Figure 9-1 the four resistance units R1, R2, R3, and R4 comprise the bridge circuit. Resistors R3 and R4 are enclosed in a metal block to insure their being at the same temperature. Resistors R1 and R2 complete the bridge circuit. Current from the battery flows through the bridge, heating the resistors, which, in turn, lose heat to the outside air through the walls of the cell. When there is no exhaust gas surrounding resistor R4, the bridge is in balance, at which time there is no deflection of meter M from the center balance position (13 on the AFR scale).

When a sample of the exhaust gas enters the gas cell at the inlet port IN and out at exhaust port OUT, the thermal conductivity of the exhaust gas as compared with air causes resistance R4 to vary in temperature and resistance, unbalancing the Wheatstone bridge. This produces a deflection of the meter M needle to the right or left of the central point. The extent of deflection is a measure of the thermal conductivity of the exhaust gas. Thermal conductivity is proportional to the air-fuel ratio. Therefore, the meter reading is a direct indication of the air-fuel ratio.

This Wheatstone bridge analyzer is accurate within the normal range of operating mixtures, up to a point of air-fuel ratios of about 14.7:1, because at that ratio sufficient air is mixed with the fuel to send an excess of oxygen through the engine, which reverses the thermal conductivity of the gases. This reversal of gas proportions above about 14.7:1 will reverse the meter reading toward the left.

On some of these analyzers there will be a SET calibration to adjust the meter to read full scale to the right, which is merely compensating for any aging of the battery. Rheostat R5 serves as a calibrating control. The meter needle position is normally set at midscale or on 13 of the gasoline fuel scale. This may be adjusted by the usual zero adjustment screw at the bottom of the meter glass.

This type of analyzer is typical of Source 1, Model E-1206, and others and may be a separate, self-contained unit, an add-on unit, or may be incorporated in a scope-type analyzer console.

9-3 Construction and Use of Typical Exhaust Gas Analyzer, Source 15

An exhaust gas analyzer typical of Source 15, Model 42–141, is shown schematically in Figure 9-2. This is an ac-dc analyzer that uses an internal battery for portable or road test use, and an ac power supply

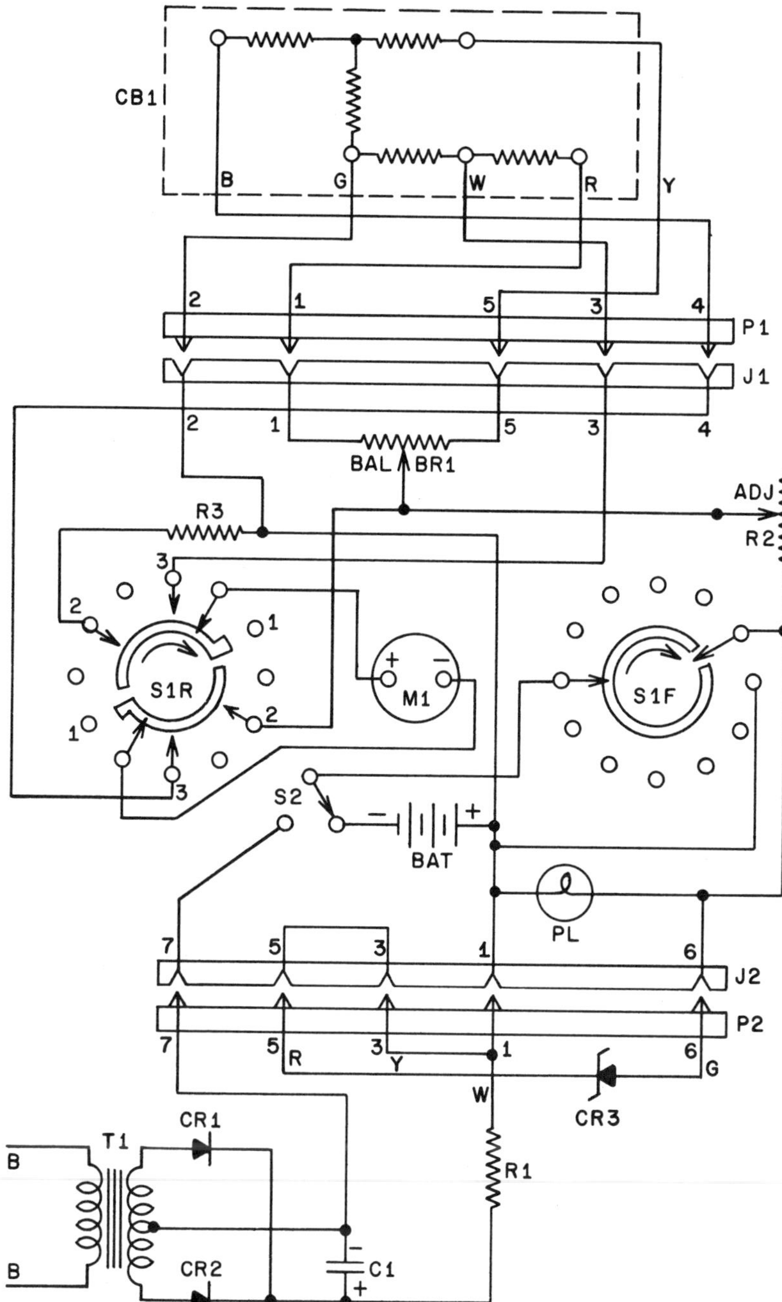

Figure 9-2 Exhaust gas analyzer, Source 15, Model 42-141

that can be plugged into jack J2. Plug P2 is connected to the ac power supply, which consists of a transformer T1 with a primary connected to 115-V ac at leads B and B, a center-tapped secondary that feeds a full-wave rectifier comprising diodes CR1 and CR2. Resistor R1 (25 ohms, 5 W) and capacitor C1 act as a filter to smooth out the rectified dc. Zener diode CR3 (4.7 V, 1 W, 5%) serves to hold the output voltage constant but is not connected until the unit is plugged in and is under load. The gas cell assembly CB1 located in the pickup unit is shown schematically connected to plug P1.

Switch S1R (rear) and S1F (front) is the function switch and is shown as viewed from the knob end and in the OFF position 1. The second position (when rotated clockwise CW) is the ADJUST position 2, moving two terminals on one indent, and the third position when rotated CW is the BALANCE AND TEST position 3, moving two terminals on one indent.

PL is a pilot light, and switch S2 changes from battery operation, as shown, to a second position for ac power supply operation. Resistor R3 is 4.8 Kohms, 1%.

Meter M1 shows the air-fuel ratio for gasoline on the top scale and for propane and butane on the bottom. The scales are calibrated from 9 to 18, with each division equaling 0.25.

The unit should be calibrated each time it is used. Before connecting the hose to the tail pipe proceed as follows. With the analyzer pickup plugged in, but out of tail pipe, set selector switch to ADJUST. Turn ADJUSTMENT TO LINE knob R2 (50 ohms, 2 W) so that the pointer aligns with the ADJ line on the left-hand side of the top scale. Then turn selector switch to BALANCE AND TEST. Turn knob of BALANCE TO LINE control BR1 (2 ohs, 2 W) so that the meter pointer aligns with the BAL line on the meter near the middle of the scale.

When switch S1 (Figure 9-2) is turned to the ADUST position the components in the circuit are as shown in Figure 9-3. The purpose of this position is to calibrate the meter and rheostat R2 so that the meter reads on the ADJ line to the far left of the scale. This is accomplished by turning the ADJUST TO LINE rheostat R2 (50 ohms, 2 W), which is in series with part of the cell assembly CB1, the meter, resistor R3 (4.8 Kohms 1%) and across the battery. The ADJUST TO LINE control R2 should not be disturbed after this adjustment.

The third position of switch S1 of Figure 9-2 is for BALANCE AND TEST. Only the components used in this position are shown in Figure 9-4.

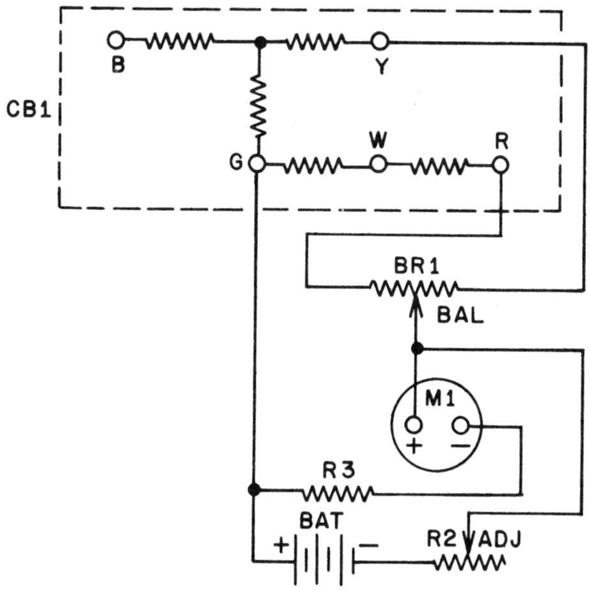

Figure 9-3 ADJUST position of switch S1, Figure 9-2

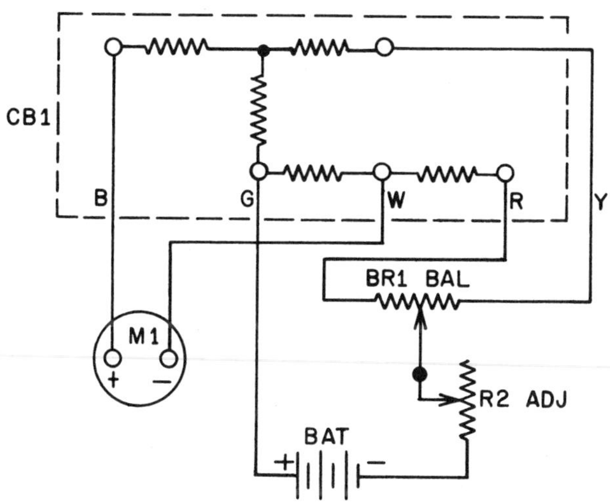

Figure 9-4 BALANCE AND TEST position of switch S1, Figure 9-2

Meter M1 is now connected across the black B and white W leads of cell assembly CB1, BALANCE TO LINE potentiometer BR1 is now across the red R and yellow Y leads of CB1, the battery positive is connected to the green G lead of CB1, and the battery negative is connected through R2 and to the middle terminal of BR1.

It should be noted that switch S1 has twice as many switch contact positions as it has knob or indent positions.

The third position of S1 forms a bridge circuit in CB1, with meter M1 connected across the bridge in the usual manner, with the battery voltage through the bridge at right angles to the meter connection. The BALANCE TO LINE control BR1 is now adjusted (with the pickup hose out of the tail pipe) to give an air-fuel ratio reading at the balance line BAL corresponding to 13 on the upper scale.

Cell CB1 is sensitive to carbon monoxide (lack of air or too rich a mixture) and also to oxygen (excessive air or too lean a mixture). As a sampling of the engine exhaust gases passes over cell CB1, the meter will read below the 13 mark for a rich mixture (carbon monoxide detected) and will read above 13 for a lean mixture (excess oxygen detected). The change in electrical characteristics of the various elements in the bridge circuit of cell CB1 unbalances the bridge and indicates on the meter.

9-4 Carbon Monoxide Analyzer, Source 15, Model 42-146

Principles of Operation

The presence of carbon monoxide in engine exhaust gases indicates incomplete combustion and the degree of air pollution. Carbon monoxide is very poisonous.

The Source 15, Model 42-146, carbon monoxide analyzer samples exhaust gases and indicates the true carbon monoxide percent concentration on the upper meter scale calibrated zero to 5% and also reads on the lower meter scale the approximate air-fuel ratio (AFR) calibrated 12.75 to 16 AFR.

Carbon monoxide (CO) present in a sample can be determined by measuring the amount of infrared energy absorbed by the gas sample. Since CO absorbs energy within the infrared spectrum, an infrared beam is directed through a sample tube and focused on a detector.

The detector (DET1 Figure 9-5) is covered by an optic interference filter which is coated with a substance that permits a band of energy of 4.7-micron (μ) wavelength to pass. Carbon monoxide will absorb energy of 4.7 micron wavelength.

Different gases in the sample tube absorb energy of different wavelengths; it is therefore necessary to pass only the band in which CO absorbs infrared. This prevents other gases from interfering with the detector.

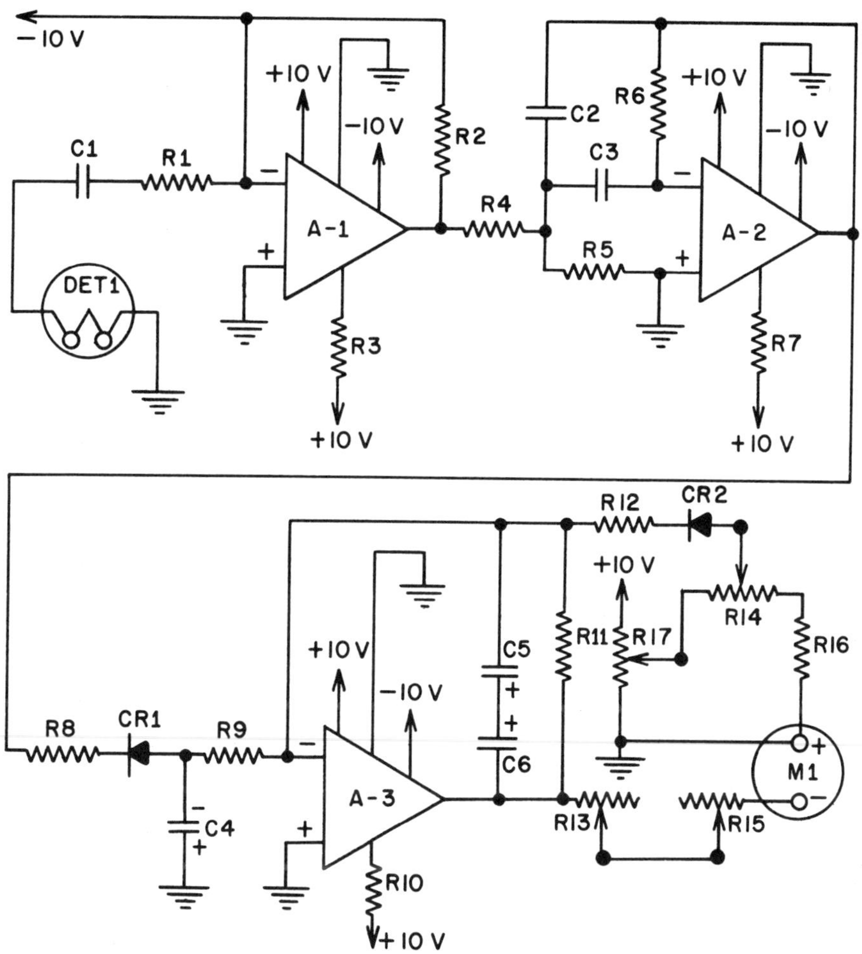

Figure 9-5 Carbon monoxide analyzer detector and amplifier circuit, Source 15, Model 42-146

Infrared energy focused on the detector is sensed as heat and is transformed into an electrical signal.

With no CO in the sample tube, a maximum signal of about 50 microvolts is sensed by the detector. It follows then that inasmuch as CO absorbs infrared, the signal becomes weaker as the amount of CO increases. This signal is therefore inversely proportional to the amount of infrared absorbing CO.

Carbon Monoxide Analyzer Detector and Amplifier

Figure 9-5 shows the schematic wiring diagram of the detector DET1, the amplifier, and the indicating meter M1.

The small dc signal from the detector DET1 is so small (about 50 microvolts) that it is very difficult to amplify because of drift: the signal is therefore converted to ac by chopping the infrared source beam with a motor-driven disc before it reaches the detector. The placement of holes on this disc and the speed at which it rotates provides a 12-Hz (cycles per second) pulsating source of voltage. This 12-Hz chopped signal is then emitted by the detector as a pulsating dc voltage which is coupled through a capacitor C1 (2000 mfd, 6 V) to provide an alternating current of 12 Hz to the first amplifier.

The amplifier board has basically three stages, the first of which is a high-gain, low-noise operating amplifier A-1 (SQ-10A). The output of A-1 will be about 1 V peak-to-peak and will probably be heavily masked by noise. This signal is then fed to A-2 (SQ-10A), which is a very sharply tuned, active electronic filter that will pass only 12-Hz signals. In this stage all unwanted noise or "hash" is blocked and the output will be about 10 V peak-to-peak 12 Hz. This signal is then rectified and filtered by CR1 (10D1) and capacitor C4 (20 mfd, 20 V dc), giving a dc level that represents the amount of energy that strikes the detector. Since the amount of energy that strikes the detector is affected by the amount of CO in the gas sample, the level of this signal across C4 is analagous to the amount of CO in the sample tube.

It follows then that, when no CO is present in the sample tube, a maximum potential will be present across C4. The greater the amount of CO in the sample tube, the smaller the amount of signal that amplifier A-3 will receive. Since the unit is adjusted with no CO in the sample tube, a maximum voltage of about 4.5 V dc is coupled to amplifier A-3 (SQ-

10A) through resistor R9 (100 Kohms, 1%). At this point an equal potential of opposite polarity is injected by adjusting potentiometer R14 (1 Kohms, 20%) AIR ADJUST POT. This in effect cancels the signal and gives a zero reading on meter M1. In operation, this canceling voltage remains unchanged, and if the CO is introduced into the sample gas, there is a change in the output of amplifier A-2 which produces less signal to amplifier A-3. The difference is compared, then amplified by A-3 and applied to meter M1 as an accurate indication of the presence, and percentage, of carbon monoxide.

Other component values in Figure 9-5 are as follows: resistors (in ohms and tolerance—unspecified are 20%); R1 (100, 1%), R2 (2 meg, 1%), R3 (22 K), R4 (13 K, 1%), R5 (681, 1%), R6 (267 K, 1%), R7 (22 K), R8 (100, 1%), R10 (22 K), R11 (150 K, 1%), R12 (100 K, 1%), R13 (10 K, 10%), R14 (1 K), R15 (1 K), R16 (10 K, 10%); capacitors in mfd, C2 and C3 (1, 50 V, 5%), C5 and C6 (20, 25 V, 20%).

Power Supply for Detector and Amplifier

Figure 9-6 shows the power supply for the Source 15, Model 42-146, carbon monoxide analyzer. This power supply provides two separate 10-V dc regulated and separately filtered dc sources. One is 10 V positive above ground, and the other is 10 V negative above ground.

In Figure 9-6 the positive dc 10-V power supply consists of one secondary of transformer T1 with two yellow outside leads and a green center-tap lead. It has diodes CR1 and CR2 (10D2), filter capacitors C1 (100 mfd, 25 V dc), capacitor C2 (0.1 mfd, 200 V dc), and resistor R1 (2.7 ohms, 10%), a voltage regulator integrated circuit board IC-1 (MC1460), filter capacitor C3 (0.1 mfd, 200 V dc), capacitor C4 (10 mfd, 35 V dc) and resistors R2 (4.7 ohms, 10%), R3 (6.8 Kohms, 1%) and R4 (13 Kohms, 1%).

The negative dc 10-V power supply consists of the other secondary of transformer T1 with two red outside leads and an orange center-tap lead, diodes CR3 and CR4 (3AF2), filter capacitor C5 (8200 mfd, 25 V dc), C6 (0.1 mfd, 200 V dc), and resistor R5 (2.7 ohms, 10%), a voltage regulator consisting of transistor Q1 (2N3055) and integrated circuit board IC-2 (MC-1460), filter capacitors C7 and C8 (0.1 mfd, 200 V dc), C9 (100 mfd, 25 V dc), resistors R6 (51 ohms, 5%), R7 (1 ohm, 10%), R8 2.7 ohms, 10%), R9 (6.8 Kohms, 1%), R10 (13 Kohms, 1%); infrared heat source IRS (2–3 ohms) and resistor R12 (3 ohms, 25 W, 10%).

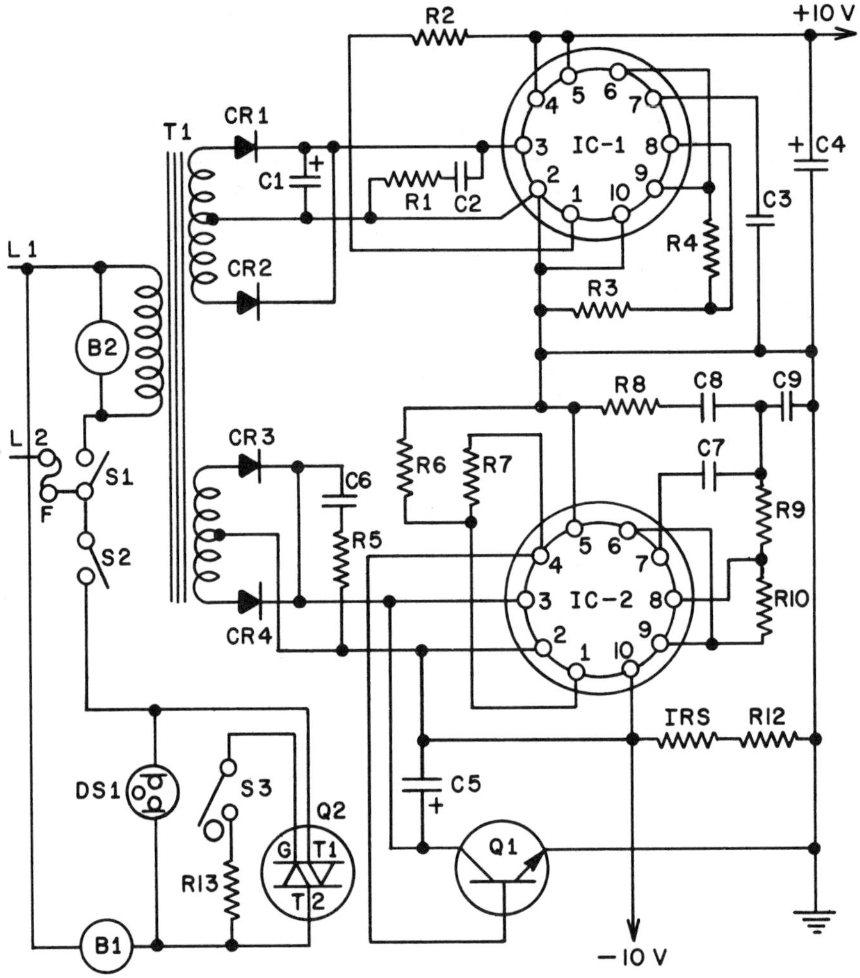

Figure 9-6 Power supply circuit for Figure 9-5, carbon monoxide analyzer detector and amplifier circuit

In the primary circuit of transformer T1, the chopper motor B2 is connected across the transformer primary to operate all the time the unit is operating.

The reference B1 is the pump motor that circulates the engine exhaust gases through the sample tube and detector. The triac Q2 is a three-terminal, five-layer NPNPN silicon structure. It is an integrated version of two silicon-controlled rectifiers (SCRs) connected inverse-parallel (back-to-back) with a common gating system G. In this CO analyzer, the triac is used to open the pump motor B1 circuit if the operator fails to empty the condensation-collecting water trap. A small magnetic reed switch S3 with float attached activates the triac. In normal operation the reed switch S3 is closed, triggering the triac into conduction, causing motor B1 to run. When the triac is conducting, the drain water pilot light (DS1 neon) is off because of the low voltage across the triac. Opening the reed switch S3 opens the triac gate G, placing ac line voltage across the triac and causing the drain water light to come on. Switch S1 turns on the unit for warm-up, and S2 operates the pump motor. Fuse F is rated at 2 amp.

Calibration of the CO Analyzer

1. Connect the power cord to a 120-V ac outlet and turn on detector switch S1 and pump motor switch S2.
2. With the exhaust pickup hose connected to the analyzer, allow unit to warm up at least three minutes.
3. Turn AIR ADJUST knob (R14, Figure 9-5) and SPAN ADJUST knob (R15, Figure 9-5) to center position. Meter should read at AIR ADJUST LINE (zero percent CO). If it does not, adjust R17 pot. (Figure 9-5) to set meter pointer at AIR ADJUST LINE (zero percent CO).
4. Turn CALIBRATING CONTROL, on the rear of the analyzer counterclockwise against the stop. This calibrating control actuates a gate or bar that is placed between the infrared source and the detector CCW for calibrating the SPAN adjustment, but the gate is removed for exhaust gas analysis by turning the calibrating control clockwise. Meter should read at the SPAN ADJ (2.5% CO) line. If it does not, then bring the meter pointer to the SPAN ADJ line by adjusting R13 (Figure 9-5) located on the chassis.
5. Turn CALIBRATING CONTROL clockwise against the stop for exhaust gas tests.
6. These calibrating and adjusting potentiometers and rheostats may bear different reference numbers on the chassis. In Figure 9-5, R17 may be marked R11, R14 may be marked R1, and R15 may be marked R2. The reason for this is that there would otherwise be a conflict of reference numbers on the power supply board, the amplifer board, and the panel controls.

7. Certain precautions in calibrating and servicing should be observed. Any voltage—static electricity or any other form—between either detector lead and its case will cause permanent damage to the detector elements. Do not touch detector leads at either end without first grounding the detector leads by connecting a jumper between both leads and case. The detector can, however, be tested "in circuit" with an ohmmeter by leaving it connected to the amplifier. First, momentarily touch one prod of the ohmmeter to the chassis to discharge any static electricity. Next, using a 10 Kohm resistor in series with the other ohmmeter lead, check detector continuity across the detector connections. If the detector has normal conductivity, the ohmmeter should read close to 10 Kohms (the value of the series resistor). If the ohmmeter reads infinity, the detector element is open. Visually inspect the detector to see that the element is black and square.

9-5 Hydrocarbon Gas Analyzer

There is another model, 42-149, similar to Model 42-146 gas analyzer that is a dual-gas analyzer indicating the carbon monoxide content on one meter and hydrocarbon (unburned gasoline) on a second meter. The circuit is the same in both except for the filter on the detector.

There is a common power supply as shown in Figure 9-6 except there are two infrared heat sources IRS.

The carbon monoxide analyzer has a separate amplifier, detector, infrared heat source, and meter, as shown in Figure 9-5.

Likewise, the hydrocarbon analyzer has its own separate detector (having a different filter), amplifier, infrared heat source, and meter, as shown in Figure 9-5.

The principle of operation and adjustments are the same for both analyzer sections, and the components are interchangeable except for the detector filter.

10
Car Start Boosters

10-1 General Principles

There are many ways to start an automobile engine with a dead battery besides pushing the vehicle.

The simplest means is to use jumper cables to connect the dead battery across a fully charged battery of the same voltage, or higher, either from another vehicle or from a portable battery. Positive terminals should be connected together and the negative terminals connected together. If there is a large spark, check the polarity. Reversed connections can damage an alternator instantly, so the "spark" method should not be used.

More elaborate systems use an engine-driven generator, or a permanently installed self-charging battery system.

If ac is available, a high-rate battery charger can be used for boosting the battery and assist in carrying the starter load. Regardless of instructions or other advice, it is always safer to put the dead battery on fast charge for at least five minutes, then, if it is adjustable, reduce the charging rate one or two notches on the charge rate control switch. If the engine does not turn over fast, or does not start within three seconds, stop and repeat the process. More battery chargers are damaged by improper use in car starting than in any other way. Some fast chargers have a ballast resistor that limits the charger output to a safe value under all conditions, and these are preferred for car starting.

If there is no rate adjustment on the charger (single rate charge), then it is even more important to take precautions about overloading. Even if the charger is equipped with circuit breakers or fuses, frequent overloading can shorten the life of the charger or damage it.

If the engine does not start immediately, it could indicate the need for a tune-up before going further.

Some road service trucks carry two batteries, each having the same voltage as the stalled vehicle battery. By connecting them in series to obtain twice the voltage, and by using long cables, a quick start can be obtained without damage. The long cables cause a voltage drop due to vehicle starting current, so that the actual voltage at the vehicle starter will be at least equal to, or slightly higher than, the normal battery voltage. These service batteries are kept charged at the service shop.

A control system permanently installed in the service truck will keep both batteries charged by the service truck charging system. They are connected in parallel for charging and in series for vehicle starting with long cables. Only one extra battery is needed in addition to the regular service truck battery.

10-2 Typical Hot Start Booster Using Batteries

One commercially available Hot Start Booster unit for permanent installation in a service truck, with a built-in alternator protector, is model 140-N2, typical of Source 10 Hot Start Boosters, shown schematically in Figure 10-1.

In Figure 10-1 the clips marked BAT 1 go to the stalled 12-V vehicle battery, BAT 2 is the service truck's regular 12-V battery, and BAT 3 is an auxillary 12-V battery. SW1 is a heavy-duty DPDT knife switch to change from CHARGE position CH, as shown, to the opposite position for BOOST AND START BS. When SW1 is on charge, to the right as shown, BAT 2 and 3 are connected in parallel, so that the two batteries in the truck are charged from the service truck charging system.

The ammeter A has a zero center scale and is of the induction type with two leads clamped to the back of the meter. The lower lead carries the charging current to BAT 3. This lead is located at the most sensitive position on the back of the meter because the charging current of 40–50 amp must give full-scale deflection to the right. The charge sector on the meter dial is green and is labeled AUX BAT CHARGE OK, shown as CH in Figure 10-1 (on ammeter A).

When switch SW1 is in the BOOST START BS position to the left, the two batteries BAT 2 and 3 are connected in series to provide 24 V. The

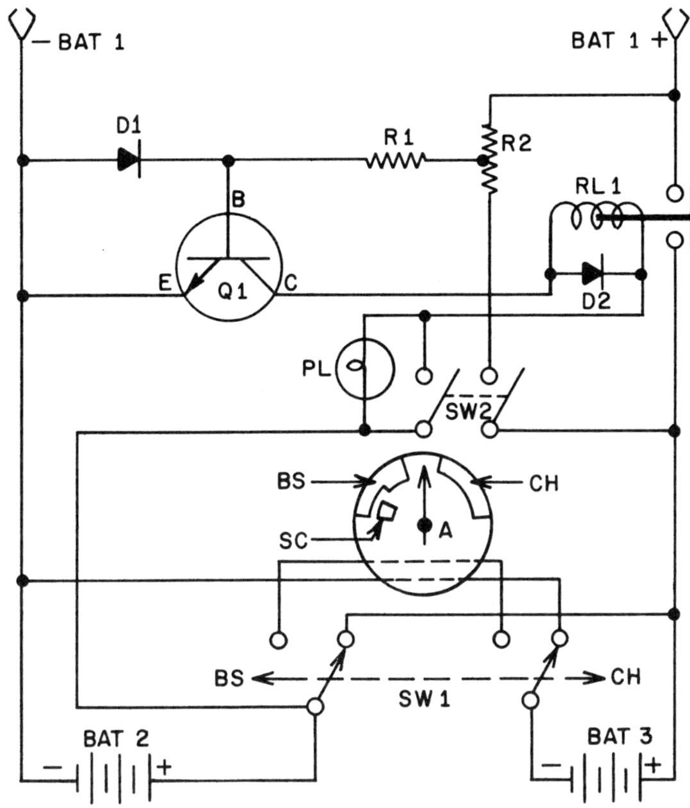

Figure 10-1 Hot start booster, typical, Source 10, Model 140N-2

boost start current goes through the upper lead in back of the meter, which is less sensitive and gives a full-scale deflection to the left with a current of 300–400 amp. The boost start sector on the meter dial is green and is labeled BOOST & START OK, shown as BS on ammeter A.

Switch SW2 is a remote control toggle switch DPST that actuates the solenoid RL1 for boosting and starting. The solenoid RL1 will not close if the polarity of BAT 1 is reversed because of the built-in alternator protector, which consists of transistor Q1, diode D1, and resistors R1 and R2. Current-limiting resistor R2 is tapped as shown. It bypasses solenoid contacts and also serves to bias the transistor during systems check. Pilot light PL is a green light that glows if the polarity connection is correct but goes out when switch SW2 is closed. Diode D2 is across the solenoid RL1 coil to protect the transistor from transient voltage when the solenoid coil discharges.

Operating Instructions

The instructions printed on the panel of the unit are as follows:

1. Connect clamps, red to positive, black to negative (green light on panel indicates proper connection). If no light, clamp connections are poor or reversed, or battery is abnormally low.
2. Set switch (SW1) to BOOST AND START and press remote toggle switch (SW2) for one minute. (This is to boost the battery (BAT 1) for one minute quick charge.)
3. Press remote toggle switch (SW2) while starting car.
4. Set switch (SW1) to CHARGE position.

> NOTE: With switch set to Boost and Start and clamps on shorting bar (a short piece of steel on the case to short the two clamps for systems check), operate remote switch (SW2). Ammeter A should read in yellow zone SC (extreme left of full scale), which is labeled SYSTEM CHECK OK. If it does not, check batteries and connections.

Precautions. The purpose of the systems check is to be sure that all connections are correct and in good working order. CAUTION: Do not attempt the systems check using shorter cables than provided with the unit, since this is a dead short across 24 V, and short cables provide insufficient resistance and would damage the unit or the cables.

The unit is left on the CHARGE position at all times except when boosting and starting. Do not leave switch SW1 in the mid position (OFF).

10-3 Engine-Driven Car Start Generators

The battery system of boosting and starting is very satisfactory for the average road service call, but in situations requiring the starting of several cars in a short time, such as in parking lots on a cold night after a football game, the gasoline engine-driven generator is a life saver and a money maker.

These systems basically are all very much alike. They consist of a governor-controlled gasoline engine, a dc generator, meters, and controls.

Some have automatic voltage regulators, alternator protectors, and a 115-V ac or dc power outlet to operate lights and electric tools.

Gasoline engine-driven dc arc welders have been successfully used for car starting when certain precautions have been observed. It is essential to have a voltmeter and a field rheostat, as well as an ammeter, built into the unit so that the voltage and current can be adjusted to a safe value. Many arc welders have an open circuit voltage as high as 70–100 volts. This can be reduced to the battery voltage with the field rheostat, by adjusting the engine speed, or both. The ampere switch should be set to give a safe booster current of 50–100 amp to the dead battery before attempting to start the car. If this system is used regularly, it is faster and safer to have a set of batteries floating across the arc welder output. Once adjusted, it is only necessary to crank up the arc welder engine.

10-4 Typical Engine-Driven Car Start Generator

In Figure 10-2 is shown a typical Source 4, Model GSI, engine-driven "Kar Start"® electrical system.

The driving engine is normal with electric starter, driving a dc generator with armature A, brushes B, and shunt field SF. Diode D1 is connected across the shunt field SF to short out surge voltages generated by the inductance of the field when the current is interrupted. These surge voltages have polarity opposite to the applied voltage; therefore, the diode D1 will conduct only negative peaks to protect the transistor Q1. A voltmeter VM and an ammeter AM are provided as standard equipment.

The voltage regulator RL is connected across the generator output through rheostat R3 and is adjusted for the desired control voltage. This operates like a regular automotive voltage regulator, except that the contacts RLC do not carry the field current but act only as a trigger to operate the transistor Q1 base B bias. The transistor carries the full field current between the positive clamp CL, transistor Q1, emitter E, collector C, through shunt field SF, and back to the negative clamp CL. When contacts RLC of relay RL are closed, the base B of transistor Q1 is connected to collector C for full bias, causing transistor Q1 to conduct. When the contacts RLC are open, the base B of transistor Q1 is connected to the emitter E by resistor R2, stopping conduction of transistor Q1. This action may take place rapidly, and the relay contacts will vibrate between opened and closed positions to maintain a constant volt-

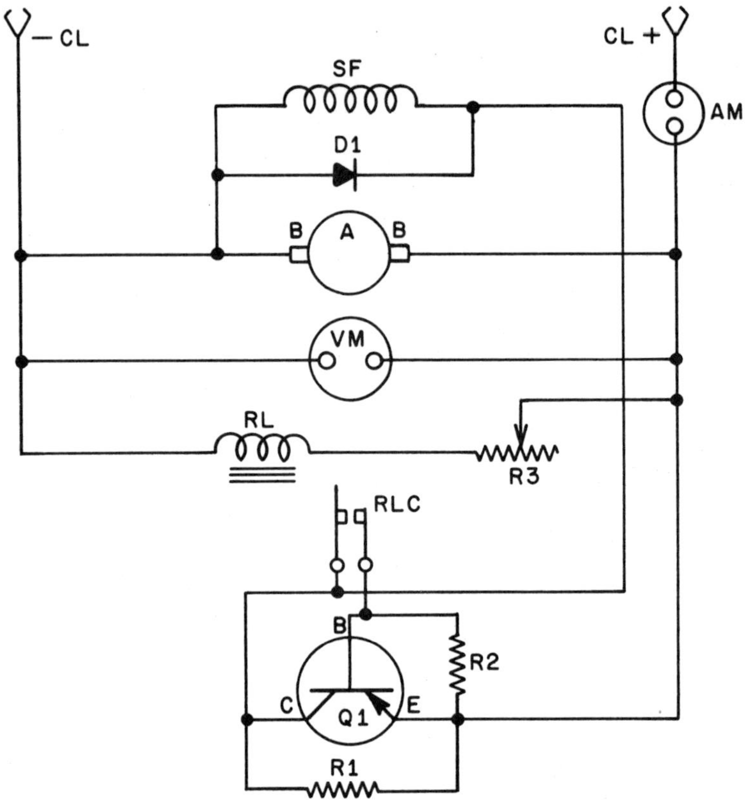

Figure 10-2 Typical engine-driven "Kar Start," Source 4, Model GSI

age. Resistor R1 is necessary to help protect the transistor Q1 against voltage surges during switching.

10-5 Typical Engine-Driven Car Start Generator with Auxilliary AC Generator

Automotive road service calls often involve more than a means of starting a stalled vehicle. A gasoline engine-driven dc generator for starting cars, incorporating an ac generator supplying 120 V, has enough power to operate impact wrenches, power tools, air compressors, and lighting. The unit shown in Figure 10-3 is typical of Source 15, Model 34-107, Heavy Duty Power Start.

The Power Start is a heavy-duty generating plant designed specifically for emergency starting service. The plant supplies direct cranking current to stalled vehicles without the need for booster batteries. In addition, the plant will supply a continuous 120 V ac for power tools of up to 2500 W (20 amp).

The generator is driven by a one-cylinder governor-controlled electric-start gasoline engine. The starter, which is also the generator, plus

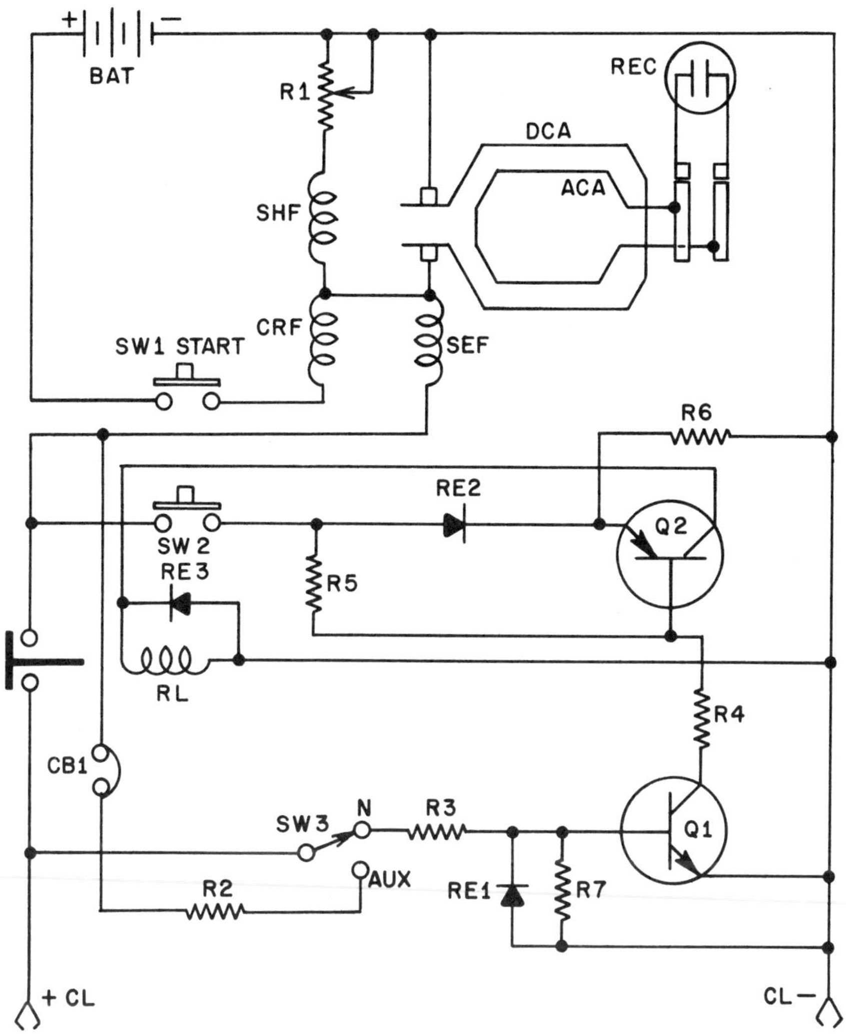

Figure 10-3 Engine-driven power start generator, typical, Source 15, Model 34-107

cranking series field CRF, can be operated from the service truck's 12-V battery by pressing starter button switch SW1.

The generator has one set of field poles on which are wound the shunt field SHF, series field SEF, and the series dc cranking field CRF. There is one armature on which is wound the dc coils DCA, and in the same slots the ac coils ACA. The ac coils are connected through collector rings and brushes to the ac receptacle REC.

Resistor R1 is a wire-wound slide unit mounted inside the generator end cover. It is in series with the shunt field to adjust the output voltage (ac and dc) of the generator.

Battery BAT is the service truck battery for starting the Power Start engine.

The remote switch SW2 applies power to the alternator protector circuit, and the two transistors Q1 and Q2 act as a switch to energize the solenoid RL on correct polarity.

Switch SW3 is shown in the normal N position. When switch SW3 is in the auxilliary position there is a small charging current through circuit breaker CB1, resistor R2 (5 ohms), and positive battery clip CL. It is enough current to boost a dead battery and to operate the alternator protector circuit.

The transistors get their power from the generator and the control signal from the battery to be boosted. With the switch SW3 in the normal N position, the control current from the battery goes through R3 (560 ohms) and the base-emitter junction of the small transistor Q1. This current is multiplied by the gain of the transistor Q1 and fed through R4 (220 ohms) to the base-emitter junction of the power transistor Q2. This current, in turn, is multiplied by the gain of the power transistor Q2 and applied to the solenoid RL to energize it, thereby connecting the generator to the battery to be boosted.

Other component values are, R5 (originally 100 ohms, changed to 560 ohms), R6 and R7 (2.2 Kohms), and R1 (3 ohms). RE1, RE2, and RE3 are small silicon diodes.

10-6 D.C. Converter Kar Start

A car starting system is available from Source 4 that also supplies 115 V dc through a converter for operating certain electric tools and lamps that will operate on direct current and will not draw over a total of 5 amp. The converter alone is available as Model A115, or as a combination Converter Kar Start®, Model B115, Figure 10-4.

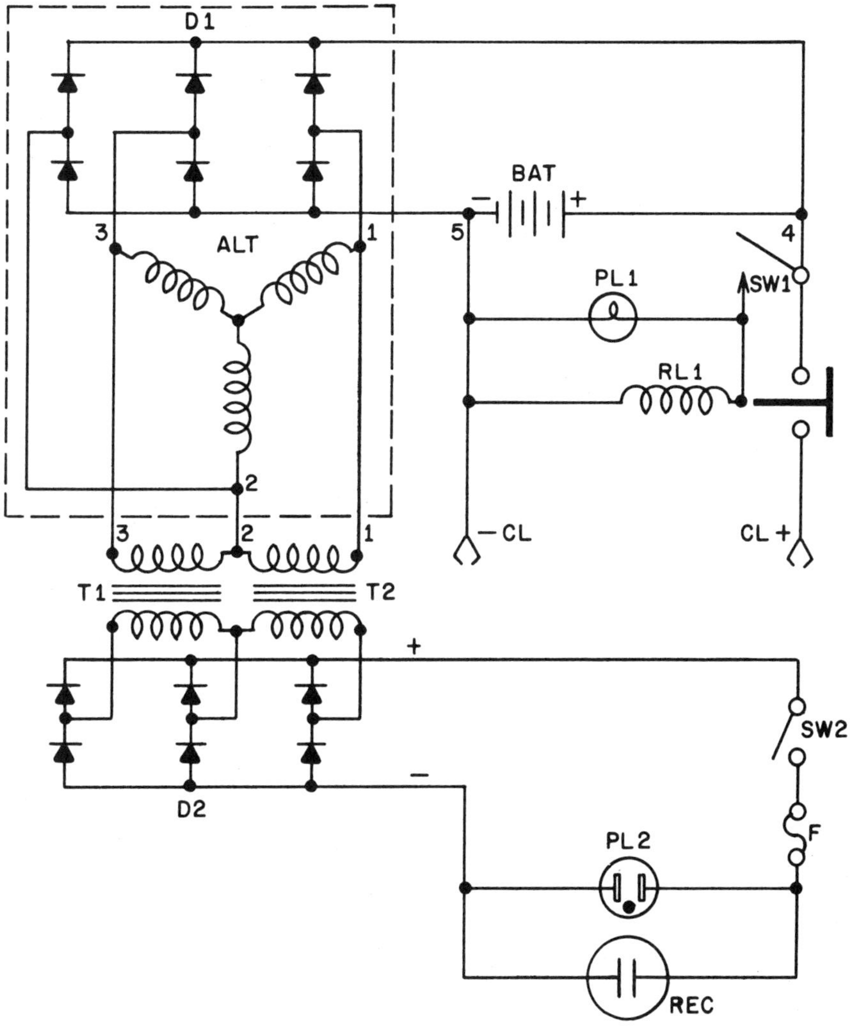

Figure 10-4 Direct current "Converter / Kar Start," Source 4, Model B115

The car-starting section consists of the service truck battery BAT, a solenoid RL1 operated by a toggle switch SW1, and pilot light PL1, which indicates when the solenoid is energized. It does not include an automatic alternator protector, but that item is available separately if desired. It does have, however, a polarity indicator light on one of the cable clamps.

The converter section consists of two identical transformers T1 and

T2, a three-phase rectifier D2, a switch SW2, a fuse F, and a neon pilot light PL2. The two transformers T1 and T2 are connected by three leads 1, 2, and 3 to the three-phase ac stator winding of the service truck alternator. A 12-V alternator of 52 amp or more is recommended. The three-phase stator winding of the alternator produces only about 15 V, but it is stepped up to 115 V, three-phase, by the transformers T1 and T2. The three-phase rectifier D2 converts this to 115 V dc, which is connected to the receptacle REC. Single-phase ac, 115 V, could be made available from one of the transformer secondaries, but the power available would be much less than this arrangement.

If the alternator has an external rectifier and leads 1, 2, and 3 are already at the alternator terminal board, then installation is simplifed. Otherwise, it is necessary to disassemble the alternator and bring out the three leads, after soldering one end to the stator winding.

There are five connections between the service truck and the "Converter Kar Start." Leads 1, 2, and 3 are connected to the alternator stator, and leads 4 and 5 are connected to the truck battery BAT.

Any tool or accessory equipped with a 115-V brush type universal ac-dc motor may be used on this dc converter as long as its maximum current draw does not exceed 5 amp. Such tools include drills, impact wrenches, and small electric saws. Incandescent lamps rated at 110–130 V may be used. Resistance load accessories, such as vulcanizers, radiant heaters, and resistance soldering irons, may be used. Remember that accessories or a combination of accessories may be used as long as the total rated current draw does not exceed 5 amp.

Power tools that have an electronic speed control must be locked to operate on high speed only.

Any tool or accessory that contains a transformer cannot be used, such as ac radios, television sets, and battery chargers. Devices equipped with "ac only" motors cannot be used.

Induction and synchronous motors or devices cannot be used, such as timers and electric fans.

10-7 Remote Starters

Remote starting devices and switches are used to crank an engine, to turn it over at cranking speed, or for "jogging" or "inching" the engine over, a small amount at a time.

Remote switches. The most commonly used remote starting system consists of a push button switch on the end of a long two-wire cable terminating at the other end in small clips. One clip is connected to the starter solenoid coil connection and the other to the ground or to the battery, depending on the position in the circuit of the vehicle's starting switch. The push button has normally open contacts and completes the circuit only when pressed. These are available commercially or can be built in the shop.

Key operated by remote control. This type of remote starter is also available commercially or can be made from readily available material. It consists of a strip of metal to 3 to 4 in. long and ½ to ¾ in. wide. A hole is drilled near each end. One hole is used to bolt the strip to the ignition key. The key and strip should be at right angles. The other hole is used to secure a light string, such as a nylon cord.

To use this device, insert the key in the ignition lock and thread the string out the vehicle door in such a direction that pulling the string will start the vehicle.

11
Timing Lights

11-1 Neon Gas Timing Lights

One of the first special tools used in automotive service was the neon timing light. It consists of a glass tube filled with neon gas, and it is formed in a helical coil to concentrate the light given off. This tube is place in an insulated holder back of a lens to focus the light in one spot.

To use the neon timing light, place it in series with the No. 1 spark plug. The high-voltage wire is removed from the spark plug and connected to one of the leads of the timing light. The other lead of the timing light is connected to the spark plug. When the spark plug fires, the high voltage breaks down the neon gas and makes it conductive, thereby conducting the high-voltage current to the spark plug. The neon tube ignites and flashes, emitting a red light, and the spark plug fires.

The timing light is focused on the timing mark located on either the flywheel or the front damper. This timing mark is established at top dead center (TDC) of the No. 1 cylinder, and it is usually marked with a few degrees on either side of the TDC mark to indicate the degree of retard or advance of the spark. The timing light flash is of such short duration that it makes the timing mark appear to be stationary.

The disadvantage of the neon timing light is that the light given off is in the red spectrum and the brilliancy is inadequate. It is usually necessary to rub white chalk on the timing mark to see it clearly.

11-2 Timing Lights Powered by Xenon Gas

The disadvantage of the neon timing lights has been overcome with the invention of the xenon-powered timing light. Whereas the neon timing light obtains all of its power from the ignition coil, the powered timing light uses the high voltage of the ignition coil to trigger the flash only. All of the power to flash the light is obtained from another power source, which may be the vehicle's battery, an enclosed battery, or a 115-V ac power line. In all cases the voltage is built up to a value of approximately 400 V dc, or about one-half of the voltage required to flash the tube. A high-voltage lead from the spark plug is placed near the tube so that, when the plug fires, it ionizes the gas in the flash tube. The gas then breaks down and conducts. A charged capacitor furnishes the power to flash the tube.

In the xenon-powered timing light, the flash tube is shorter and is U-shaped. It is filled with xenon gas, which gives off a brilliant white light of shorter duration than the neon. This makes the timing mark easier to read, even in daylight; it is brighter, whiter, and sharper in detail because of the shorter duration of the flash.

11-3 Vibrator Power Supplies for Timing Lights Powered by Direct Current

The vibrator type of power supply is used to obtain high ac and dc voltages from a battery for operating an automotive timing light or other test equipment. It consists of a special step-up transformer combined with a vibrating interrupter called a "vibrator." When the unit is connected to a battery, the transformer secondary output power is obtained by passing current from the battery alternately through each half of the transformer primary. The circuit is made and reversed rapidly by the vibrator contacts interrupting the current at the rate of approximately 100 to 200, or even up to 400 Hz. This produces a changing magnetic field which induces a voltage in the secondary of the transformer.

An appreciable time is required for the vibrator reed to move from one side to the other, so that the contacts are closed, one way or the other, for only about 80% of the time. The transformer is connected across the battery only that percent of the time. This percentage is known as the "time efficiency" of the vibrator.

The resulting square-wave dc pulses in the transformer primary cause an alternating voltage to be developed in the secondary.

This high-voltage ac, in turn, is rectified either by rectifiers or by an additional synchronized pair of vibrator contacts. The rectified output is pulsating dc, which can be filtered by the usual means, if desired.

Figure 11-1 shows the two basic types of circuits. Figure 11-1a is one form of nonsynchronous vibrator VIB1. When the battery is not connected, the vibrator reed VR is midway between the two stationary contacts, touching neither. On connecting the battery, the magnet coil MC pulls the reed VR into contact with the lower contact point (called the pull interrupter), causing current to flow through the lower half of the transformer T1 primary winding. Simultaneously, the magnet coil is short-circuited, de-energizing it, and the reed swings back. Inertia carries the reed VR into contact with the upper contact point (called the inertia interrupter), causing current to flow through the upper half of the transformer primary. The magnet coil, in the meantime, is again energized, and the cycle repeats itself at the natural period of vibration, or frequency, of the vibrating reed. Diodes D1 and D2 are the rectifiers, and C1 is a buffing capacitor.

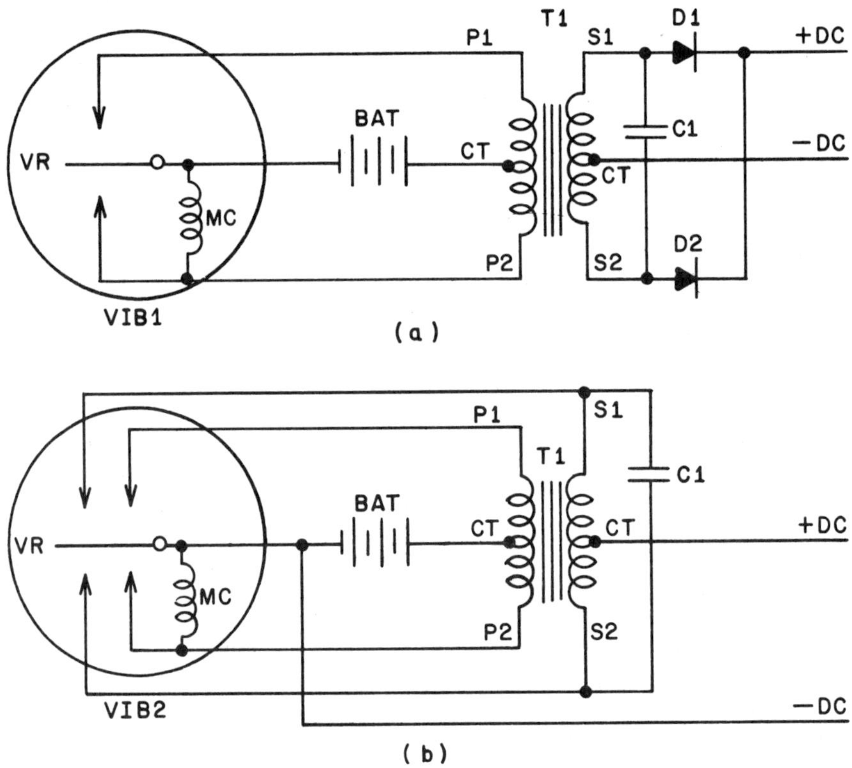

Figure 11-1 Basic types of vibrator power supply circuits. (a) Nonsynchronous (b) Synchronous

The synchronous circuit shown in Figure 11-1b has a vibrator VIB2 provided with an extra pair of points. The points, connected to the secondary of the transformer, rectify the secondary output of the transformer, thus eliminating the need for a separate rectifier. These two stationary contacts are called the "pull rectifier" and "inertia rectifier" contacts. The secondary center tap furnishes the positive dc output terminal when the relative polarities of the primary windings are correct.

The synchronous vibrator was popular before the high-voltage solid state rectifiers appeared, but it is giving way to the simpler and more rugged nonsynchronous type. The development of a solid state vibrator, or oscillator, replacement will eventually make the mechanical vibrator obsolete. The solid state device consists of an oscillator which has no limit on frequency and can therefore use a smaller transformer at higher frequency.

The buffer capacitor C1 across the transformer secondary of both circuits absorbs the surges that occur when the current is broken, which collapses the magnetic field practically instantaneously and hence causes very high peak voltages to be induced in the secondary. Without this buffer capacitor, excessive sparking occurs at the vibrator contacts, shortening the life of the vibrator. Correct values usually lie between 0.005 and 0.03 mfd, and for 250–300 V (under load) supplies, the capacitor should be rated at least 1500 to 2000 V DCWV. In some cases a resistor is placed in series with the buffer capacitor to prevent a short in case the capacitor is shorted. The resistor also improves the buffer action. The exact capacitance is critical, and any defective units should be replaced with the exact value. The correct value can best be determined by observing the patterns on an oscilloscope.

The buffer capacitor can be effectively placed across the primary winding instead of the secondary winding, but it would require a larger value of capacitance and would have the disadvantages of larger size and higher cost.

Ideal performance is reached when the battery current is at a minumum for maximum output and the sparking at the vibrator contacts is at a minimum. Since most commercial vibrators are sealed at the factory, the sparking of the contacts cannot be observed. Therefore the oscilloscope will reveal any irregular sparking of the vibrator, point spacing, point resistance (from pitting or burning) and other defects.

With an oscilloscope connected between primary point P1 and the center tap CT, and also between P2 and CT, a uniform and symmetrical square wave pattern should be observed. A slight difference in the wave form is to be expected between the two primary sections, because magnet coil MC is shunted across one-half of the primary P2 to CT.

The secondary wave forms between S1 and CT and between S2 and CT, and it should form a symmetrical square wave pattern.

Although it is beyond the scope of this book to show typical patterns of all possible irregularities and defects that can possibly occur in the vibrator, transformer, or buffer capacitor, any deviation from a normal pattern indicates trouble that must be analyzed. Points to consider are —contact condition, contact spacing and centering, coil resistance, vibrator frequency, transformer design, buffer capacitor value, and minimum starting voltage.

For service in dc-powered timing lights, the normal output dc voltage should be approximately 400 to 500 V and be capable of delivering at least 10 to 15 ma at that voltage.

The voltage across the common timing light flash tube is about 400 to 500 V, or about one-half the voltage at which the tube will flash and conduct. The ignition electrode then fires this gas tube to conduct at 400 to 500 V. It will continue to glow and conduct until the voltage reaches the extinguishing voltage below about 50 V.

In a powered timing light circuit the power supply must be capable of building up to 400–500 V and holding steady until the tube is fired. Then, it must be capable of dropping quickly to almost zero volts (or to the point where the tube is extinguished), and then recover quickly to 400–500 V. Some circuits have inherently poor voltage regulation to accomplish this purpose, with the filter capacitor discharging through the tube. Others may use a resistor in series with the capacitor connected across the flash tube electrodes. When the capacitor discharges through the tube, the voltage across the capacitor drops to nearly zero and the tube is extinguished. The voltage across the capacitor then builds up to 400–500 V for the next flash at the time rate of the R-C combination.

At slow engine speeds, or idling, the capacitor will build up to almost full voltage, say 350 V, because of the time delay constant of the resistor-capacitor combination. The higher the resistance, or the larger the capacitance, or both, the time to charge the capacitor is increased. At higher engine speeds the voltage may be built up to only about 200 V or less between flashes. This results in a less brilliant flash, but since these flashes occur more frequently, the average light output is calculated to be about the same as at low speed.

If the engine speed is too high, the tube may skip some flashes because there is not sufficient time for the capacitor to build up to the required voltage. Any attempt to hold this voltage up at high speeds may result in overheating and destroying the flash tube.

Any timing light designed for timing automotive engines depends on the high-voltage pulse found at the spark plugs. This voltage varies widely in cars of different manufacture, cars of the same make, and even at the plugs on the same car.

If the flash tube electrode in the timing light is set for maximum sensitivity to assure proper operation on unusually low voltages, double tripping may occasionally take place. This is caused by the proximity of adjacent plugs or wiring when the light is used on cars having very high voltage at the plug.

If the flash tube electrode is adjusted to function properly on unusually high voltages, the timing light may occasionally skip or misfire on very low voltages, frequently caused by a defective ignition system.

Occasional misfiring, or double tripping is not necessarily indicative of trouble in the timing light, neither does it interfere with the timing procedure.

Nonsynchronous Vibrator, Source 14, Model 1621

Figure 11-2 shows a vibrator of Source 14, Model 1621, or equivalent having three leads out instead of a four-prong socket, and can be used as a universal replacement for units having these connections. The vibrator coil VC has a dc resistance of 42 ohms. The terminals numbered 1, 2, and 3 correspond to the terminals on a four-prong base as shown in Figure 11-3 and the leads are colored Y for yellow going to terminal 1, G for green to terminal 2, and B or BR for black or brown to terminal 3.

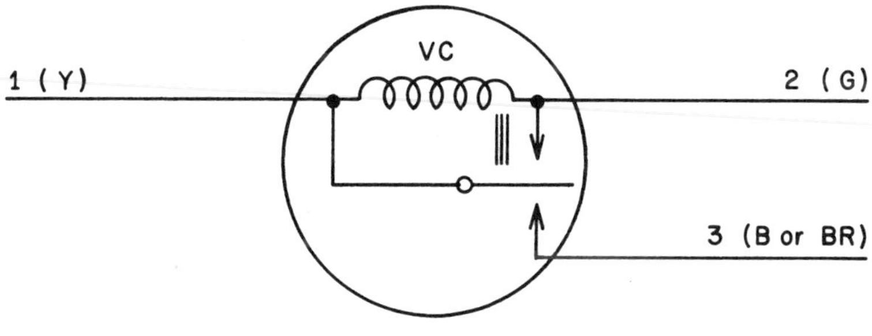

Figure 11-2 Nonsynchronous vibrator, Source 14, Model 1621

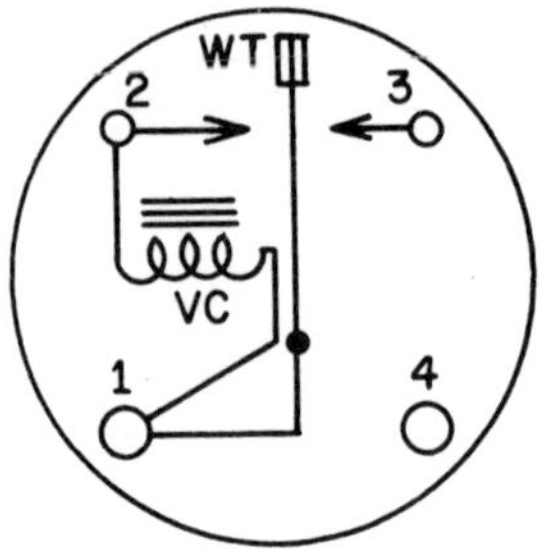

Figure 11-3 Common nonsynchronous vibrator

Common Nonsynchronous Vibrator

The common nonsynchronous vibrator shown in Figure 11-3 operates the same as the one of Figures 11-1a and 2 except that it is mounted on a four-prong base.

Nonsynchronous Vibrator, Source 14, Model 1610

Figure 11-4 shows a Source 14, Model 1610, vibrator (or equivalent). It has a four-prong base like the one in Figure 11-3, but the internal connection is different. Any conventional secondary circuit may be used on transformer T. Here, the vibrator VIB coil VC (40 ohms) is connected between prongs 1 and 4, and a ballast lamp BL (No. 51 bulb) is connected externally to prongs 2 and 4.

In Figure 11-4 the action is the same as explained for Figure 11-1a except that the ballast lamp BL allows about the same current to flow

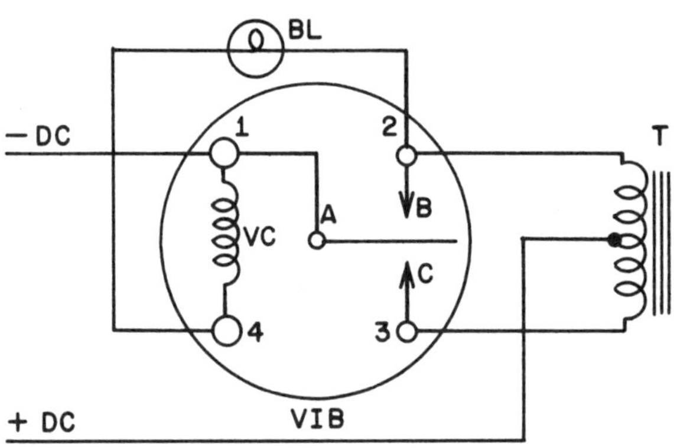

Figure 11-4 Nonsynchronous vibrator, Source 14, Model 1610

through coil VC on 6 V as on 12 V because of the cold resistance characteristics of ballast lamp BL. On 6 V the current is lower and the bulb has a lower resistance than it does when burning more brilliantly on 12 V.

Synchronous Rectifier Vibrator, Source 14, Model 1802

The vibrator of Source 14, Model 1802, (or equivalent), is a synchronous rectifier vibrator, Figure 11-5. Viewing the five-prong tube base from the prong-side, VC is the vibrator coil and R is a vibrating reed with a weight WT to give it inertia. Contacts pull interrupter PI, inertia interrupter II, pull rectifier PR, and inertia rectifier IR. This operates the same as explained for Figure 11-1b.

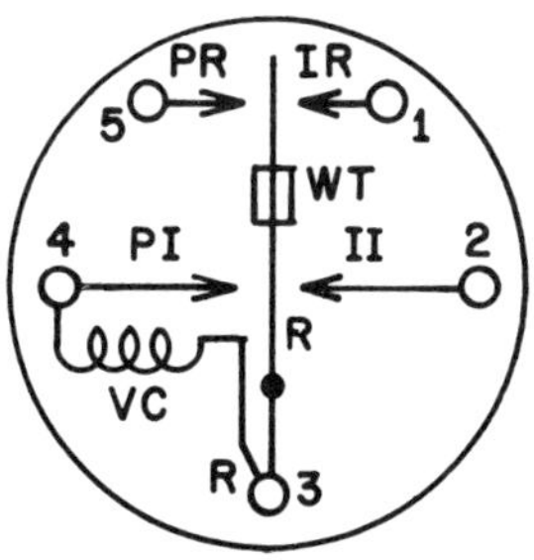

Figure 11-5 Synchronous rectifier vibrator, Source 14, Model 1802

11-4 Vibrator Transformer Design

Transformers used in vibrator (or solid state equivalent) power supplies require special consideration due to the direct current in the primary circuit that tends to saturate the core if too small. The core must be larger than for an ac transformer having the same output. The "shell" type of core is preferred. If the "core" type is used, one-half of the primary and one-half of the secondary winding must be wound on each leg to avoid saturation.

As battery voltage varies widely, the flux density must be kept low. The primary winding is almost twice as bulky as that of an equivalent ac transformer. The current is flowing only part of the time, making a larger-size wire necessary for the same average current. On the other hand, if the frequency of the vibrator is much higher than 60 Hz, such as 150 Hz, the higher frequency will reduce the size of the vibrator transformer. It will still be somewhat larger than an equivalent 60-Hz transformer.

In the design of small transformers it is difficult to calculate all of the values precisely because of the many unknown variables, but the following formulas should give a good first approximation.

Flux Density and Primary Turns

The flux density B max. in a vibrator transformer core is given by

$$B \text{ max.} = \frac{(1+P)(El \times 10^8)}{4\ Np\ A\ k\ f} = \text{lines per square inch.}$$

where P is time efficiency, El is normal battery voltage, Np is the total number of primary turns, A is the cross section of the core in square inches, k is the stacking factor of the core, and f is the vibator frequency in hertz. Flux density B max. should be kept below 40,000 lines per square inch at normal battery voltage. A still lower figure may be desirable at high frequencies because of core loss.

Secondary Turns

As a first approximation use

$$\text{Ns} = 1.33 \times \text{Np} \times \frac{\text{desired dc output volts}}{\text{battery volts}}$$

where Ns is the total number of secondary turns.

Current and Wire Size

The rms current in the secondary winding is

$$\text{Is} = \frac{0.707 \times Idc}{\sqrt{\text{P}}}$$

The rms current in the primary winding is

$$\text{Ip} = \frac{0.707 \times Idc}{\sqrt{\text{P}}} \times \frac{Ns}{\text{Np}}$$

in which *Idc* is the dc load current of the secondary. Wire sizes should be chosen to have about 800 circular mils/amp. Usually the primary has relatively few turns because of the relatively low-input battery voltage as compared with the output voltage. The primary wire size should be chosen to give 2, 4, 6, or 8 even layers. This will place the center tap at the end of a layer, which is convenient for multiple winding.

Output Voltage

The total resistance r_T in ohms of the transformer and associated circuits, referred to the working half of the secondary, is

$$r_T = \left(\frac{r_1}{2} + r_V + r_L\right)\left(\frac{Ns}{Np}\right)^2 + \frac{r_2}{2} + r_3$$

in which r_1 is total primary resistance, r_V is vibrator contact resistance, r_L is the battery lead resistance, r_2 is the total secondary resistance, and r_3 is the rectifier resistance, if any.

During the time that the vibrator contacts are closed, the secondary current equals *Idc/P*. Then the *IR* drop expressed in volts on the secondary is $r_T \times Idc/P$.

The no-load dc voltage across the first filter capacitor is

$$E_0 = aE1 \times \frac{Ns}{Np}$$

in which a is a constant which takes care of imperfect contacts, primary and secondary contacts not exactly synchronized etc. This constant equals 0.94 for a synchronous rectifier and 0.98 for a separate rectifier, such as a tube or diode.

Then the full-load dc voltage across the first filter capacitor is

$$E_2 = E_0 - \frac{r_T \mathrm{Idc}}{P}$$

With the use of transistors or other solid state oscillators instead of the mechanical vibrator, there is no limit to the frequency used, and therefore the transformers can be correspondingly smaller.

11-5 Voltage Multiplier and Rectifier Circuits

The powered timing light requires a relatively small amount of power to charge the capacitor that ultimately fires the flash tube. Therefore an elaborate rectifier and power supply having good voltage regulation is not required. Usually some form of voltage multiplier circuit is used. From an initial 115 V ac supplied from the power line, or 200 V from a battery-operated vibrator, and small transformer, the voltage is doubled, tripled, or quadrupled by one of the circuits shown in Figures 11-6 and 11-13. These circuits raise the ac voltage to a higher dc voltage.

There are two diodes and two electrolytic capacitors used in the voltage doubler circuit of Figure 11-6a. At low current drain, the doubler circuit will deliver a dc voltage equal to approximately 2.8 times the rms value of the ac input voltage. During the positive half-cycle of the ac voltage, when AC1 is positive and AC2 is negative, diode D2 is blocking and D1 is conducting, which charges capacitor C1 at the polarity shown. During the negative half-cycle, when AC1 is negative and AC2 is positive, diode D1 is blocking and diode D2 is conducting, which charges capacitor C2 at the polarity shown. At 60 Hz (and even higher when a dc vibrator is used) this process repeats itself so rapidly that both capacitors remain charged at approximately the rms voltage of the ac. Since they are in series, and adding polarity, the dc voltage output will be more than twice the ac rms voltage at low current drain, but the dc voltage drops rapidly as the load is increased. This is desirable in timing light applications.

There are three diodes and three electrolytic capacitors used in the voltage tripler circuit of Figure 11-6b. At low current drains the tripler circuit will deliver a dc voltage equal to approximately 4.2 times the rms value of the ac input voltage. During the positive half-cycle of the ac voltage, when AC1 is positive and AC2 is negative, diodes D2 and D3 are blocking, and diode D1 is conducting. Capacitor C1 is charged through capacitor C3 and diode D1. During the negative half-cycle, when AC1 is negative and AC2 is positive, diode D1 is blocking but D2 and D3 are conducting, which charges capacitors C2 and C3. During the second positive half-cycle (and thereafter), when AC1 is positive and AC2 is negative, the voltage of charged capacitor C3 is added to the ac rms voltage and therefore approximately doubles the voltage charge across capacitor C1. With C1 at double voltage in series with ca-

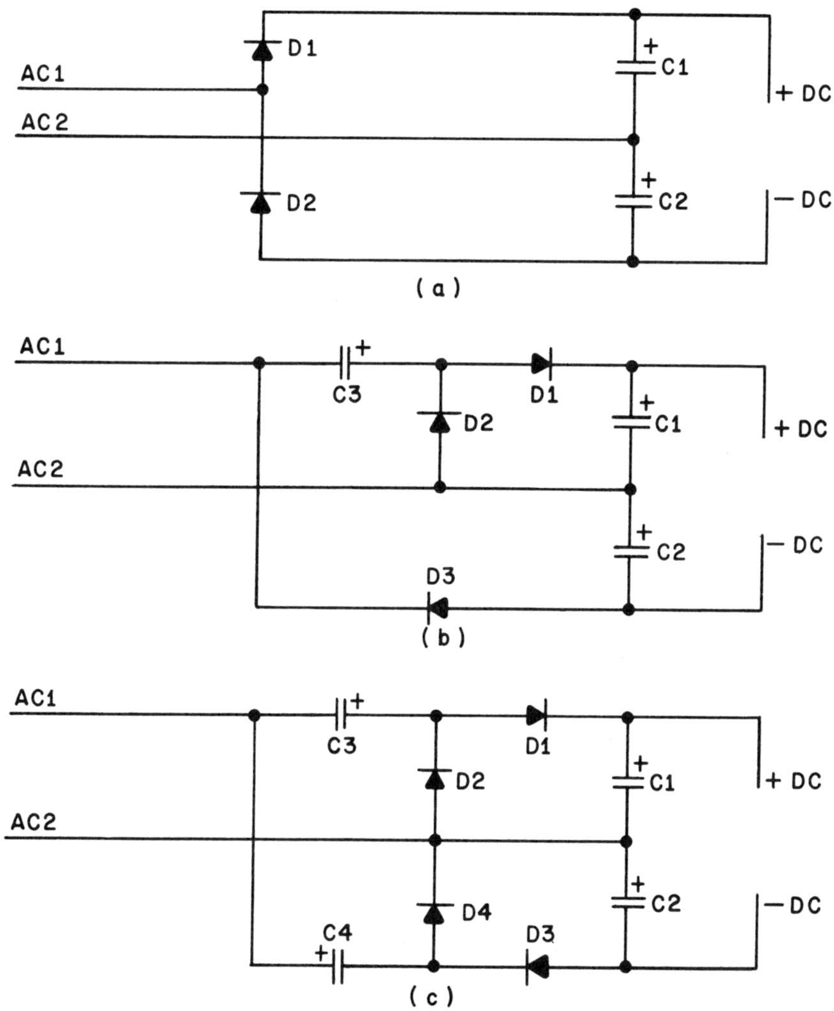

Figure 11-6 Voltage multiplier and rectifier circuits. (a) Doubler. (b) Tripler. (c) Quadrupler

pacitor C2 at single voltage, there is delivered a dc voltage of approximately three times the ac rms voltage.

There are four diodes and four electrolytic capacitors used in the voltage quadrupler circuit of Figure 11-6c. At low current drains, the quadrupler circuit delivers a dc voltage equal to approximately 5.6

times the rms value of the ac input voltage. The explanation for the tripler circuit shown in Figure 11-6b applies to the quadrupler circuit shown in Figure 11-6c, except that the additional capacitor C4 and diode D4 places a double voltage charge on capacitor C2 in the same manner. With capacitors C1 and C2 each charged at double voltage and in series adding, the dc output voltage is approximately four times the rms ac voltage.

For frequencies between 60 Hz and 10,000 Hz, use 8-mfd electrolytic capacitors throughout. Lower values are usually used, depending on the voltage regulation required. The diodes are all the same and are rated according to the current and voltages encountered. For the voltage doubler circuit shown in Figure 11-6a, which is common in timing lights, the diodes are sometimes in the form of a stack of selenium plates with three leads out—one outer lead the anode, the other the cathode, and a middle lead common to an anode and cathode.

11-6 Common Nonsynchronous Timing Light Powered by Direct Current Vibrator

Figure 11-7 shows a common timing light circuit for use on either 6- or 12-V batteries. The relay RL remains in the de-energized position as shown, with armature A in contact with the 6-V contact. When the positive dc and negative dc clips are connected to the battery, the voltage is applied across the relay coil RL (200 ohms). If the voltage is below 9 V, the contacts remain in the 6-V position. If the voltage is above 9 V (12-V system), then armature A moves to the 12-V contact. This relay selects the proper tap on the secondary of the transformer T1 to give approximately the same voltage output whether on 6 or 12 V.

The vibrator VIB is a four-prong Oak No. 449612-034 (249-6639), or equivalent, with coil VC (55 ohms) connected internally to prongs 1 and 2. This is not interchangeable with other four-prong vibrators, for example, those that have the coil VC across prongs 1 and 4.

The positive dc is fed to the center tap on primary P of transformer T1 and through the upper half of the primary to terminal 2, through coil VC and back to the negative dc. This causes coil VC to attract the movable armature A, making contact with B. This momentarily shorts out coil VC, which releases the armature A, but because of spring action it flies over to make contact with C, causing current to flow through the lower half of the primary. Armature A is again attracted away from con-

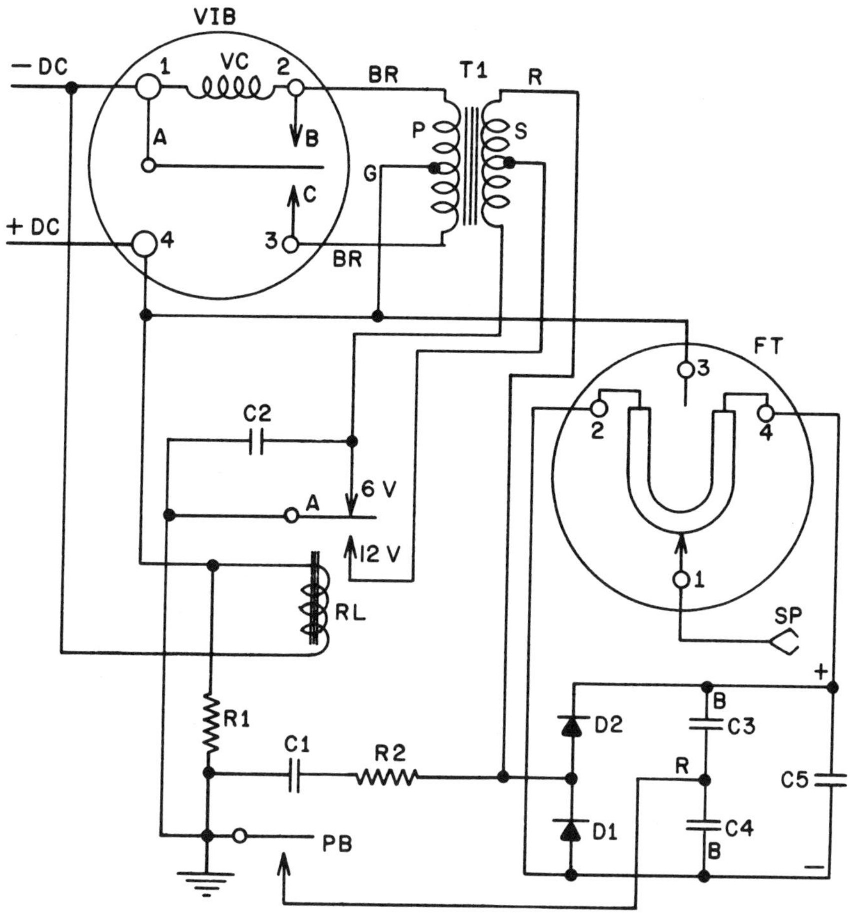

Figure 11-7 Direct current timing light, common, nonsynchronous, 6/12-volt

tact C to contact B. This cycle repeats itself at a high frequency of approximately 100 to 400 Hz, causing a reversal of current in the primary P of transformer T1. This reversing current (square wave alternating current) induces a higher voltage of about 220 V in the secondary S of transformer T1. This ac voltage is fed to the voltage doubler circuit. This circuit consists of a selenium stack represented by diodes D1 and D2 and capacitors C3 and C4 (0.5 mfd) with capacitor C5 (1.25 mfd) acting as additional capacitance to discharge and fire the flash tube FT. Terminal 1 of the flash tube socket connects to the ignition electrode of the flash tube FT and to clip SP, which is connected to the No. 1 spark plug. Terminal 3 has a 3/8-in. extension of the lead to cause a corona from high voltage lead terminal 1.

Other values are, R2 (8.2 Kohms), R1 (330 ohms), and C1 and C2 (0.01 mfd). The ground is the die cast case of the timer.

The vibrator VIB is designed to operate efficiently on either 6 or 12 V. When connected to the battery the vibrator operates continuously, but the flash tube does not flash until the high voltage dc circuit is completed by depressing the push button switch PB located in the trigger position of the pistol grip handle of the timing light case.

The voltage regulation of the power supply must of necessity be "poor," or a series resistor used, to allow the voltage to drop to a low value when the capacitors C1, C2 and C3 discharge through the flash tube. The capacitors are then recharged and the voltage builds up again. Capacitors C1 (0.01 mfd) and series resistor R2 (8.2 Kohms) serve as a buffer circuit to prevent vibrator sparking.

On most powered timing lights there is actually no metallic return path to ground for the ignition voltage from the No. 1 spark plug that fires the flash tube. Sufficient current is conducted back to ground through the capacitance between the secondary and primary of the transformer. Also the small mass of the timing light acts as a small earth ground, and due to the static "inertia" it acts as a ground.

Figure 11-8 shows the schematic of a timing light that is similar to Figure 11-7 except for the vibrator internal connections, and the use of a Source 14, Model 1610, vibrator VIB. Other values are, bulb B1 (No. 51), C1 (0.047 mfd, 600 V), C2 (0.01 mfd, 1.6 KV), C3 and C4 (4 mfd), R1 (470 ohms), RL (6/12-V relay), diodes D1 and D2, flash tube FT, and a switch for turning the battery off and on (not shown).

11-7 DC-Powered Timing Light, Nonsynchronous Half-Wave Vibrator, Full-Wave Bridge Rectifier, Source 15, Model 41-117

Figure 11-9 shows a dc-powered timing light that uses a nonsynchronous vibrator VIB operating on half-wave by interrupting the primary P circuit of transformer T1 using vibrator reed contacts 1 and 3. Contacts 1 and 2 are normally closed, but when the vibrator coil VC is energized, it disconnects contact 2, and after it has made contact with 3, it bounces back to start a new cycle.

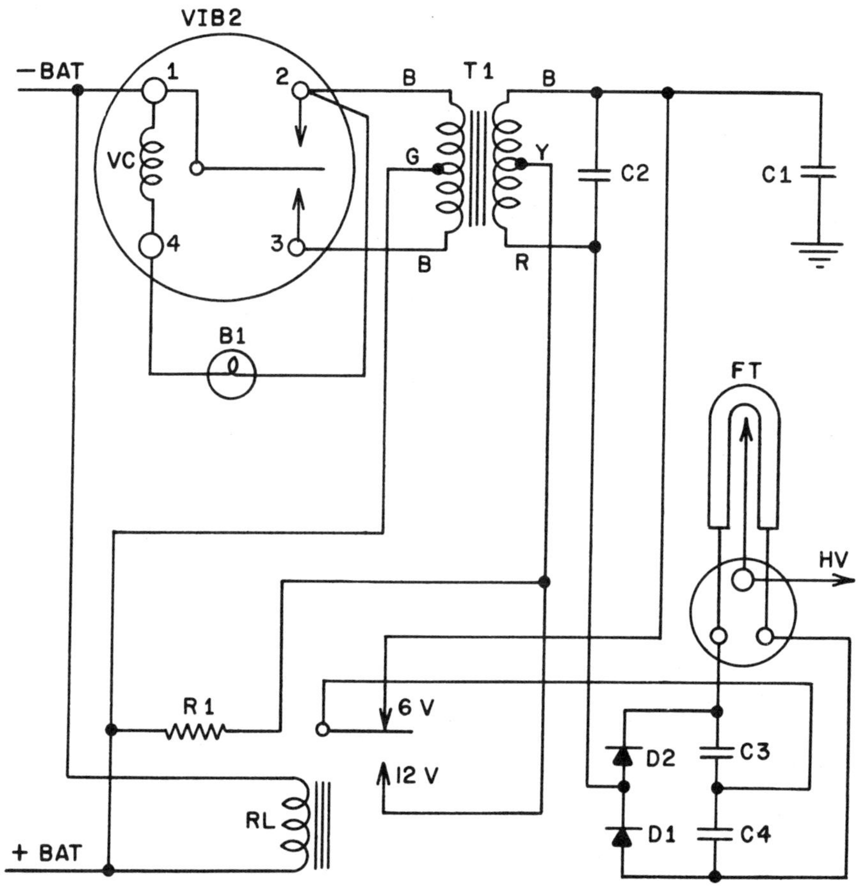

Figure 11-8 Direct current-powered timing light, nonsynchronous vibrator, 6/12-volt

Voltage-sensitive relay RL coil actuates the three-pole double-throw set of contacts RLC1, RLC2, and RLC3 to select a different tap on the primary of the transformer T1 for 6 or 12 V, and it also inserts a resistor R1 (22 ohms) for 12-V operation to be in series with the vibrator coil VC. Normally open switch SW1 is closed to energize the timing light.

The high-voltage secondary circuit shown in the lower half of Figure 11-9 consists of a selenium stack bridge rectifier represented by diodes D1, D2, D3, and D4. Condenser C1 (0.018 mfd, 1000 V) is a buffer, and capacitors C2 and C3 (1 mfd, 400 V) are the flash tube firing capacitors.

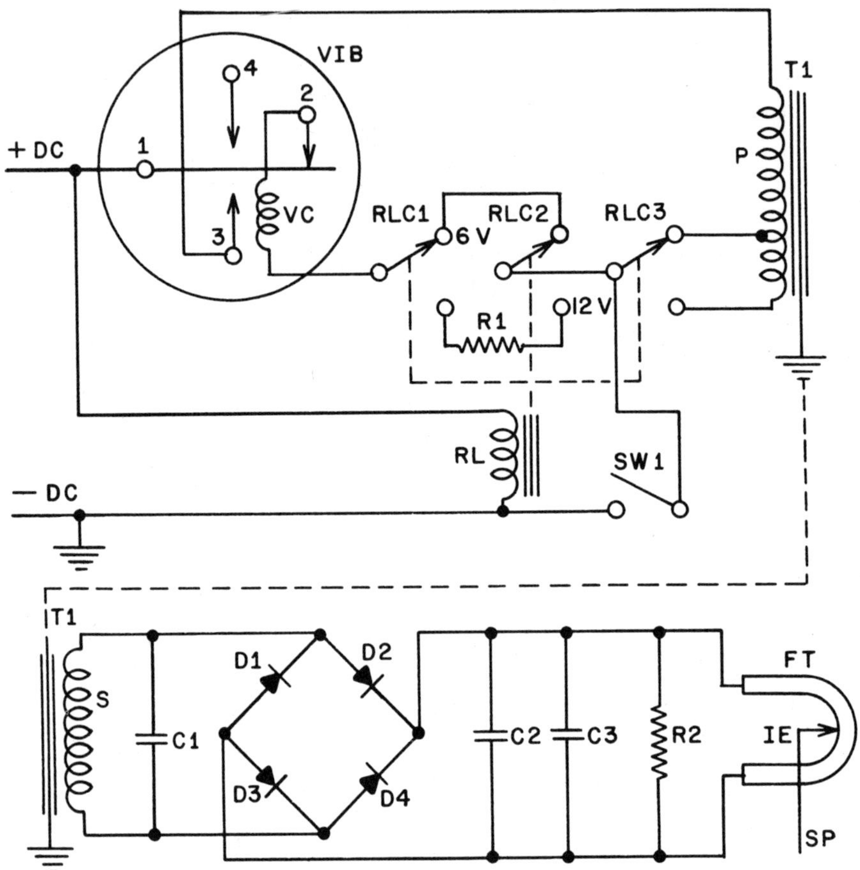

Figure 11-9 Direct current-powered timing light, typical, nonsynchronous half-wave vibrator, full-wave bridge rectifier, Source 15, Model 41-117

Resistor R2 (220 Kohms) is a bleeder resistor. The ignition electrode IE of flash tube FT goes to the No. 1 cylinder spark plug through a lead terminating in a clip at SP.

11-8 Typical DC-Powered Timing Light with Synchronous Half-Wave Vibrator and Rectifier Source 15, Model M804D

Figure 11-10 shows a dc-powered timing light using a synchronous vibrator VIB having only half-wave primary interrupter and half-wave secondary synchronous rectifier.

Vibrator reed 1 is normally closed to contact 2, which energizes the vibrator coil VC. This opens contact 2 from reed 1, which flies over to make contact with contacts 3 and 4. Contact 5 is not used. Contact 3 is in the primary P circuit of step-up transformer T1, and contact 4 is in the secondary S circuit of transformer T1. The primary and secondary voltages are interrupted in unison to provide 300 to 400 dc volts across the flash tube FT and capacitor C2 (2 mfd, 600 V). Resistors R2 (4700 ohms) and R3 (220 Kohms) with capacitor C1 (0.007) provide a secondary buffer network. Capacitor C2 (2 mfd, 600 V) is the flash tube FT discharge, or firing capacitor.

Relay RL coil is connected across the battery voltage, either 6 or 12 V. This relay RL is normally in the position shown for 6 to 9 V (6-V operation), but at voltages above 9 V (12-V operation) the relay contacts move to the 12-V position to change a tap on the primary P of trans-

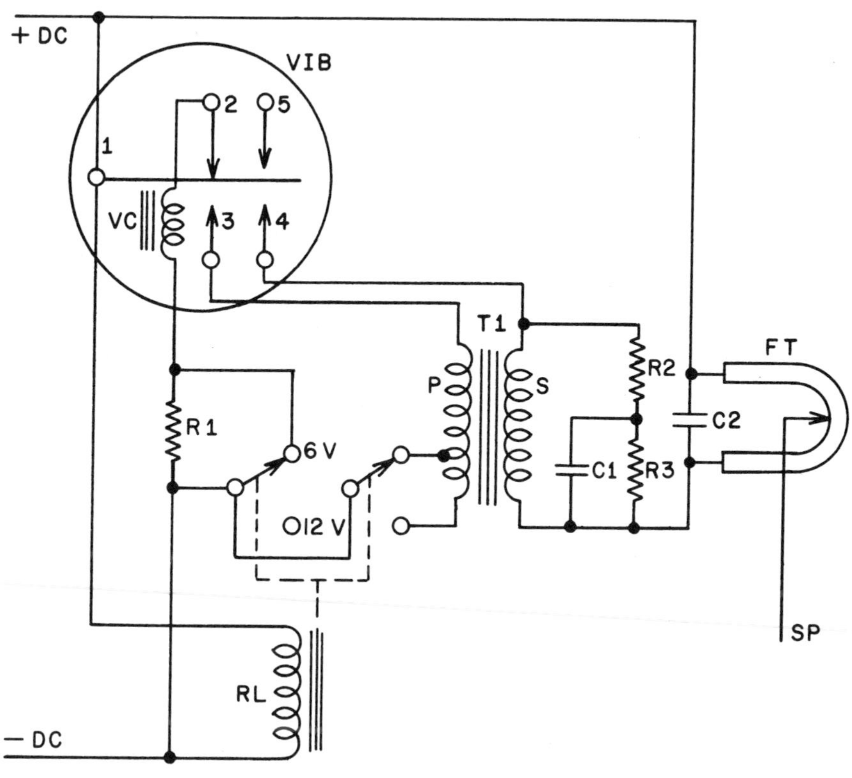

Figure 11-10 Typical dc-powered timing light having synchronous half-wave vibrator and rectifier, Source 15, Model M804D

former T1, and also to insert a resistor R1 (33 ohms) in series with the vibrator coil VC so that the dc output will be approximately the same on either 6 or 12 V.

A circuit very similar to this is used by Source 12 and others, but the 6/12-V relay RL is eliminated. Instead, a 6/12-V ballast resistor is substituted in series with the battery input to hold the output substantially the same on 6- or 12-V operation.

11-9 Transistorized, 6/12-Volt-Powered Timing Light, Source 15, Model 317B

Figure 11-11 shows a transistor Q1 (2N1540) functioning in an oscillator circuit in place of a mechanical vibrator, and is typical of Source 15, Model 317B. The ac output voltage of transistor Q1 is fed into the primary of transformer T1, where it is stepped up to a higher voltage and rectified by the high voltage silicon diode D2, and direct current is fed to flash tube FT. Capacitor C1 (0.001 mfd, 3KV) acts as a buffer, and C2 (2 mfd, 500 V) is the discharge capacitor for firing the flash tube. Resistor R4 (390 ohms) is a current-limiting resistor for charging capacitor C2.

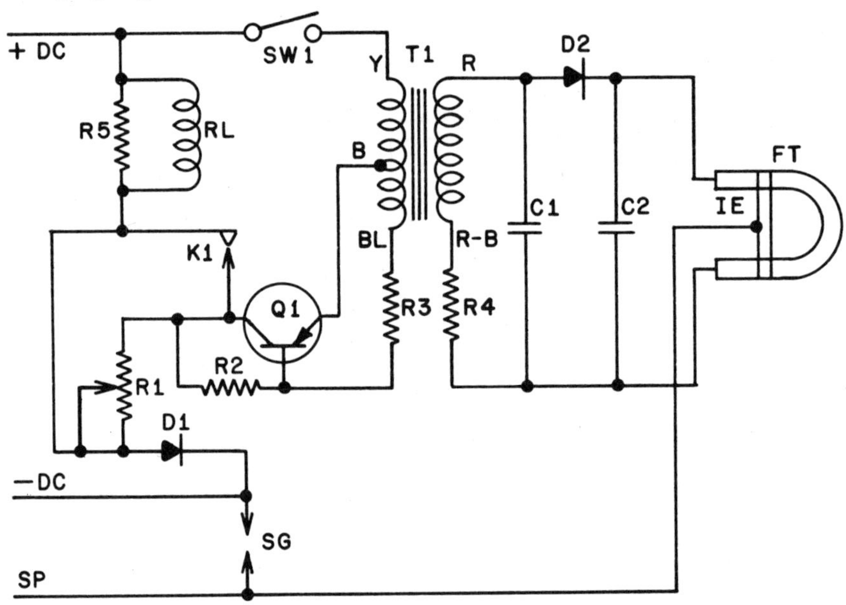

Figure 11-11 Transistorized dc-powered timing light, Source 15, Model 317B

The ignition electrode IE of the flash tube FT is connected to the No. 1 spark plug by clip SP. The spark gap SG is connected between this lead and the battery negative and is adjusted to 15 KV peak voltage, or to about ½-in. gap to limit the ignition voltage to the flash tube FT for vehicle ignition voltages above 15 KV.

The positive and negative dc input leads are connected to a 6- or 12-Volt battery. Relay RL is voltage sensitive and is adjusted to open the contacts K1 of relay RL at about 9 V. Therefore, on 6 V the contacts K1 remain closed, and on 12 V they are open to insert rheostat R1 (3 ohms) to limit the input current on 12 V. Rheostat R1 is adjusted to give the same voltage across the transistor Q1 collector to emitter on 12 V as on 6 V. Resistor R5 (100 ohms) is connected across relay coil RL to act as a damper to prevent relay chattering and prevent transient voltages from damaging diode D1. Diode D1 is a small selenium plate (1.3×1.3 in.) that acts as a reverse polarity protector. No current will flow to the circuit if the battery connections are reversed.

Other values are, resistor R2 (470 ohms) and R3 (18 ohms). Switch SW1 is a N.O. push button switch for energizing the timing light.

11-10 Typical Transistorized Timing Light 12 Volt Only, Powered by Direct Current

Figure 11-12 shows the schematic wiring diagram of a battery-powered timing light for 12 V only, that uses a PNP transistor Q1 (2N3612) acting as a chopper, or oscillator, to change the vehicle battery dc voltage into ac. The transformer T1 primary P steps up the voltage of the oscillating current in the secondary S. A half-wave diode D1 rectifies this ac and applies a dc voltage of approximately 300 V across the flash tube FT and firing capacitor C1 (2 mfd, 600 V).

The ignition electrode IE fires the flash tube from the spark plug connection SP. Resistor R1 (120 ohms) is connected between the battery negative terminal and the metal case of the timing light to offer a ground return. This prevents a short circuit should the negative clip be accidentally connected to the hot battery and the case touched to ground. Resistors R2 (390 ohms) and R3 (3.3 Kohms) are transistor Q1 base B bias resistors. Diode D2 protects the transistor against high reverse voltage peaks.

The initial start of oscillation depends on capacitor C2 momentarily holding the base of the transistor Q1 at approximately the same voltage

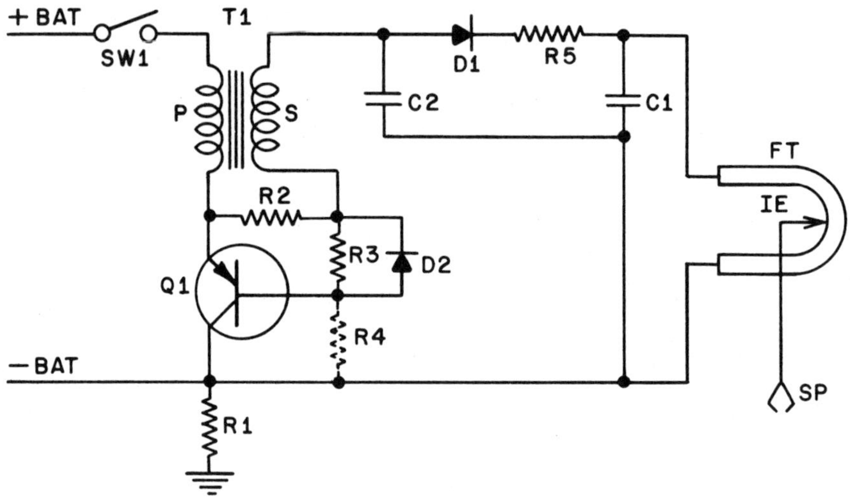

Figure 11-12 Transistorized dc-powered timing light, 12-volt only

as the collector, which causes transistor Q1 to conduct between the collector and emitter and through the transformer T1 primary P. When this conducts, the bias of the base is now at the emitter potential, and the transistor Q1 ceases to conduct. This causes the flux to collapse, inducing a high voltage in the secondary, at the same time conducting some current through resistor R2 and the secondary. This again biases the transistor base to conduct, and the cycle repeats itself at a frequency that depends on the constants in the circuit, such as the inductance of the transformer primary, capacitor, and the resistors.

In this circuit, if transistor Q1 has lower than normal gain, it may not start to oscillate. To correct this, either replace the transistor, or add a resistor R4 (34 Kohms), which places a more sure bias on the base.

Resistor R5 (120 ohms) is located in the flash tube base and serves as a filter and current-limiting resistor.

11-11 Cordless Timing Light

A timing light is available that has a built-in rechargeable battery and a charging stand. The primary circuit uses a high-frequency transistorized oscillator instead of a vibrator, and it is designed to require a minimum of battery power, which is not an important consideration when the power is obtained from the vehicle battery. The secondary circuit may have any conventional arrangement.

Only one connection, to the No. 1 spark plug, is required. The timing light is encased in a plastic shock-proof case. There is no actual metallic return path to ground from the No. 1 spark plug high voltage. This high voltage ignites the flash tube, however, because the timing light acts as a small earth ground, since it has some capacitance coupling to ground, but mainly because its small mass has a static inertia at ground potential.

11-12 Timing Light, Powered by 115-Volt, Alternating Current

The ac-powered timing light is by far the simplest and least expensive of all powered timing lights, but it has the disadvantage of re-

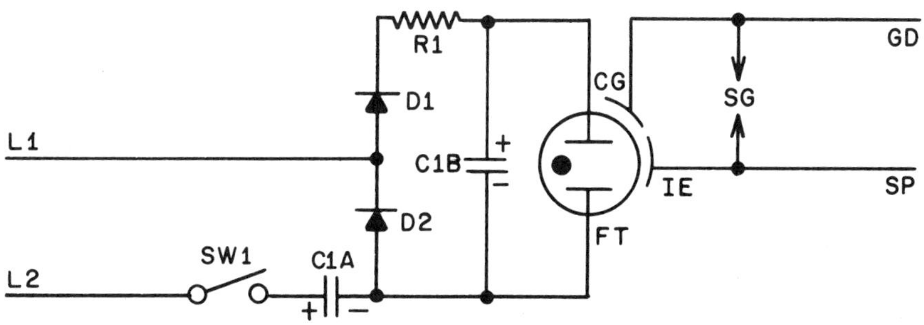

Figure 11-13 Alternating current-powered timing light, Source 15, Model 319

quiring a 115-V ac electric outlet near its point of use, or a long extension cord. The current drawn is very low, so the lightest available extension cord may be used.

Figure 11-13 shows an ac-powered timing light typical of Source 15, Model 319, which uses a unique voltage doubler and rectifier circuit. When the ac line lead L1 is negative and L2 is positive, capacitor C1A (10 mfd, 150 V) is charged through diode D2 to approximately 125–150 V, and diode D1 is blocking. When L1 is positive and L2 is negative, D2 is blocking and D1 is conducting a positive charging current through resistor R1 (120 ohms) to the positive of capacitor C1B (10 mfd, 300 V). Since capacitor C1A is already charged to 125–150 V, it adds to the line voltage at that instant, thereby placing a voltage of approximately 250–300 V across capacitor C1B and the flash tube FT.

This cycle repeats itself 60 times per second, restoring the voltage across the flash tube after each flash that discharges capacitor C1B.

The ignition electrode IE is connected to clip SP, which is connected to the No. 1 spark plug. A ground lead GD terminates near the flash tube at CG, causing a corona discharge as a return to the engine ground, but it does not flash over.

The spark gap SG is placed between the spark plug wire SP and ground to suppress vehicle ignition voltages above 15 KV, and it is adjusted to spark over at 15 KV peak, or a spacing of about ½ in., depending on the sharpness of the air gap points.

Switch SW1 is a N.O. push button switch to energize the timing light.

In this circuit the two capacitors C1A and C1B are in the same enclosure. The circuit will not function properly if the internal resistance of C1B is higher than normal, that is, no greater than 1.3 ohms when measured at 50 KC.

11-13 Cantonwine Timing Light for Measuring Dwell Angle

The author has invented, and officially filed a patent application for, a dual-flash timing light that measures the dwell angle precisely.

Disclosure of the details here is intended only to provide advance information for the individual student and technician, and does not imply nor grant rights that the inventor may have in this device. Any commercial use of this information should be preceded by a thorough patent search and with full knowledge that a patent application has been filed covering the specifications and drawings of this device. The patent application was filed substantially as follows.

A. Abstract of the Disclosure

The conventional timing light used in automotive engine adjustments, flashes a stroboscopic light on a timing mark, usually located on the flywheel or front damper, indicating the firing of cylinder number 1, as caused by the *opening* of the distributor points. This invention teaches how the timing light can be made to also flash when the distributor points *close*. If an additional "points closing" mark is located on the same flywheel or damper as the "points opening" mark in an advanced position of the "points opening" timing mark, at the point where the distributor points should close to achieve the correct dwell angle, then the distributor can be adjusted to the precise dwell angle by aligning the two marks so they appear to coincide. This will eliminate the need to look up, or even know the correct dwell angle in each case as is required using the present system. This invention will also teach a mode of construction to provide for measuring the dwell angle on engines not yet provided with the additional "points closing" mark.

B. Background of the Invention

1. *FIELD OF INVENTION* This invention relates to stroboscopes in general and more particularly to improved means of using a stroboscopic timing light for indicating and determining the dwell angle of internal-combustion engine distributor points, as well as calibrating dwell meters and tachometers.

2. *DESCRIPTION OF PRIOR ART* In the past, it has been the practice to measure the dwell angle with a calibrated voltmeter which, at best, is subject to error due to rough handling and constant use. Some are even affected by point resistance and battery voltage.

C. Summary of the Invention

The present invention overcomes these objectionable features by providing a relatively simple dwell angle measuring device that can never lose calibration, and the accuracy is limited only by the ability to read a dial calibrated to one-tenth of a degree, or better, which is far more accurate than required. The accuracy is not affected by point resistance, battery voltage, or any other known conditions.

This invention will serve two distinct functions. Function (1), to indicate the correct dwell angle, and other purposes, on engines marked with a "points closing" mark located in advance position of the present "points opening" timing mark on top dead center (TDC). Function (2), to indicate the dwell angle, and other purposes, on engines not marked with a "points closing" mark.

It is therefore a principle object of this invention to provide a means to adjust accurately and dependably the distributor points of an internal-combustion engine according to the manufacturer's specifications without referring to charts or service manuals. Other objects of this invention are as follows: To provide,

1. Means to accurately calibrate existing dwell meters to within one-tenth of a degree, or better.
2. Means of accurately calibrating tachometers.
3. An adjustable speed stroboscope accurately calibrated in revolutions per minute (rpm) and for stop-action study of engine valves, springs, and the like.
4. A means of accurately adjusting distributors in a test stand for "off the engine" adjustments and tests.
5. A means of measuring the dwell angle of any individual cylinder lobe, the two opposite lobes, and the average of all cylinder lobes, or any combination thereof.
6. A timing light that will fire a flash tube at the instant the distributor points close, or fire at the instant the points open, or fire at both the closing and opening instant.
7. Improved means of indicating the amount and position of imbalance when used on a dynamic balancing machine, by firing a stroboscopic light at both the closing and opening of the contacts on the dynamic balancer.
8. An internal-combustion engine that fires on the closing of the distributor points instead of firing on the opening of the distributor points which has been the practice in the past. The ignition system to be within the scope of this invention using any applicable solid state electronic devices or mechanical devices.

These and other objects and advantages of the present invention will become apparent after considering the following detailed specifications in conjunction with the accompanying drawings.

D. Brief Description of the Drawings

In the drawings—

1. Figure 11-14 is a schematic wiring diagram showing the ignition system of a typical internal combustion engine and the electronic circuit embodying the present invention for function (1) and function (2).
2. Figure 11-15 is a schematic drawing showing a calibrating device embodying the present invention to be used in conjunction with the circuit shown in Figure 11-14 for function (2).
3. Figure 11-16 is a detailed assembly drawing of rotating dial 57 shown in Figure 11-15.

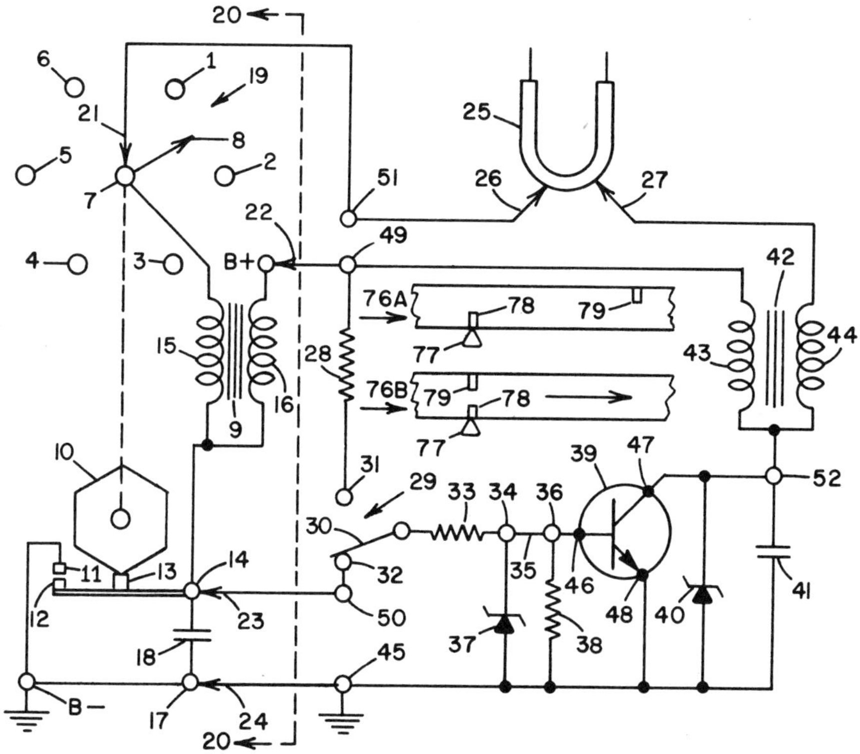

Figure 11-14 Cantonwine timing light for measuring dwell angle

E. Description of Preferred Embodiment

Referring to the drawings by reference numbers, the number 20 in Figure 11-14 refers to a typical automative ignition system which has a distributor 19, and ignition coil 9 and a battery B plus and B minus.

The distributor 19 consists of a 6-cylinder distributor cap designated 1-7 having a rotor 8 driven by a 6-lobe cam 10. This invention is not restricted to 6 cylinders, but applies equally to any number of cylinders. The cam 10 is driven by the crankshaft at one-half the crankshaft speed, on a 4-cycle engine. Cam 10, in contact with rubbing block 13, opens and closes the contact points 11 and 12. Condenser 18 is connected across points 11 and 12. Coil 9 consists of an iron core having a low-resistance primary winding 16 and a high-resistance, high-voltage secondary winding 15 which is connected to the center terminal 7 of the distributor cap 19. All reference numbers above 20 represent the present

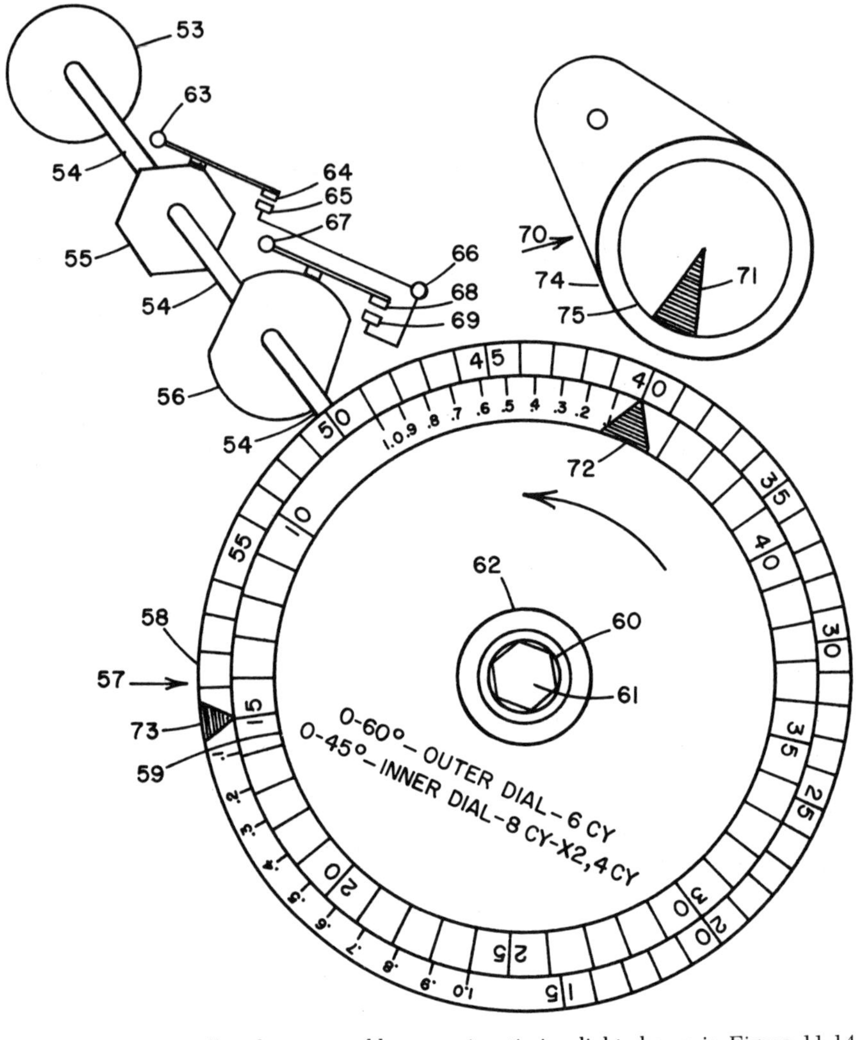

Figure 11-15 Dwell tachometer calibrator using timing light shown in Figure 11-14

invention. The basic circuit as shown in Figure 11-14 requires only four connections to the ignition system. Clip 21 (arrow) is connected to the center terminal 7 of the distributor cap 19, clip 22 (arrow) is connected to the positive battery terminal B plus, clip 23 (arrow) is connected to the distributor points 14, and clip 24 (arrow) is connected to the battery negative B minus and ground through terminal 17.

The reference number 25 in Figure 11-14 refers to a flash tube of a typical powered timing light, but has two ignition electrodes 26 and 27 instead of the usual one. It will be shown that electrode 27 flashes the tube 25 when the contact points 11 and 12 close, and electrode 26 flashes the tube 25 when the contact points 11 and 12 open. Although only one flash tube 25 is shown with the two ignition electrodes 26 and 27, this light can consist of a single flash tube having a single ignition electrode, by isolating the two high voltages from each other by a double spark gap, for example, but each high voltage would feed the common ignition electrode; or it can have two separate flash tubes having one ignition electrode each and powered by a common power supply of ample capacity, which is the preferred embodiment; or there can be two completely independent timing lights. The dc power supply, or amplifier, is not shown, as any conventional system can be used. Even plain neon timing lights can be used. All of these modes are within the scope of this invention. Any flash tube that flashes a light at short time intervals and duration may be synonomously referred to as a timing light,

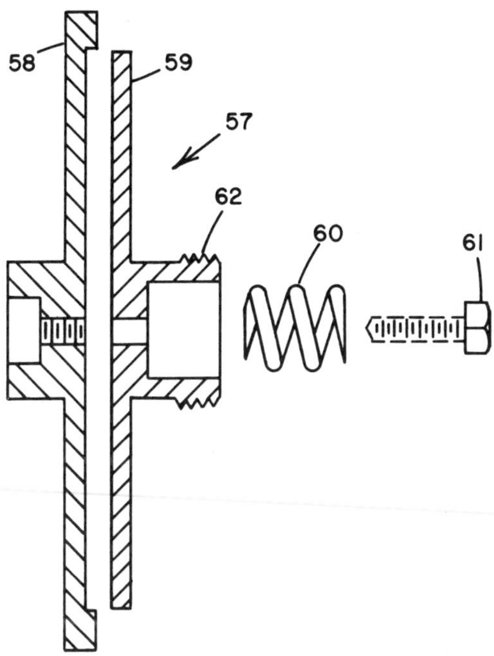

Figure 11-16 Details of dial 57 of Figure 11-15

stroboscope, stroboscopic flash light, strobe light, neon light, or the like.

In Figure 11-14, with contact points 11 and 12 open as shown, the full battery positive voltage is present at terminal 14. With clip 23 attached to terminal 14, a positive dc voltage is fed to the base 46 of transistor 39 through switch 29 stationary contact 32, movable contact 30, resistor 33, terminal 34, lead 35, terminal 36, resistor 38, and back to ground and battery negative B minus. Resistor 33 and resistor 38 form a voltage divider network, and zener diode 37 holds the transistor base 46 bias within safe limits. With the base 46 of NPN transistor 39 positively biased, the transistor 39 is conducting between collector 47 and emitter 48. With clip 22 connected to the battery positive B plus, current now flows through a standard ignition coil 42 primary 43, through terminal 52 to transistor 39 collector 47 to emitter 48 and back to the battery negative B minus and ground. This current builds up the ignition coil 42 in the usual manner. Condenser 41 is the usual ignition condenser and zener diode 40 is rated to protect the transistor 39 from the high voltage surges when the coil 42 fires. When the contacts 11 and 12 close, the positive bias is removed from the base 46 of transistor 39 and it ceases to conduct, which causes the coil 42 to generate a high voltage in the usual manner, igniting flash tube 25 through ignition electrode 27 from coil 42 secondary 44.

With the contact points 11 and 12 closed, the coil 9 now builds up in the usual manner. When the contact points 11 and 12 open, the coil 9 secondary 15 generates a high voltage and ignites flash tube 25 through ignition electrode 26, and the cycle repeats itself.

In Figure 11-14, the reference number 76A represents the flywheel, damper, or any other moving part of the engine, when not running. Reference number 77 represents the pointer or indicator located on a stationary member of the engine. Reference number 78 represents the "points opening" mark, and number 79 the "points closing" mark. Reference 76B, having an arrow pointing in the direction of rotation when running, and shows two marks, 78 and 79 in line for correct dwell angle, as they appear under the flash tube light. This is the preferred embodiment. The existing timing mark alone, could also be used by adding another stationary index pointer 77, but would have the disadvantage of requiring two separate observations at each pointer 77 and also allow for the degrees of advance or retard. The preferred embodiment will show the correct dwell angle at all degrees of the spark advance or retard, providing of course the dwell angle does not change; and marks 78 and 79 will line up with each other, but not necessarily line up with the stationary pointer 77, unless the spark is on TDC. Therefore pointer 77 is not required to read the dwell angle.

The flash tube fires twelve times during one revolution of the cam 10, six times when the contact points 11 and 12 close and six times when they open (for 6 cylinders).

When used in function (1) each revolution of the crankshaft will show six flashes, three opening and three closing for cylinders 1, 2, and 3 in the firing order of 153624 (cylinders 1, 5, and 3) and will show one exposure of the point closing mark 79 and the points opening timing mark 78 at the indicator 77 for cylinder number 1. For the second revolution of the crankshaft there will be six flashes, three opening and three closing, for cylinders 4, 5, and 6 in the firing order of 153624 (cylinders 6, 2, and 4), and will show one exposure of the points closing mark 79 and the points opening mark 78 at the indicator 77 for cylinder number 6 (4th in firing order). If the distributor is perfect, these two exposures will appear as one and be superimposed. Since these two flashes are from the opposite sides of the cam, any eccentricity of the cam will show up. This may be a desirable feature.

If it is desired to read the dwell angle on only one cylinder at a time, such as number 1, then connect clip 21 to the number 1 cylinder spark plug or distributor cap terminal number 1. This will cause the opening of the points to fire flash tube 25 from electrode 26 on the number 1 cylinder only, however the closing of the contact points will still cause the flash tube 25 to fire on each cylinder from ignition electrode 27. Provision can be made to cause the closing of the points to fire only on the number 1 (or any other selected) cylinder if desired. This can be accomplished in several ways that are well within the capabilities of one skilled in the art. A suggested method is to use a silicon controlled rectifier (SCR), or similar device to activate the transistor 39 only between the firing of the last cylinder, on contact point opening and the firing of the first cylinder on contact point closing position. To do this, remove lead 35 from between terminals 34 and 36. Connect the anode of a SCR to terminal 34, and the cathode of the SCR to terminal 36. The gate of the SCR can be connected to a voltage divider network between distributor terminal 6 and ground so that when the 6th cylinder fires on the opening of the points, it triggers the SCR into conduction and builds up the coil 42. When the points 11 and 12 close on number 1 cylinder the SCR holding current drops to zero and deactivates the transistor 39 which fires electrode 27 and remains inactive until the gate of the SCR is again triggered by the firing of the number 6 cylinder. Another more elaborate system could comprise a delay circuit similar to those used in DYNAMIC COMPRESSION and POWER CHECK circuits (described elsewhere herein), except here the selective firing of one cylinder at a time would be used instead of "shorting out" one cylinder at a time as is done in DYNAMIC COMPRESSION and POWER CHECK circuits.

From these suggestions and explanations, anyone skilled in the art can apply a SCR or similar device in the primary 16 circuit of coil 9, or in the primary 43 of coil 42 to activate or deactivate the firing on any group of cylinders desired.

The circuit shown in Figure 11-14 is for a single battery voltage and for a negative ground. For 6, 12, and 24 volts, for example, it is only necessary to insert the proper ballast resistor in series with the primary 43 of coil 42 by switching means. For a positive ground system, the NPN transistor 39 requires replacing, by switching means with the PNP transistor. The two zener diodes 37 and 40 require reversing, by switching means, or two additional zener diodes could be attached to the PNP transistor to simplify switching.

The components used in a successful working model, although not necessarily the best values to use ultimately, are as follows; resistor 33 (50 ohms), resistor 38 (100 ohms), resistor 28 (10 ohms not critical), zener diode 37, (4.7 V), zener diode 40 (60 V), transistor 39 (NPN Sylvania ECG-130), switch 29 (SPDT with OFF position), contact points 11 and 12, coils 9 and 42, condenser 18, and condenser 41 (not necessarily required) are all standard automotive parts.

The circuit shown in Figure 11-14 is all that is required for function (1) where the engine has been marked with a points closing mark.

Until such time as this invention is universally adopted it has been necessary to devise means of reading the dwell angle on engines not so marked. All conceivable methods have been tried or considered, such as applying precalibrated magnetic or adhesive tape, or marker to the engine damper, or drive a calibrated disc from the moving engine parts, or synchronized therewith, but it has been concluded that the best and simplest all-around mode of operation is to use an existing, or built-in, dwell meter and tachometer which can be periodically calibrated accurately by the calibrating device shown in Figure 11-15 and 11-16 using the present invention shown in Figure 11-14.

The circuit shown in Figure 11-14 can be further simplified if a transistor 39 is available that will withstand the high-voltage surges, by eliminating zener diodes 37 and 40, resistors 33 and 38, and condenser 41.

The calibrator shown in Figure 11-15 consists of an adjustable speed (200–2000 rpm) 115 volt ac motor 53 driving a shaft 54, on which is secured a six-lobe cam 55, a one-lobe cam 56, and the calibrated dial 57. A set of adjustable contact points 64 and 65 are actuated to open and close by the six-lobe cam 55. A set of adjustable (while running) contact points 68 and 69 are actuated to open and close by the one-lobe cam 56.

The rotating calibrated dial 57 is shown in separated cross-sectional detail in Figure 11-16 where dial 58 is fastened to the end of shaft 54 and calibrated 0–60 degrees for 6-cylinder engines as shown for a CCW rotation on the outer rim of the dial, having triangular pointer 73 located at zero and 60 degrees. The inner dial 59 is calibrated 0–45 degrees for an 8-cylinder engine (multiply the angle by two for a 4-cylinder engine) as shown for a CCW rotation on the inner dial having triangular pointer 72 located at zero and 45 degrees. Dial 59 is held in friction contact with outer dial 58 by a spring 60 and screw 61. Knob 62 is knurled and fastened to dial 59. To adjust the dials, pull out on knob 62 against spring 60 to turn freely to the desired degree setting, and then release slowly to the precise angle. Each dial has a vernier scale to enable adjustment in increments of one-tenth of a degree. The dial is shown set at 40 degrees on the outer dial by pointer 72. For most calibrating purposes, it is better to set the dial on a whole number, rather than a fraction. Instead of the knurled knob 62 and spring 60, any standard planetary geared down dial drive can be used for precise vernier adjustment.

The single-lobe cam and points allow the 60 degrees to be expanded over the entire circumference of 360 degrees. If the six-lobe cam is used, the scale would be compressed six times.

The calibrator shown in Figures 11-15 and 11-16 may be driven from the engine, or from the distributor while cranking and geared up 6 to 1, or may be driven by a synchronous motor through a converter deriving power from the engine battery through the distributor points and synchronized therewith, or by a sending device such as a Selsyn generator and motor, all of which are well within the scope of this invention.

In Figure 11-15 the reference number 70 shows a 115-volt, 60 Hz, 450 rpm (200 preferred) synchronous clock or timer motor having a stator 74 and an exposed rotor 75 having a triangular pointer 71. Clock motor 70 is only used to calibrate tachometers against 60 Hz frequency which is considered a very accurate source for such purposes, if obtained from the regular Utility Company outlet.

The switching operation means has purposely been omitted to simplify the drawings and permit a clearer understanding of the principles of operation. The connections required, and the results are stated under each of the following operations. Anyone skilled in the art can arrange the switching to suit.

The present invention can very well be used for other purposes to fire in reverse order, that is, fire first on the opening of the switching means and second on the closing of the switching means, or vice versa.

A switching device, or switching means, is defined herein as any mechanical, electrical, or other means of providing a relatively low impedance referred to as the closing position, or activated, and a relatively high impedance, referred to as the open position, or deactivated, or the application of a generated wave shape, or any form of pick-up or sending device. Anyone skilled in the art can adapt this invention to any ignition system such as, pointless, transistorized, condenser discharge (C-D), magnetos, balancing machine senders and the like, all of which are within the scope of this invention.

Dwell angle is herein defined as the percentage of time, expressed in degrees, that the distributor points remain in a closed position relative to the total time between the first and second opening position of the distributor points. The average dwell angle for engines is approximately two-thirds of the total time, or 67% of full dwell.

F. Operations That Can Be Performed with the Present Invention

1. *TIMING LIGHT ONLY.* For a conventional single flash timing light, connect clip 21 to number 1 on the distributor cap 19 or number 1 spark plug. Turn off switch 29 to place movable contact 30 midway between stationary contacts 31 and 32. Proceed to use the tim-

ing light flash tube 25 in the usual manner with the engine running. The dwell meter and tachometer can be connected to terminal 50 and ground terminal 45 (or the recommended connection) and left in this position, unless otherwise specified.

2. *FUNCTION (1).* To adjust the dwell angle on engines having a TDC or points opening mark and a points closing or dwell angle mark, connect clip 21 to the center terminal 7 of the distributor 19. Leave switch 29 in the position shown in Figure 11-14 with movable contact 30 engaging stationary contact 32. With the engine running, use the dual-flash timing light 25 and adjust the distributor dwell angle until the marks on the flywheel or damper are alongside of each other or superimposed, as shown in member 76B Figure 11-14.

3. *CALIBRATE A DWELL ANGLE METER.* In the past, dwell angle meters have been calibrated by using a square wave generator, which determined only one point of calibration at 50%, whereas the important point is approximately 67%. This invention can calibrate precisely at any angle between zero and full dwell. To calibrate, connect dwell meter to terminal 50 and terminal 45. With ignition ON, turn engine over by hand until points 11 and 12 open, as indicated by a zero dwell angle reading. Leave ignition on, but do not start engine. This operation can be performed independently of an engine if an ignition coil and condenser be built into the unit. Connect terminal 67 (contact point 68) to terminal 50, and connect terminal 66 (contact point 69) to ground terminal 45. Switch 29 and clip 21 should be in the position shown in Figure 11-14.

For 0–60 degrees on 6-cylinder engines, set pointer 72 for the desired dwell angle on the outer dial 58. For 0–45 degrees on 8-cylinder engines, set pointer 73 to desired dwell angle on inner dial 59. For 4-cylinders use 8-cylinder dial and multiply by 2. Connect motor 53 to power and adjust speed to just fast enough to show a steady reading on the dwell meter. Focus the twin timing light 25 on the rotating dial 57 near the area where the two triangular pointers 72 and 73 are near each other. Adjust the contact points 68 and 69 until the two pointers 72 and 73 are exactly opposite each other, and read the dwell meter. If it does not read the same as the degrees set on dial 57, then readjust the dwell meter if possible, if not, make a chart showing the amount to add or subtract. Usually, only one point in the commonly used range is sufficient. There will be two other pointers appearing on the dial 57. Pointer

73 will appear as shown and pointer 72 will appear at the 20–30 degree point. The reason for this is that when the light flashes on the closing of the points, they are in the position shown, 40 degrees before the points open, and when the points open, the dial 57 has rotated 40 degrees CCW placing pointer 73 exactly where pointer 72 was. Due to retentivity of the eye, it appears to still be there, but it actually has moved around to the 20–30 degree point and will appear there. This is not objectionable nor confusing, and cannot give an erroneous alignment. This could be prevented by mounting two flash tubes in back of the dial, but it would complicate the construction. The cam 56 and the bearings carrying the shaft 54 require careful design in order to repeat readings to within a tenth of a degree. The cam 56 must open and close the contacts rapidly to avoid flutter of the observed pointers. The dial 57 should be adjustable on shaft 54 to initially set it in relation to cam 56 so that the pointers line up in the area at the top near motor 70 so that a single mounting bracket can hold the timing light for dwell and tachometer calibrations.

The dwell angle calibrator is even more precise than the timing light itself. It is well known that there is an insignificant, but measurable, time delay between the opening of the contact, and the time the light is seen by the eye. This delay may be in the order of about one-thousandth of a degree. If the first timing light that flashes on the closing of the contacts has the same delay as the light that flashes on the opening of the contacts, then these two equal delays offset each other. It can therefore be said that the interval between the two flashes (dwell angle) is so precise that the present invention can appropriately be classified as a PRIMARY STANDARD DWELL ANGLE CALIBRATOR. It can never lose calibration, and there is no calibration of the instrument itself, all calibration consists of mechanical location of timing marks or calibrated dials with the accuracy limited only by readability. Anyone skilled in the art can devise means of measuring the dwell angle to any degree of accuracy desired, even to one-millionth of a degree or better.

4. *TO CALIBRATE A TACHOMETER.* Starting from scratch as shown in Figure 11-14 and 11-15, with the points 11 and 12 still open and ignition ON, but the engine not running, disconnect clip 21 from the center of the distributor 19 terminal 7. Connect terminal 63 (contact 64) to terminal 50, and connect terminal 66 (contacts 65 and 69) to the ground terminal 45. Connect terminal 67 (contact 68) to stationary con-

tact 31 of switch 29 and turn switch 29 movable contact 30 to engage stationary contact 31. Connect tachometer between terminal 50 and ground terminal 45, and set to "6 cylinders." Connect synchronous motor 70 to 115-V ac, 60 Hz frequency-regulated power supply source, such as any Utility Company power outlet. Connect adjustable speed motor 53 to power source. Turn the ignition ON (points 11 and 12 open), but engine not running. This calibration can be made independent of the engine if a separate coil and condenser be built into the calibrator. Focus the timing light 25 on clock 70 rotor 75 and adjust the speed of motor 53 to show approximately 450 rpm on the tachometer. Adjust the speed of motor 53 slightly until the triangular pointer 71 is standing perfectly still. The tachometer should read exactly 450 rpm (or the speed of motor 70). Only one triangle should appear in half exposure because it is running twice as fast as the calibrator motor 53. The cam 55 runs only one-half the crankshaft speed as shown on the tachometer. Switch the tachometer to "8 cylinders" and read 337½ rpm, and to "4 cylinders" and read 675 rpm with triangle 71 still stationary. Switch tachometer to "6 cylinders" and increase the speed of motor 53 to read approximately 900 rpm on the tachometer, and adjust motor 53 speed so pointer 71 in single exposure is perfectly stationary. That is 900 rpm on "6 cylinders," 675 rpm on "8 cylinders," and 1350 rpm on "4 cylinders." Two triangles 71 will appear at 1800, 1350 and 2700 rpm respectively and so forth.

If the tachometer is adjustable, correct the calibration, otherwise make a chart to show the amount, or percentage, to add or subtract at each range or reading.

Another simple method to calibrate a tachometer is to use the motor 70 in the same way, but use a conventional timing light in the same manner and connections used to time an engine. Focus the timing light on motor 70 rotor 75 and follow the instructions above, changing the engine speed to obtain stop-action. In this case, however, the flash tube fires on the *opening* of the contacts.

5. *STROBOSCOPE.* With the tachometer accurately calibrated, this calibrator can be used to measure rpm, belt speed, and the like, at a distance and without any mechanical connection. It is valuable in automotive service to study valve spring action, for example, by stop-motion observations. Use the same connections as No. 4 above, *TO CALIBRATE A TACHOMETER.*

6. *DYNAMIC BALANCERS.* The present invention provides im-

proved means of stopping action in a dynamic balancing machine. Any machine operated by mechanical, magnetic, inductive or electronic switching means can flash a strobe light at the closing as well as the opening of the contacts (just before and just after the "run out" point) providing a more accurate means of determining the exact point and amount of lateral movement or imbalance. For example, the dwell angle will indicate the amount of imbalance by holding the contacts closed for a longer period of time.

7. *DISTRIBUTOR TEST STAND.* The conventional distributor test stand, for testing and adjusting distributors off the engine, uses a single timing light usually rotating with the disc and mounted in back of the disc with a dial calibrated in degrees. This timing light, or flash tube, fires on the opening of the distributor points and shows the angle of firing for each cylinder. A second flash tube using the present invention, could be mounted adjacent to the existing flash tube to indicate the dwell angle, thus eliminating the dwell meter built into the unit, and providing a more accurate adjustment of the dwell angle.

12
Dwell Tachometers

12-1 Simple Transistorless Dwell Tachometer

Dwell tachometers are combination instruments to measure the dwell angle, and the tachometer section is to measure the revolutions per minute, rpm, of the engine. These two functions are combined for reasons of economy because some of the components, such as the meter and resistors, are common to both circuits. They are usually connected to the engine distributor points and to ground (or the "tach block") with correct polarity observed.

Figure 12-1 shows a slight modification of a Source 13, Model 11-0101, dwell tachometer. Switch SW1 is shown as viewed from the back and in the dwell position. The second position in the direction of the arrow is the 0–1000 rpm tachometer position. The third position in the direction of the arrow is the 0–5000 rpm tachometer position. Switch SW2 is shown in the first position for a 4-cylinder engine, second position 6-cylinder, and third position 8-cylinder. Switch SW2 is in the circuit only when used as a tachometer. The meter reads down-scale with zero dwell at the top of the scale (right hand) for dwell angle readings, but reads up-scale for tachometer rpm readings.

An explanation of the principle of operation of the dwell meter section only is shown in Figure 12-2 and accompanying description. The explanation accompanying Figure 12-3 gives the principle of operation of the tachometer section only.

This type of circuit has the disadvantage of introducing error depend-

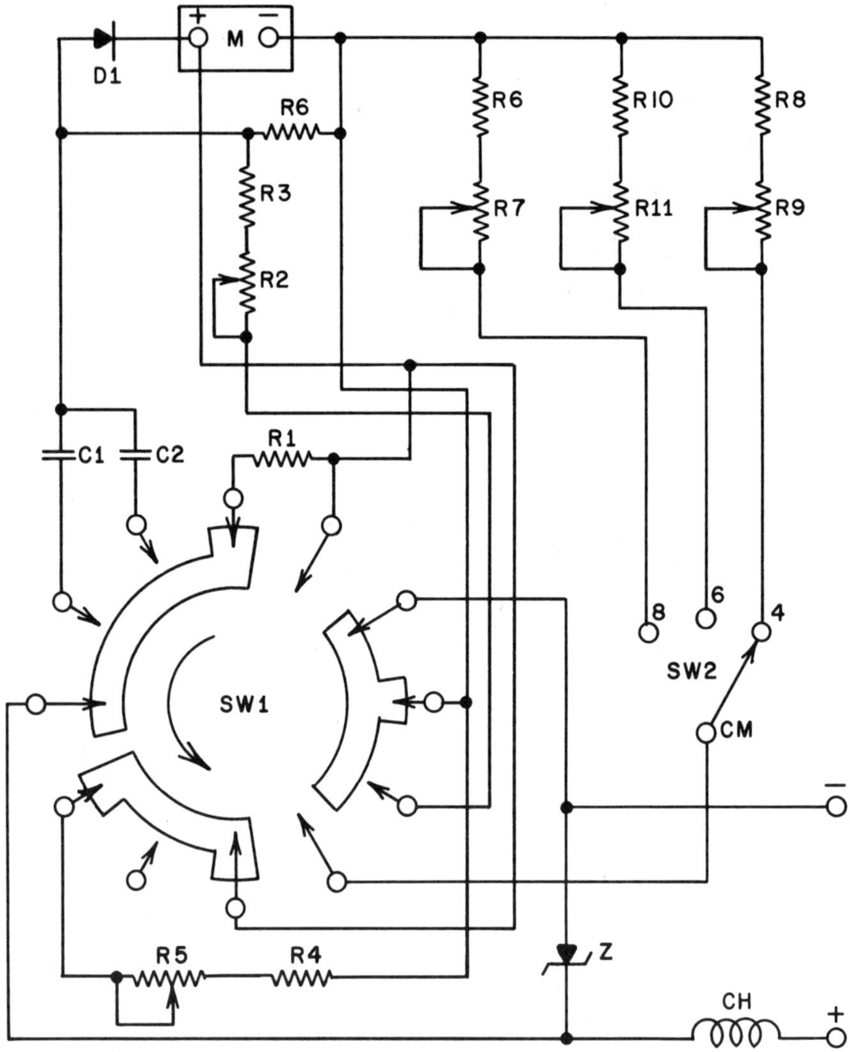

Figure 12-1 Dwell tachometer, Source 13, Model 11-0101

ing on the distributor point resistance, and, somewhat, on the battery voltage of the vehicle.

The values of the components in Figures 12-1, 12-2 and 12-3 are; D1 (diode), M (meter 200 μa, 500 ohms), C1 (0.27 mfd), C2 (1.0 mfd), Z (zener diode 5.5 V), CH (100 ohm choke), R1 (20 Kohms), R2 (high rpm

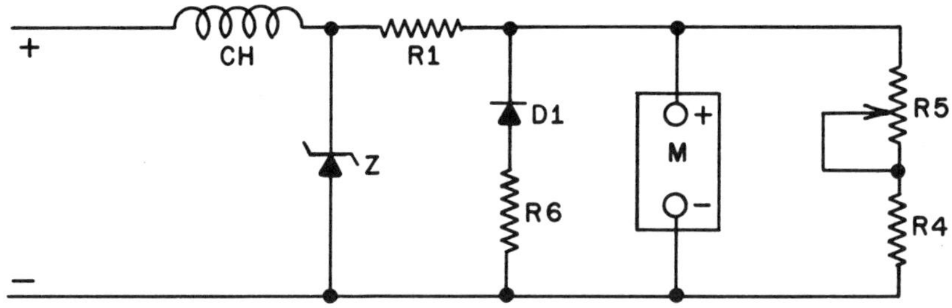

Figure 12-2 Dwell angle meter section of Figure 12-1

adjust 5 Kohms), R3 (calibrating resistor if needed), R4 (680 ohms), R5 (dwell adjust 500 ohms), R6 (180 ohms), R7 (8 cy. low range tach. adj. 500 ohms), R8 (680 ohms), R9 (4 cy. low range tach. adj. 500 ohms), R10 (220 ohms), R11 (6 cy. low range tach. adj. 500 ohms).

A. Dwell Angle Meter Section Only of Figure 12-1

Figure 12-1 shows a dwell tachometer; Figure 12-2 shows only the dwell angle section of Figure 12-1 and uses the same reference numbers and values.

With the positive and negative terminals connected across the engine distributor points of either a 6- or 12-volt system, the choke CH sup-

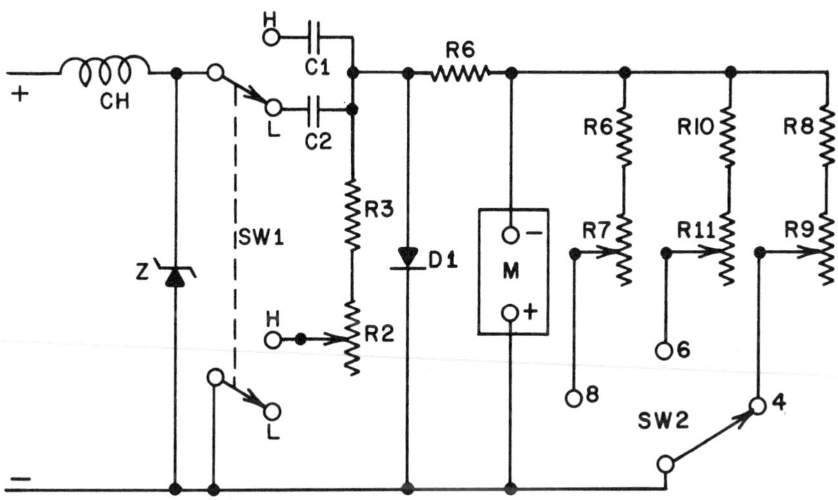

Figure 12-3 Tachometer section of Figure 12-1

presses the high frequency voltage spikes and the zener diode Z maintains a practically constant voltage of 5.5 V across the zener diode. The negative pulses are conducted to ground through resistor R1, diode D1 and resistor R6, but the positive pulses pass through to the meter M positive terminal and register the average on the dwell meter M. Rheostat R5 and resistor R4 comprise the dwell angle calibration circuit. Rheostat R5 is adjusted to read zero dwell (4500 on tachometer scale) with the distributor points open. Better yet, adjust to zero across battery under same conditions as taking dwell readings, such as cranking, or idle speed, allowing for change in voltage.

This circuit does not compensate for the voltage drop (points resistance) that registers on the meter when the distributor points are closed at 100% dwell reading. About two-thirds of this small voltage drop, in degrees, should be added to the meter reading to get the correct dwell angle.

B. Tachometer Section Only for Figure 12-1

Figure 12-1 shows a combination dwell tachometer; Figure 12-3 shows the tachometer section only of Figure 12-1 and uses the same reference numbers and values.

With the positive and negative terminals connected between the engine distributor points and ground of either a 6- or 12-V system, the choke CH and zener diode Z will maintain a constant voltage of approximately 5.5 V across the zener diode. The peak voltage, on the opening of the distributor will pass through capacitor C2, as shown for the low-rpm range of 0–1000 rpm, or through capacitor C1 for the high-rpm range of 0–5000 rpm. The positive peaks are shorted back to the negative terminal by diode D1 but will allow the negative peaks to pass through to the negative terminal of meter M and register a pulse reading on the meter. The higher the speed, the more pulses and, therefore, the higher the reading on meter M. Meter M is calibrated in rpm, but actually functions as a frequency meter.

When switch SW1 is in the low rpm range position, the rpm is calibrated by adjusting R9, R11, and R7 for 4-, 6-, and 8-cylinder engines respectively. When switch SW1 is in the high rpm range (0–5000), R2 is adjusted for calibration of all number of cylinders. The usual zero adjustment is made from the face of the meter M, and is adjusted only when the tachometer is disconnected from the engine.

The calibrating rheostats R9, R11, R7, and R2 place a shunt load across the meter to compensate for variations in commercial tolerance of capacitors C1 and C2.

12-2 Transistorless Dwell Tachometer Having Self-Contained Battery, Source 1, Model E314

Figure 12-4 shows a typical dwell tachometer circuit of Source 1, Model E314, using an internal battery, but no transistors—only two germanium diodes in series as a rectifier. This circuit is not affected by the distributor point resistance or change in vehicle battery voltage. This unit will operate on either 6 or 12 V because the relay RL, which is the tachometer vibrator, will operate efficiently on either 6 or 12 V.

Selector switch SW1 consists of a wafer switch having two decks and four sections in all, designated as SW1-1, SW1-2, SW1-3, and SW1-4 which are ganged to act in unison to select five positions in the order of; OFF, SET, 8 CYLINDER, 6 CYLINDER, and 4 CYLINDER. The off position turns the internal battery BAT of 6 volts (4, 1.5 V dry cells) off and should always be left in this position when the instrument is not being used to prevent drawing the batteries down. All other positions of SW1 connect the battery to the circuit.

The SET position is for adjusting the meter M (0.9 ma, 50 ohms) to the set line to compensate for any change in battery voltage. The set adjust rheostat is R4 (200 ohms). When the meter cannot be adjusted to the set line, the batteries should be replaced. The other three positions of switch SW1 are used for tachometer readings and dwell angle readings. It is turned to the position corresponding to the number of cylinders of the vehicle being tested. When testing 2-cylinder engines, turn the cylinder selector switch SW1 to the 4 CYLINDER position and double the indicated cam angle and engine rpm readings to obtain the correct values.

Selector switch SW2 consists of a wafer switch having two decks and four sections in all, designated as SW2-1, SW2-2, SW2-3, and SW2-4 which are ganged together to act in unison to select four positions in the order of; S (start), R (run), H (0–4500 rpm) and L (0–900 rpm).

The start S position is for reading the cam angle at very low speeds when turning the motor over by the starter only (cranking speed). The run R position is used for reading the cam angle at running speeds when the engine is running under its own power. The 0–4500 rpm H position is for reading engine rpm from zero to 4500 rpm. The 0–900 rpm L posi-

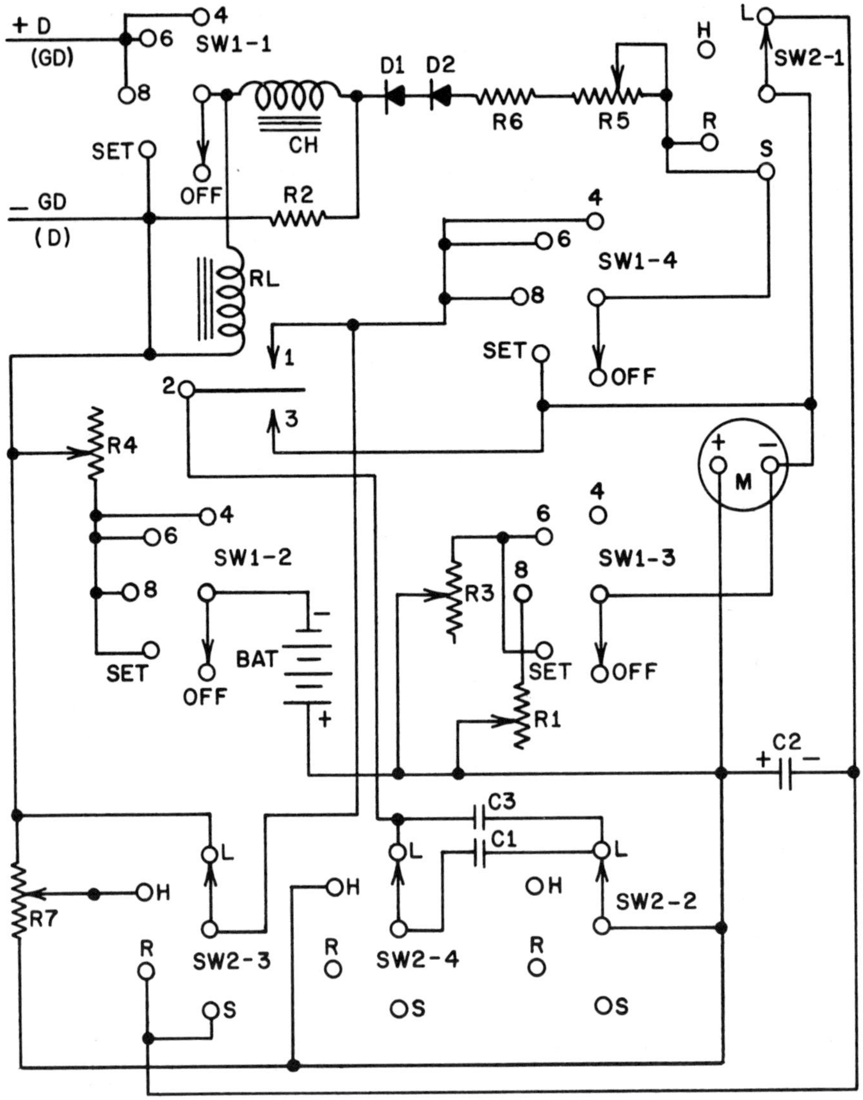

Figure 12-4 Dwell tachometer, typical, Source 1, Model E314

tion is for reading engine rpm from zero to 900 rpm which is the engine idle adjustment range.

The adjustable resistors R1 (100 ohms), R3 (200 ohms), R5 (2 Kohms), and R7 (200 ohms) are located on the chassis for calibrating purposes. Other component values are; choke CH (5.5 mh, 30 ohms), diodes D1 and D2 (1N34 germanium), resistors R6 (2.4 Kohms), R2 (120 ohms), capacitors C1 and C3 (3 mfd), capacitor C2 (5000 mfd).

The instrument can be used for either positive or negative ground systems by connecting the red or positive clip always to the positive, and the black or negative clip to the negative. For a negative ground system, connect the red positive clip D to the distributor points and connect the black negative clip GD to the ground of the vehicle. For a positive ground system, connect the red positive lead (GD) to the ground, and connect the black negative clip (D) to the distributor points.

This instrument meter is calibrated in rpm from zero-to-9 and zero-to-45 when multiplied by 100 gives a range of zero-to-900 and zero-to-4500 rpm reading up-scale from the left zero point. A third scale for cam angle measurement reads up-scale from zero-to-70 degrees. At the left zero point is zero dwell angle with the distributor points open, and the full vehicle battery voltage across the instrument leads. The full dwell angle (points closed) for an 8-cylinder engine would read 45°, for a 6-cylinder 60°. For a 4-cylinder engine, 90° would go off scale, but the usual dwell angle setting for a 4-cylinder engine is near 60°.

A. *"Set" Position of Figure 12-4*

The SET position of switch SW1 in Figure 12-4 is shown in simplified schematic form in Figure 12-5. The "set adjust" knob on the front panel operates rheostat R4 to set the meter pointer on a set mark located at 5.8 on the zero-to-9 low rpm (0–900) scale. Rheostats R3 and R5, for dwell and tachometer adjustments, are located inside the cabinet on the chassis. This "set adjust" simply compensates for variations in the internal battery BAT voltage and should be checked every time it is used.

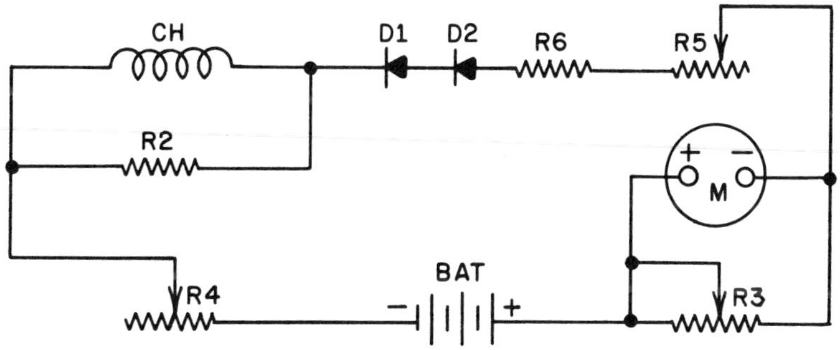

Figure 12-5 SET position schematic of Figure 12-4

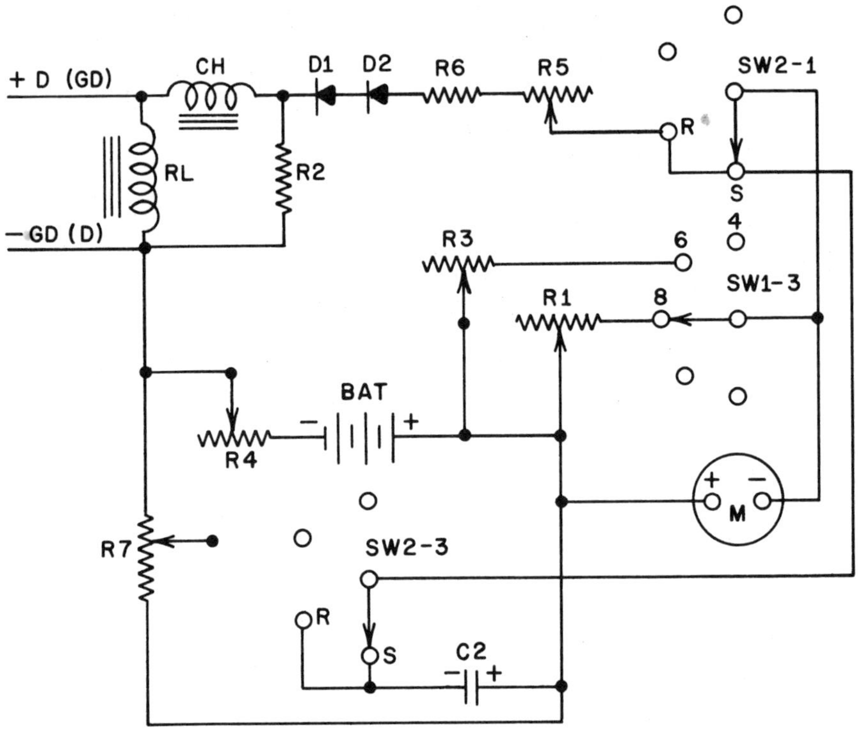

Figure 12-6 Dwell meter section only of Figure 12-4

B. Dwell Meter Section of Figure 12-4

Only the dwell meter section of Figure 12-4 is shown is simplified schematic form in Figure 12-6.

To function as a dwell meter, or cam angle meter, the selector switch SW2 must be in the run R or start S position. In the model shown in Figure 12-4 the run R or start S positions are the same. However, some other models having the same panel layout may have two different values of capacitor C2 switched in the circuit to give a steadier reading on the start S position. The cylinder selector switch SW1 must be in the position of 8, 6, or 4 cylinders corresponding to the number of cylinders on the engine being tested.

Regardless of the polarity of the ground to the input leads, the diodes D1 and D2 will pass only the negative peaks of the pulses developed across the points when they are open. When the points close, there is no voltage across the input leads. These negative pulses are present during the time the distributor points are open. The choke CH suppresses the extremely high voltage pulses to protect the diodes D1 and D2. The negative pulses pass through the choke CH, diodes D1 and D2, resistors R6 and R5 to the negative terminal of the meter M and return to ground lead through resistor R7. Capacitor C2 is now shunted across the meter and builds up a charge through resistors R6 and R5 at a time constant depending on the duration (time the points are open) of the pulses.

Rheostat R1 is for calibrating on 8 cylinders and R3 is for calibrating on 6 cylinders. There is no separate calibration for 4 cylinders.

C. Tachometer Section for Figure 12-4

Only the tachometer section for Figure 12-4 is shown schematically in simplified form in Figure 12-7.

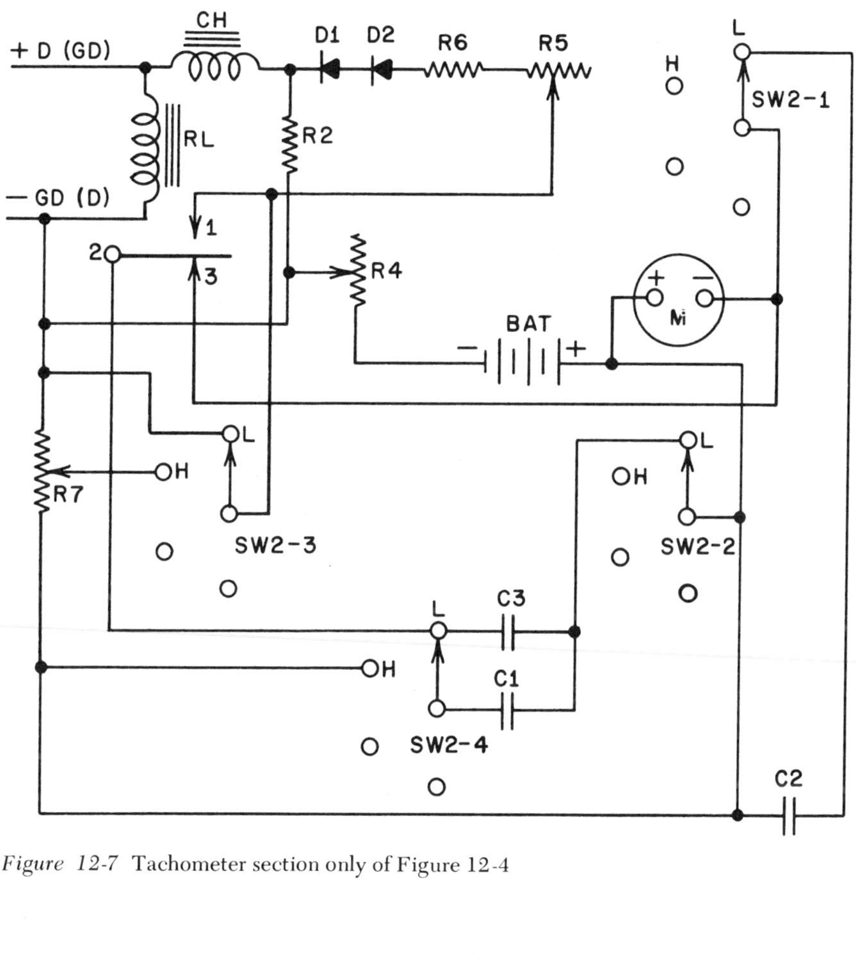

Figure 12-7 Tachometer section only of Figure 12-4

Figure 12-7 is shown with switch SW2 in the L (0–900 rpm) position. The relay RL coil is across the distributor points and is energized by the vehicle battery when the distributor points are open, and de-energized when the points are closed. This relay RL is a millisecond relay and will vibrate in unison with the distributor points even at relatively high speeds.

When relay RL contacts 1 and 2 are engaged (coil energized) the battery BAT places a charge on capacitors C1 and C3 (in parallel) through resistor R4. When the relay contacts 1 and 2 are disengaged (coil de-energized) contacts 2 and 3 are engaged, which connects the partially charged capacitors C1 and C3 across the meter M terminals and causes an upward reading. As the speed of the engine increases, the capacitors are charged faster and faster and cause a higher meter reading. Capacitor C2 is connected across the meter M at the L (0–900 rpm) position to steady the meter readings at the lower speeds. There is no calibration on the low L speed, the factory selectively chooses a combination of C1 and C3 to give an accurate calibration.

On the high speed H (0–4500 rpm) position of switch SW2, a calibrating resistor R7 is switched into the circuit, capacitors C1 and C3 are now connected in series, and C2 is switched out of the circuit. The unit functions on the same principles as explained for the low L speed position.

12-3 Dwell Angle Defined

When the engine distributor shaft is turning, usually at one-half the crankshaft rpm, the lobes on the cam close and open the distributor points to fire the plugs. For best performance of the engine, the points should remain closed approximately twice as long as they are open, or two-thirds of the time between plug firings, unless otherwise specified.

The dwell angle, or cam angle, is the percentage of time that the points are closed relative to the total time between firings of the plugs. This dwell angle is expressed in degrees. The degree of dwell is determined by the number of cylinders in the engine.

For a 4-cylinder engine, the distributor is shown in Figure 12-8a where the four-lobe cam R has four, 90° spaced lobes. The points consist of a grounded contact C, a movable contact D, an insulated rubbing block E, and a stationary pivot point F which connects to the coil and capacitor.

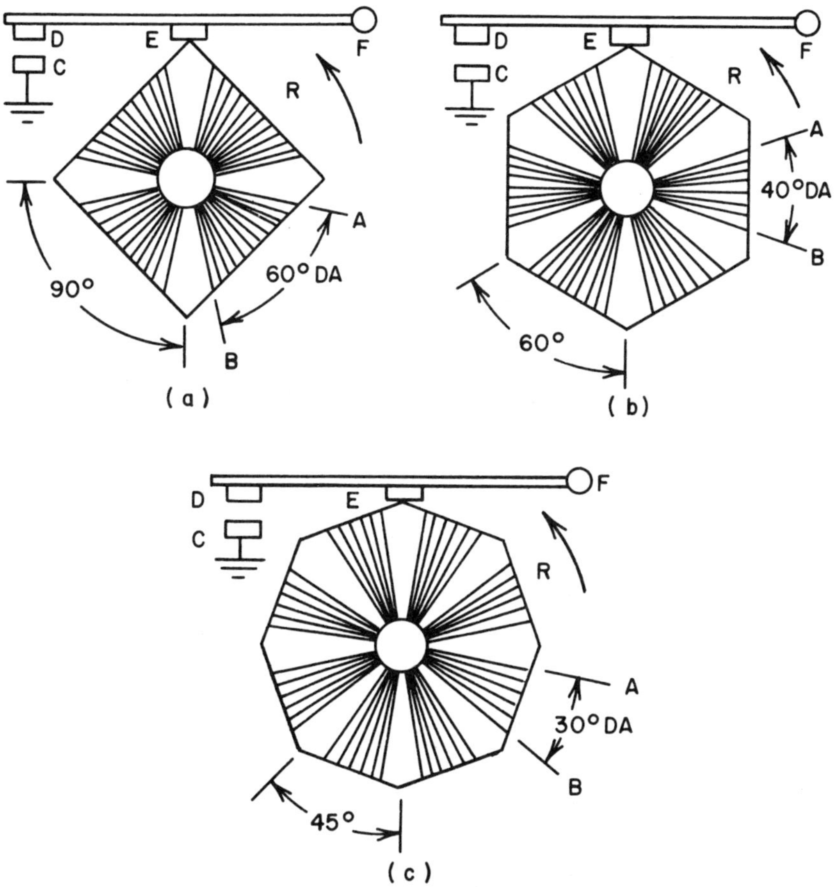

Figure 12-8 Dwell angle. (a) Four-cylinder. (b) Six-cylinder. (c) Eight-cylinder

As the rotary cam R turns in the direction of the arrow, the rubbing block E causes contacts C and D to close at point A and open at point B. The contacts C and D are closed for 60° (shaded area) of each 90° of rotation of R and open for 30°, reaching maximum opening at the position shown. The dwell angle DA is 60 degrees.

Figure 12-8b shows a six-cylinder engine where the six-lobed cam R has six 60° spaced lobes. The contacts C and D are closed for 40° of each 60° of rotation of cam R and are open for 20°. The dwell angle DA is 40 degrees.

Figure 12-8c shows an eight-cylinder engine where the eight-lobed cam R has eight lobes 45° apart. The contacts C and D are closed for 30° of each 45° of rotation of cam R, and are open for 15°. The dwell angle DA is 30 degrees.

For a one-cylinder engine, the dwell angle is, therefore, approximately two-thirds of 360°, or 240°, four times that of a four-cylinder engine.

For a two-cylinder engine the dwell angle is two-thirds of 180°, or 120°, twice that of a four-cylinder engine.

For a three-cylinder engine the dwell angle is two-thirds of 120°, or 80°, twice that of a six-cylinder engine.

The complete engine cycle takes place in 360° rotation of the distributor regardless of the number of cylinders or number of cycles (2- or 4-cycle).

12-4 To Calibrate a Dwell Angle Meter

The simplest procedure to follow in calibrating a dwell angle meter is to compare it with a standard, or master dwell angle calibrator.

Another method is to use a square wave generator that is known to have a true square wave. That is, positive for one-half of the cycle and negative for the other half of the cycle. By connecting a half-wave diode in series with the generator and the dwell meter, only the positive (or negative) half of the wave is used to simulate the opening and closing of the distributor points. The meter can be calibrated at only one point of 50% dwell, which is not in the commonly used range of approximately 67% of full dwell.

The ignition simulator shown in Figure 3-1 can be used to calibrate a dwell angle meter accurately. Connect a light bulb, or better yet a buzzer to the distributor points and to ground. The buzzer should sound when the points open and be silent when the points are closed. As the disc wheel W is turned in the direction of the arrow, adjust spark advance SA until the buzzer sounds when the No. 1 mark is in line with the 60° mark on the dial DL. The No. 6 mark should also be in line with zero on dial DL. Turn the disc wheel W very slowly by hand in the direction of the arrow. When the buzzer stops at the closing of the points, read the dwell angle on dial DL opposite the No. 1 mark. Readjust the points, if necessary, to be at 40° dwell. Repeat the operation several times to be sure. Remove the buzzer, and connect the dwell angle meter. With the simulator running at a low speed, just fast enough to give a steady reading, the dwell angle meter should read the same dwell angle as shown by the buzzer when the points closed.

The most precise method to use in calibrating a dwell meter is the author's invention described elsewhere under the heading, "Cantonwine Timing Light to Measure Dwell Angle." This timing light can also be used with any ignition simulator having a calibrated dial showing cam angle of the distributor.

12-5 Single-Range Electric Tachometer Circuit Using Self-Contained Battery

Figure 12-9 shows a typical Source 17 electric tachometer that uses a microsecond, 150 ohm, dc relay RL1 across the distributor points. Capacitor C4 (0.003 mfd, 1600 V) is a transient voltage suppressor. The relay RL1 vibrates the movable contact RLC1 in unison with the dis-

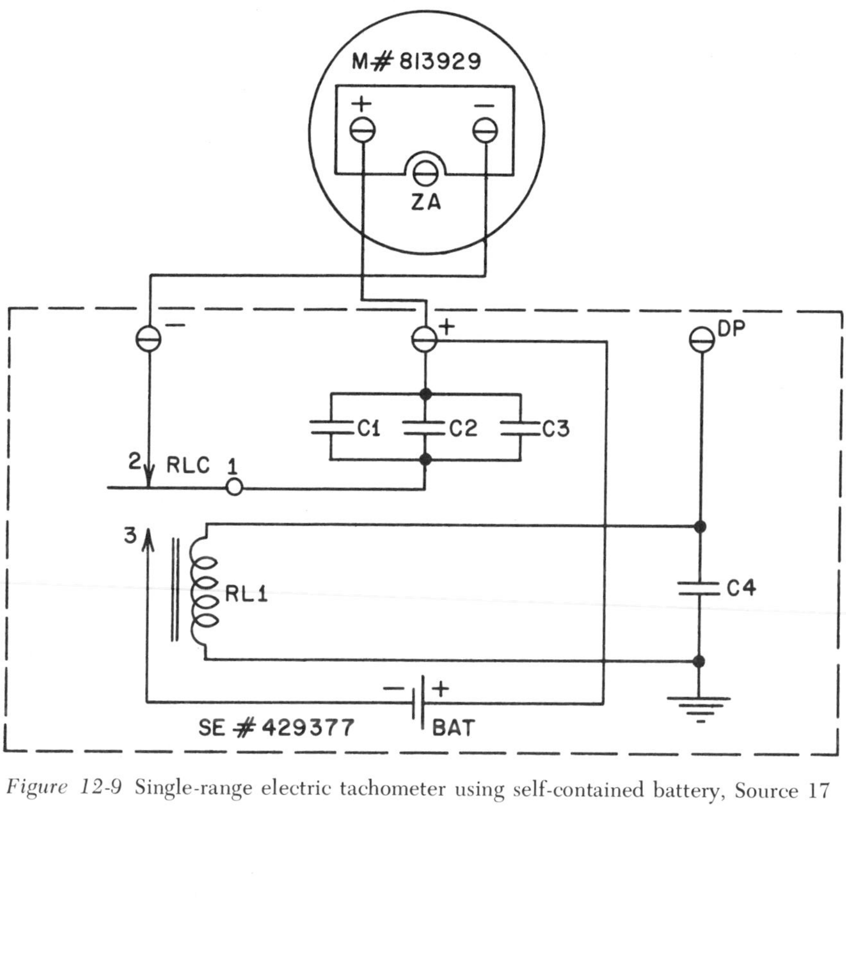

Figure 12-9 Single-range electric tachometer using self-contained battery, Source 17

tributor points opening and closing. When the distributor points are open, the relay RL1 is energized which closes RLC 1 and 3, placing the 1.5-V battery BAT across the capacitors C1, C2, and C3 thereby charging them. When the distributor points close, the relay RL1 is de-energized, closing the NC contacts RLC 1 and 2, which places the partially charged capacitors C1, C2, and C3 across the meter M.

As the speed of the engine increases, the distributor points open and close faster, and the capacitors C1, C2, and C3 discharge at the increased rate of speed, which builds up the meter M reading. The dial of meter M is calibrated in rpm (0–8000 for the #813929 meter). The only calibration is in the value of the parallel capacitors C1, C2, and C3. These are selected at the factory to add up to the correct value to give an accurate reading of the precalibrated meter dial. The same procedure should be followed in replacing the capacitors. Capacitors C1 and C2 have a nominal value of .94 mfd, 400 V, and C3 is 0.1 mfd, 400 V. The nominal sum is 1.98 mfd total, but the actual value may be an even 2.0 mfd, for example, due to selective assembly.

This circuit has the advantage of being able to operate on either positive or negative ground.

The meter M has a zero adjustment ZA located between the two terminals on the back of the meter. The companion "sender" in dotted lines is #429377.

12-6 Multirange Tachometer Using a Magnetic Pulse Sender

The tachometer circuit shown in Figure 12-10 is usually found in permanent installations such as on a generator-regulator test bench. This one is used on a Source 1, Model E241, where the magnet M is driven by the tester shaft, which induces a voltage in tachometer pick-up coil TC. The voltage induced is dependent on the shaft speed.

This voltage is fed to the tachometer full-wave bridge circuit consisting of R6 (500 ohms), R7 (500 ohms), diode D1 and D2, and meter TM. Capacitor C1 (100 mfd) in series with resistor R5 (1 Kohms) is connected across the meter M to give a steady reading, especially at low speeds.

The switch SW4 is a speed range selector switch. In the high-speed H position, the series resistor R10 (220 ohms) and rheostat R9 (300 ohms)

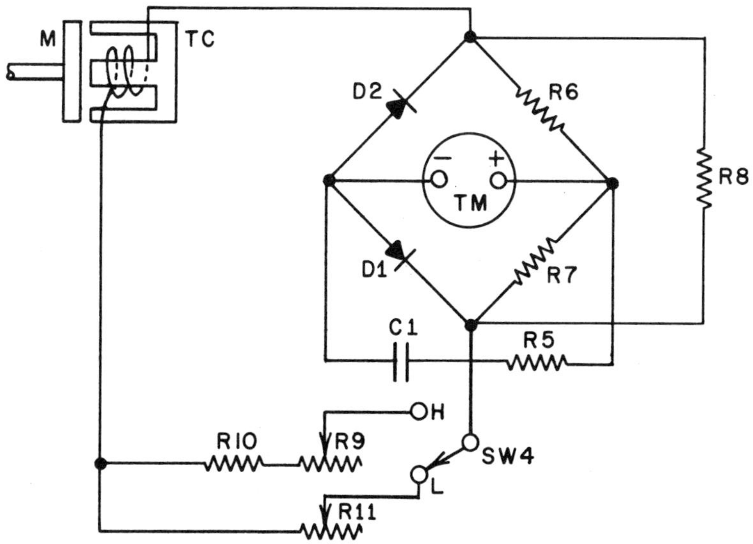

Figure 12-10 Multirange tachometer using magnetic pulse-type sender, Source 1, Model E241

are in series with the pick-up coil TC and the meter TM to register on the 0–8000 rpm scale. Rheostat R9 is the 0–8000 rpm adjustment, R8 (200 ohms).

When switch SW4 is in the low L position, it connects calibrating rheostat R11 (300 ohms) for the 0–4000 rpm meter scale.

This circuit can be adapted to more than two ranges by adding any number of selector switch positions and corresponding resistors, within limits.

12-7 Stroboscope Tachometer and Frequency Meter

A stroboscope is a very useful instrument in automotive service, or in any activity where it is desired to measure the rpm, belt speed, stop-motion study, and the like, without mechanical or electrical connection.

A stroboscope is a device consisting of a flash tube triggered by an adjustable frequency oscillator that indicates, on an electric calibrated meter, the number of flashes per second (frequency) or revolutions per minute (rpm). By adjusting the frequency of the oscillator until the flash tube light makes the rotating or moving object appear to be standing still, the rpm or frequency can then be read directly from the calibrated meter. The stroboscope meter can be calibrated by focusing the light on a revolving disc, having a white spot, driven by a synchronous motor of known speed. The 60-Hz ac power from the utility lines is held to such a close tolerance that it is accepted as a very reliable and accurate calibration reference. The synchronous motor can be an electric clock, or timer motor that has an exposed rotor so that a white spot can be attached thereto. A common speed for these timer motors is 450 rpm, and with one spot on the disc, it can cover a wide range of calibrating points.

For example, when the stroboscope is flashing at 225 rpm and the disc turning 450 rpm, one spot will appear in two revolutions. At 450 rpm of the stroboscope there will be one spot seen for each revolution. When the stroboscope is flashing at 900 rpm there will be two spots diametrically opposite. This can continue to any high range desired.

The stroboscope has many novel uses limited only by the ingenuity of the user. Some of the more common uses are given below:

A. *To measure rpm* of any low-power motor without loading it down with a mechanical tachometer, or revolution counter, and thereby not reading its true rpm.

B. *To measure belt speed* in feet per minute, or slippage of multiple V-belts, or slippage of a single belt.

C. *Study stop-motion action* of engine valve springs, tappets, and the like.

D. *To measure rpm* of any rotating shaft, pulley, or the like.

E. *Stop—Motion Study* of fan blade flutter.

F. *Check the accuracy* of an engine tachometer.

G. *Check* for bent shafts.

H. *Determine* pulley ratio, gear ratio, or differential gear ratio.

I. *To measure* the frequency of vibrations, and determine the resonant frequency of vibrating members.

J. *Measure* road speed of vehicles.

The procedure in measuring rpm of an unmarked rotating shaft is to place two marks opposite each other, one mark having a different shape than the other, such as a line parallel to the shaft, *and* the other mark a line at right angles to the first mark. Start the stroboscope at the highest speed, or at least a speed higher than the rpm of the rotating part, and gradually reduce the stroboscope speed until the marks appear stationary for the first time. Read rpm on the proper meter scale. The number of marks will appear decreased as the speed control is decreased. When both marks appear superimposed, making a cross, the meter reading is double the actual speed. Reduce the stroboscope speed control slowly until only one of the marks appears stationary and read the correct rpm on the meter. Turning the speed control still lower will also show a single mark but at one-half speed.

To measure belt speed, determine the belt length, place a white chalk mark on one spot on the belt. When the belt appears to be standing still and only one mark is visible, multiply the rpm by the belt length in feet to get the feet per minute.

To determine belt slippage in single or multiple belts, place a chalk mark on the pulley and on each of the belts. When the motion is stopped, belt slippage will be indicated by a gradual separation, or creeping, of the chalk marks.

To measure the ratio of pulleys, gears, transmissions, differentials, and the like, take the rpm of one pulley, and then the other. The ratio is determined by dividing the lower rpm into the higher rpm.

To measure the road speed of vehicles, jack up one rear wheel, measure the circumference of the tire, and place a white chalk mark on the tire. (Use the same procedure for measuring belt speed except divide by two because with one wheel blocked, the other turns twice as fast due to differential action.) This will give the feet per minute $\left(\frac{\text{rpm}}{2} \times \text{circumference in feet}\right)$, multiplied by 60 gives the feet per hour. Divide the feet per hour by 5280 (feet per mile) to get the road speed in miles per hour.

With the rpm of one wheel known, with the other stationary, use Table 12-1 to convert rpm into miles per hour for most common tire sizes. Table 12-1 shows tire sizes from 4.00–15 to 14.00–24, inclusive. This table allows for the fact that one wheel is turning twice as fast with the other blocked, due to differential action. The accuracy of the vehicle speedometer can be checked on this road speed measurement.

Miles Per Hour Road Speed for Various Tire Sizes at Known RPM

	Miles Per Hour								
Tire Size	*10*	*20*	*30*	*40*	*50*	*60*	*70*	*80*	*90*
4.00-15	147	294	441	588	735	882	1029	1176	1323
4.50-12	165	331	496	661	827	992	1157	1323	1488
4.75-19	124	247	371	495	618	742	866	989	1113
5.00-15	137	274	412	548	685	823	960	1097	1234
5.25-17	121	242	363	484	605	732	847	968	1089
5.25-18	120	240	360	479	599	719	839	959	1079
5.50-15	137	275	412	549	687	824	963	1101	1238
5.50-16	128	257	385	513	642	770	898	1026	1155
6.00-15	124	247	371	495	618	742	866	989	1113
6.00-16	127	254	382	509	636	763	890	1017	1145
6.00-17	117	234	351	467	584	701	818	935	1051
6.00-18	115	231	346	461	578	692	807	923	1038
6.00-20	105	211	316	421	527	633	738	844	948
6.40-15	130	259	389	519	648	778	908	1037	1167
6.50-15	127	253	380	507	633	760	887	1013	1139
6.50-16	121	242	363	484	605	727	847	968	1089
6.50-20	103	205	308	410	513	615	718	820	923
6.70-15	128	257	385	513	643	770	898	1027	1155
6.70-16	124	247	371	495	618	742	866	989	1113
7.00-15	121	242	363	484	605	727	847	968	1089
7.00-16	117	234	351	467	584	701	818	935	1052
7.00-17	108	216	324	431	539	646	754	861	969
7.00-20	98	197	295	393	491	590	688	786	885
7.00-24	88	177	265	353	441	530	619	706	795
7.10-15	126	251	377	502	627	753	879	1004	1130
7.50-15	117	235	343	471	588	706	824	941	1058

Table 12-1

Miles Per Hour Road Speed for Various Tire Sizes at Known RPM

	Miles Per Hour								
Tire Size	*10*	*20*	*30*	*40*	*50*	*60*	*70*	*80*	*90*
7.50-16	115	229	344	459	573	688	803	917	1032
7.50-17	104	208	312	415	519	622	726	830	935
7.50-18	100	200	300	400	500	600	700	800	900
7.50-20	95	190	285	380	475	570	665	760	855
7.50-24	85	171	256	342	427	513	599	684	719
7.60-15	124	247	371	495	618	742	866	989	1113
7.60-16	119	238	358	477	596	715	834	953	1073
8.00-15	121	242	363	484	605	726	847	968	1089
8.20-15	120	240	360	480	600	720	840	960	1080
8.20-16	117	234	351	467	584	695	812	929	1046
8.25-16	105	209	314	418	523	627	732	836	941
8.25-18	96	192	288	384	480	576	672	769	865
8.25-20	92	184	276	368	460	552	644	736	838
8.25-22	88	177	265	353	442	530	618	707	795
8.25-24	83	166	250	333	416	499	582	665	749
8.90-15	113	227	340	453	567	680	793	907	1020
9.00-16	94	187	281	375	468	562	656	749	842
9.00-18	92	184	276	368	460	552	644	736	828
9.00-20	87	174	262	348	436	523	610	697	784
9.00-22	83	166	248	331	414	496	579	661	745
9.00-24	80	161	242	322	402	482	563	643	724
10.00-18	88	177	265	353	441	530	618	706	795
10.00-20	85	170	254	339	424	509	594	678	763
10.00-22	80	160	240	321	401	481	561	641	721
10.00-24	77	154	231	308	385	462	539	606	694
11.00-20	81	163	244	326	406	489	570	651	734

Table 12-1 (continued)

Miles Per Hour Road Speed for Various Tire Sizes at Known RPM

	Miles Per Hour								
Tire Size	*10*	*20*	*30*	*40*	*50*	*60*	*70*	*80*	*90*
11.00-22	78	156	234	312	390	468	546	624	701
11.00-24	74	148	222	296	370	445	519	593	667
12.00-20	80	160	240	320	400	480	560	640	720
12.00-24	73	146	218	292	364	437	510	513	656
13.00-20	77	153	230	307	383	460	536	613	690
13.00-24	70	140	211	281	351	421	491	561	632
14.00-20	74	148	222	296	370	444	517	591	666
14.00-24	68	135	203	271	331	406	474	542	610

Table 12-1 (continued)

12-8 Strobemeter (Stroboscope) Operating from 12-Volt Battery, Source 15, Model 41-192

Figure 12-11 shows a stroboscope circuit typical of Source 15, Model 41-192, strobemeter which is portable and derives its power from a 12-V storage battery BAT.

A dc-to-dc converter to change the 12 V dc to a dc voltage of 450 V for the flash tube V1, and two dc voltages of 20 and 40 V for the oscillator circuit, is provided by transformers T1 and T2, transistors Q1 and Q2 (T13027), 450 V full-wave bridge rectifier BR1 using four diodes (5A6) and low-voltage full-wave bridge rectifier BR2 using four diodes (5A1 or SLA-12C), and associated components.

The adjustable frequency oscillator comprises a unijunction transistor UJT Q3 (2N2646), a pulse trigger stage silicon-controlled rectifier SCR12 (106-32), and associated components.

The meter M1 (272 ohms, 0.5 ma) is calibrated in two ranges 0–1500 rpm and 0–6000 rpm.

The speed control knob of potentiometer R10 (75 Kohms) is geared down to give full range on about 5½ turns. This control should always be turned fully CW at the start of the operation.

The stroboscope flash tube V1 (xenon FX6A) is triggered by the SCR12 through pulse transformer T3.

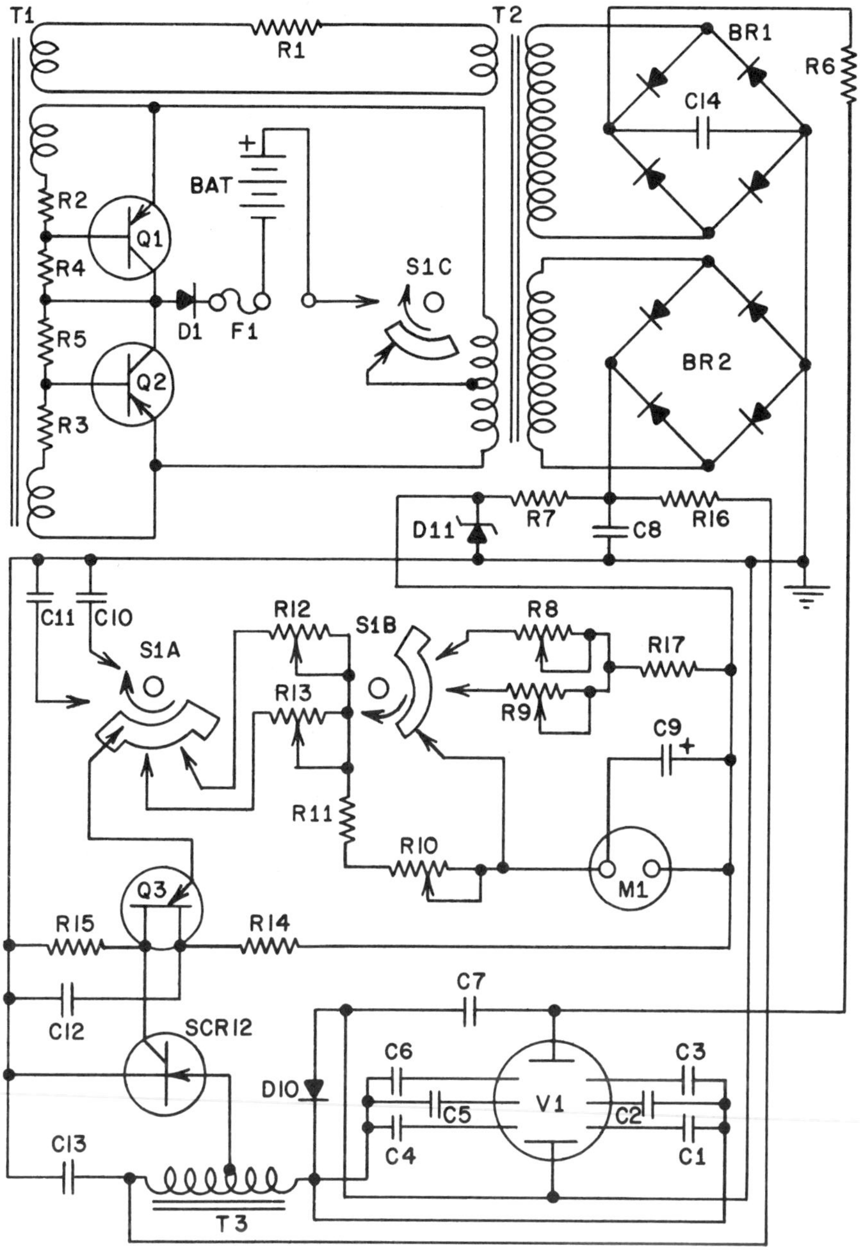

Figure 12-11 Strobemeter (stroboscope) operating from 12-volt battery, Source 15, Model 41-192

The ganged selector switch S1 (S1A, S1B, and S1C) is shown in the OFF position, the second position in CW direction is 0–1500 rpm, and the third position in CW direction is 0–6000 rpm.

Potentiometers R12 (7 Kohms, 2 W, 10%) and R8 (300 ohm, 2 W) are the 0–1500 rpm adjustments. Potentiometers R13 (7 Kohms, 2 W, 10%) and R9 (300 ohm, 2 W) are the 0–6000 rpm adjustments.

Other component values are; capacitors in mfd C1 to C6 (4.7 mmfd, 4 KV), C7 (0.68, 600 V), C8 (8, 150 V), C9 (1000, 6 V), C10 (1.0, 35 V, 10%), C11 (3.9, 35 V, 10%), C12 (0.0005), C13 (0.9, 150 V), C14 (0.25, 600 V), D1 (silicon diode 1N4719), D10 (diode 6RS 18PH110BQB1), D11 (zener diode ZA20B, 20 V, 1 W), F1 (fuse 3 amp), resistors in ohms R1 (10, 0.5 W), R2 and R3 (39, 0.5 W), R4 and R5 (2.2 K, 0.5 W), R6 (3.3 K, 5 W), R7 (2.2 K, 1 W 10%), R11 (7.5, 0.5 W), R14 (680, 0.5 W 10%), R15 (33, 10%), R16 (2.2 K, 0.5 W-10%), R17 (100, 0.5 W wire wound 10%).

12-9 Calibrating a Tachometer

Electric tachometers operating from the ignition system of an engine can be calibrated by using the simulator shown in Figure 3-1 or any engine ignition system can be used if the speed can be controlled closely by using a finely adjustable throttle control.

In Figure 3-1 motor M2 is a 60 Hz synchronous clock or timer motor having an exposed rotor running at 450 rpm (200 rpm is preferred if available). The exposed rotor has a white marker WM. To calibrate a tachometer using the 6-cylinder simulator, adjust the speed of the simulator to approximately 450 rpm as shown on thc tachometer set for the "6 cylinder" position. With a timing light connected to the No. 1 cylinder, or any one cylinder, focus the timing light on the motor M2. Adjust the speed until the white mark WM appears absolutely stationary. The tachometer should read exactly 450 rpm. Turn the tachometer switch to "4 cylinders" and read 675 rpm, and to "8 cylinders" and read 337½ rpm. The white mark WM will appear in half exposure. This is caused by the fact that the timing light is making only 225 flashes per minute (every two revolutions of the crankshaft) so the motor M2 will make exactly two revolutions per flash.

On increasing the speed to 900 rpm, there will still be only one exposure of the mark WM because there are only 450 flashes of the timing light at 900 rpm. Turn the tachometer switch to "4 cylinders" and read 1350, and to "8 cylinders" and read 675 rpm.

Table 12-2 shows other speeds and numbers of spots for a 450 rpm motor M2.

Tachometer Calibration Chart

Spots	*1*	*1*	*3*	*2*	*3*	*4*	*5*
8-CY (rpm)	337½	675	1012	1350	2025	2700	3037
6-CY (rpm)	450	900	1350	1800	2700	3600	4050
4-CY (rpm)	675	1350	2025	2700	4050	5400	6075

Table 12-2

12-10 Simple Solid State Dwell Angle Meter

Figure 12-12 shows a solid state dwell meter that operates from the vehicle battery and is connected across the ignition coil primary (and the ballast resistor) instead of across the distributor points. Zero dwell angle is at the left end of the meter scale and reads up to full dwell of 45° or 60° at the right end of the scale.

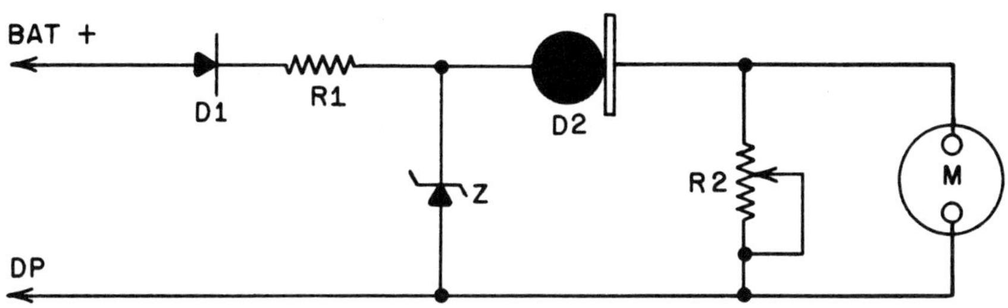

Figure 12-12 Simple solid state dwell meter

In Figure 12-12, silicon diode D1 (1N4004) protects the other components against reverse polarity and large negative peak surges, zener diode Z (6.2 V, 500 MW, 1N5234), in conjunction with resistor R1 (820 ohms, 5%) maintains constant voltage at the input to current regulator (constant current) diode D2 (1N5299), rheostat R2 (1 Kohms) and meter M (0–1 ma).

To use the meter, connect the clip BAT positive to the battery positive and clip DP to the distributor points. Turn over the engine by hand to close the distributor points, and adjust R2 until the meter reads full scale.

This dwell meter can be used on 4-, 6-, and 8-cylinder engines, and on battery voltages of 6, 12 and 24 without any switching. It may be necessary to recalibrate by R2 slightly for each of the above voltages. This unit is not materially affected by point resistance nor battery voltage changes. However more accurate results can be obtained if the meter is calibrated to full scale across the battery while the engine is running at idle speed. Then switch the ground lead to the distributor points to read dwell angle. The engine should be operated at the same speed as for calibration to have the same system voltage.

13
Wheel Balancers

13-1 General Principles of Balancing

Vehicle wheels, especially at high speeds, must be balanced and run true to prevent vibration.

The wheels, brake drums, hub caps, and tires may not always be perfectly balanced. When these parts are all assembled together and ready to mount on the vehicle (or after mounting), they must be balanced.

There are two kinds of balance—or imbalance—in any rotating piece of machinery; static balance (stationary) and dynamic balance (rotating).

A piece can be in static balance without being in dynamic balance. However, if it is in dynamic balance, it is also in static balance.

The principle of static and dynamic balance is illustrated in Figure 13-1. Figure 13-1a shows a bar ST fastened at an angle to the shaft axis AX. If the bar ST is of equal length and weight on each side of the axis AX it will be in perfect static balance, just like a teeter board. If the axis AX is rotated, then the bar ST, due to centrifugal force, tries to run at right angles to the original axis AX, and assumes a position R in dotted lines. In doing this the shaft AX now wobbles in a circle as shown in dotted lines AX1, causing vibration or wobble. To balance this dynamically, a bar of equal length and weight must be placed on an identical angle in the opposite direction as shown in Figure 13-1b.

Now, when the piece is rotated around axis AX there is no wobble—or out of balance—because the two centrifugal forces offset each other and it is dynamically balanced.

A thin disc can be statically balanced and it will also be dynamically balanced if it runs true at right angles to the axis.

A vehicle wheel is relatively thin in relation to the diameter so a fairly good balance can be achieved on the spirit level bubble-type of static balancer provided one-half of the total weight is applied to each side of the rim. If a wheel runs true and does not wobble or shimmy, any imbalance on a dynamic balancer can be equalized by adding equal weights on each side of the rim. However if there is a wobble, it is usually necessary to add more, or all of the weight to the outside of the rim at the "run-out" point. If the wobble is due to a bent wheel, the wheel should be replaced instead of trying to balance it.

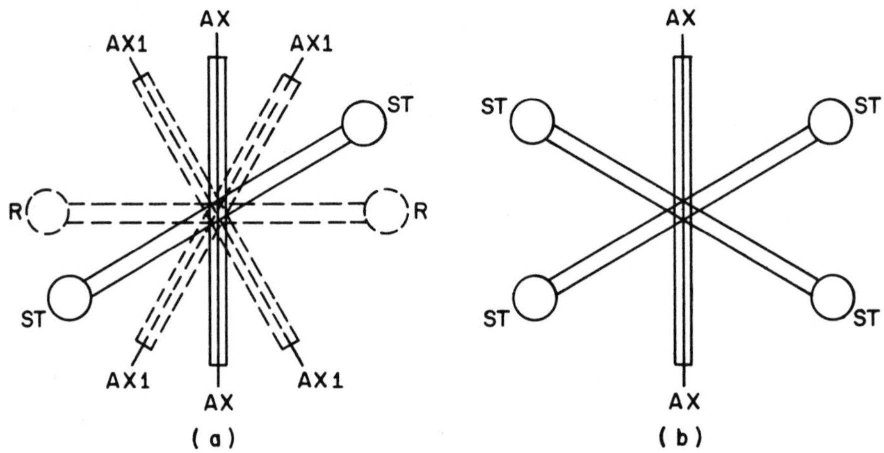

Figure 13-1 Principle of static and dynamic balance

There are many types of dynamic wheel balancers available. There are machines for balancing the wheel and tire only after removing it from the vehicle. Others balance the entire assembly of wheel, tire, hub cap, and brake drum while mounted on the vehicle. In either type, the principles of operation are essentially the same using a stroboscope flashing light.

There is another type known as the mechanical wheel balancer which has a calibrated weight and position system bolted to the spinning wheel, and by pushing or pulling knobs, the correct balance is achieved, and the amount of weight as well as the position of the weight can be determined.

13-2 Static Wheel Balancers

One common type of static wheel balancer consists of a spirit level bubble that indicates wheel balance by centering the bubble in a small marked circle. These are "off the car" balancers and can be fairly accurate if care is exercised in centering the bubble. This indicates that the wheel is horizontal in all directions. Any imbalance will throw the wheel off the horizontal and the bubble will not be centered.

This type is perhaps the quickest way to get a good balance, and if used properly can produce good results for slow to medium speed driving.

The spirit level bubble-type is also used on a motor-driven spindle to get a balance that is even more accurate than the stationary type.

13-3 Mechanical Wheel Balancers

Mechanical wheel balancers, for dynamic balancing on the vehicle, are so called because they are purely mechanical with no electronics or flash tubes involved. An electric motor-driven wheel spinner, however, is required for spinning the front wheels. The rear wheels can be engine driven. A remotely-controlled throttle control, or accelerator depressor, is helpful in controlling the engine speed during balancing.

A typical mechanical wheel balancer of Source 17 consists of a self-contained device that is clamped to the wheel being balanced. Internal weights can be shifted as to position and amount by pushing or pulling on two knobs while the wheel is spinning. These knobs are adjusted until the wheel is balanced, indicated by no vibration. The weight placement position is shown by a pointer and the amount of weight to add is found on the calibrated dial. This amount is divided by two, with one-half of the indicated weight placed on the inside of the rim and the other half placed on the outside of the rim.

13-4 Electronic Wheel Balancer for "On the Car" Dynamic Balancing

To balance vehicle wheels dynamically while still on the vehicle, an electric motor-driven wheel spinner is required, at least for the front wheels. The rear wheels may be driven by the engine, with one wheel blocked while the other is being balanced.

A remotely-controlled throttle control, or accelerator depressor, is almost a necessity for one-man operation of balancing the rear wheels from engine power.

The rotating wheel, if not balanced, sets up vibrations in the suspension system corresponding to the degree of imbalance in the wheel. At certain speeds, the rotating wheel and the suspension system vibrate at the same natural (tuned) rate to produce the highest vibrations in the suspension system. These are points of resonance. The imbalance of a wheel can be most accurately determined at the lowest speed at which resonance occurs where the movement is the greatest.

One balancer, in particular Source 17, operates in general as described below.

An electronic pick-up unit is attached to the wheel suspension system by a self-contained permanent magnet attached to the external end of an adjustable arm.

The vibrations are transferred by the pick-up arm into the pick-up head. Inside the head, the other end of the pick-up arm is fastened to another permanent magnet. This magnet moves over a coil in response to the mechanical vibrations. This movement generates an electric current of the same magnitude and frequency as the mechanical vibrations. This electric current is then fed into an amplifier, and into a calibrated meter.

The current sent to the amplifier is built up to the voltage required to flash a strobe light. The strobe light is electronically synchronized so that it flashes once with each revolution of the wheel at the point of highest vibration during resonance. This makes the wheel appear to be standing still with the imbalance remaining in the same relative position.

The meter measures the current from the pick-up. The degree of imbalance is indicated on the meter. If the meter reads in the green section of the meter scale, the wheel is balanced; if the meter reads in the red section, the wheel is out of balance. The higher the pointer is on the red scale, the higher the degree of imbalance.

The principle of operation of the pick-up, amplifier, flash tube, and indicating meter can better be understood by studying Figure 13-2.

This circuit is, in general, the same on all Source 17 balancers, differing only between models in the values and components used.

The pick-up PU consists of a magnet MA1 that is attached to the vehicle suspension system. On the other end of the rod is another magnet

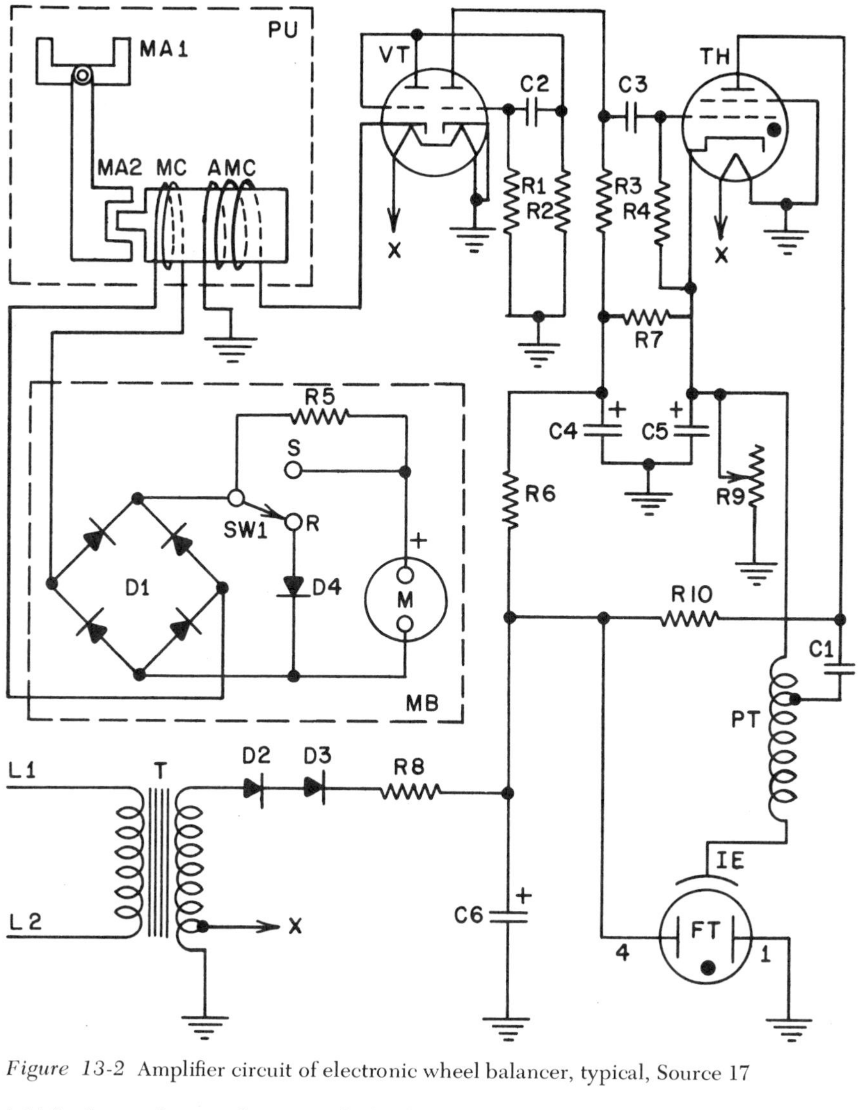

Figure 13-2 Amplifier circuit of electronic wheel balancer, typical, Source 17

MA2 that vibrates back and forth across an iron core on which are wound two separate coils. One coil MC (750 ohms) is the meter coil and feeds ac to a full-wave bridge rectifier D1 on meter board MB. This feeds dc to the 400 μa meter M through switch SW1. Position R of switch SW1 is the less sensitive (regular) position for initial balancing. Position S is the sensitive position for final check of balance.

Coil AMC (1800 ohms) on the pick-up unit PU is connected to the amplifier tube VT (12AX7) first cathode circuit, connected as a diode. The signal rectifier and amplifier consists of two triodes in the same envelope, and may be a type 6SL7, 12AX7, etc. The first triode is connected as a diode to rectify the signal and the second triode amplifies the signal.

The amplified signal is fed to the grid of a thyratron TH (2D21) trigger stage, which may be a 2D21, 2050, etc., which conducts on the signal peak, discharging capacitor C1 (0.05 mfd) through the primary of pulse transformer PT. The secondary of pulse transformer PT develops a voltage high enough to ignite the flash tube FT through ignition electrode IE. A high positive dc voltage of approximately 800 V (not high enough to fire the flash tube FT by itself) is already between terminal 4 and terminal 1 of flash tube FT.

On each excursion of the magnet MA2, there is a flash of flash tube FT, which occurs on the end of the upward excursion.

The dc power supply furnished by transformer T, may have a diode (D2 and D3 in series) rectifier as shown, or in some models this rectifier may be a vacuum tube such as type 5Y3 etc.

A vibration simulator, or "wheel balancer tester," is available from Source 17 for actuating the pick-up head at a specified rate of vibration and distance of movement to accurately check the meter circuit as well as the complete amplifier and unit. This tester consists of a back-geared motor with an eccentric moving element which is attached to the magnet of the pick-up unit. It will also indicate if the flash tube flashes in phase with the movement of the pick-up head. In addition to the usual electronic test equipment, this vibrator-simulator is the only other piece of equipment needed to service these wheel balancers properly.

A simple wheel balancer tester can also be made, or incorporated in the ignition simulator as shown in Figure 3-1, cam part WBT, and described in the explanation accompanying Figure 3-1.

The component values of a typical model are; VT (12AX7 vacuum tube), TH (2D21 thyratron tube), resistors in ohms R1 (150 K), R2 (3.3 meg), R3 (680 K), R4 (10 meg), R5 (1 K), R6 (20 K), R7 (680 K), R8 (500), R9 (10 K), R10 (150 K), capacitors in mfd C1 (0.05), C2 (0.47), C3 (0.005), C4 and C5 (8), C6 (16).

13-5 Remote Throttle Control

A means of remotely-controlling the engine speed in automotive service enables one man to perform operations that would otherwise require an extra man operating the accelerator.

These devices are commercially available under various names such as "accelerator depressor," "throttle control," and similarly meaningful names. With a little ingenuity, one can be made in the shop to fit special needs.

A. *Accelerator depressor.* One type of remote throttle control consists of a system of levers and clamping devices to depress the foot accelerator, in the driver's compartment, by a flexible cable operated remotely. The other end of the flexible cable is operated by a pistol-grip handle equipped with a means of locking it in any position. Some may have an additional "fine" adjustment for making a slight change in engine rpm after it is locked.

B. *Carburetor linkage throttle control.* Another type operates the carburetor linkage from the engine compartment. The simplest of these uses a pair of hinged jaws that fit over the idle screw between the carburetor throttle arm and the idle stop. The jaws are then expanded by means of a remotely adjustable screw. The jaws can be opened or closed to set the desired engine speed.

14
Engine Analyzers

14-1 Scopeless Engine Analyzers

The simplest engine analyzers consist of one or more multi-range and multipurpose indicating meters, most of which will perform the following functions.

A. Dwell angle meter.
B. Tachometer for engine rpm.
C. Direct Current voltmeter for reading voltages of batteries, generators, alternators, regulators, and the like.
D. Spark plug voltage meter, calibrated in peak volts.
E. Points resistance meter, millivoltmeter.
F. Capacitor tester.
G. Ohmmeter.
H. Diode tester.
I. Ignition coil tester.

14-2 Scope-Type Engine Analyzers, General

The scope-type engine analyzer consists basically of a standard cathode ray oscilloscope designed for the special needs of analyzing the voltage wave forms and magnitudes in the primary and secondary circuits of the engine ignition system. The wave form of the engine alternator system can also be analyzed, as on any standard scope.

For analyzing the ignition system, it is necessary to have a calibrated vertical input, and a triggered horizontal sweep that is linear. These modifications can also be added to any standard oscilloscope by connecting a triggered horizontal sweep circuit externally to any general purpose scope.

The standard oscilloscope, without triggered horizontal sweep, uses an internal oscillator for the horizontal sweep which must be synchronized with the horizontal input signal.

There has been so much written on the oscilloscope that the fundamental principles of operation will not be repeated here. Only the special features required for engine analyzers will be explained. Smaller CRTs use static deflection plates, larger CRTs use electromagnetic yoke deflection systems.

The triggered horizontal sweep requires no internal synchronization with the horizontal input signal. The horizontal input signal is derived from the spark plug firing of the No. 1 cylinder of the engine, or from the distributor points pulses.

By using thyratron tubes, or equivalent solid state devices, the firing of the No. 1 spark plug causes one linear sweep of the cathode-ray beam horizontally, regardless of the speed of the engine, for one complete ignition cycle. A connection across the distributor points will cause one horizontal sweep of the cathode-ray beam for each plug firing.

The vertical input to the scope is derived from one of three sources.

A. For analysis of the primary circuit of the ignition system, a voltage is obtained across the distributor points. There will be a number of vertical patterns corresponding to the number of cylinders, in parade formation. They can be arranged to be superimposed, or separated to view individual cylinder firings. For primary circuit analysis, the horizontal scope screen is calibrated to read dwell angle in degrees. It is very important that the horizontal sweep be linear. Linearity and drive adjustments are provided in the horizontal section to correct for any nonlinearity. These are adjusted to give equal spacing of the paraded patterns, using the dwell angle scale as a guide.

B. For analysis of the secondary circuit in paraded patterns of the ignition system, a voltage is obtained at the ignition coil secondary output to the center of the distributor cap. There will be a number of vertical patterns corresponding to the number of cylinders in parade formation. They can be arranged to be superimposed, or separated to read individual cylinder firings.

For secondary circuit analysis, the vertical scope screen is calibrated to read the spark plug voltages in peak volts.

C. To analyze the pattern of the alternator ripple, it is necessary to synchronize this circuit internally, because these ripples do not have a fixed frequency relationship to the firing of the ignition system.

14-3 Console-Type Scope and Engine Analyzers

The scope-type engine analyzer is available as a single-scope unit without any indicating meters, and also in a console assembly consisting of a scope, with or without a slave or monitoring scope, and an array of other instruments necessary to make complete analysis and adjustment of the entire engine and vehicle system. There is economy in the console type, as against several individual test units because many of the circuits and components are in common with more than one meter or test unit.

Appearing in a large variety of combined meter scales and circuit arrangements, the console type engine analyzer may provide the following instruments and tests in addition to the oscilloscope. Some of these may be add-ons or furnished as separate plug-in units.

A. Tachometer.
B. Ampere/dynamic compression/carburetor adjustment meter.
C. Volts/points resistance/ignition advance meter.
D. Air-fuel ratio meter, or combustion analyzer.
E. Road performance meter to measure hp output.
F. Road speed meter to measure mph.
G. Vacuum and pressure gauges.
H. Dwell angle meter, in addition to scope screen calibration.
I. Ohmmeter.
J. Timing light with spark advance control.
K. Ignition advance meter, calibrated in degrees.
L. Ammeter.
M. Volts-amperes meter.

14-4 Typical Console Scope-Type Engine Analyzer, Source 1, Model 28-01

Engine Harness Connections for Engine Analyzer Inputs

Figure 14-1 shows the connections of the engine analyzer input cable harness to a typical ignition system of a vehicle, or ignition simulator. This is a typical, Source 1, engine harness.

The vehicle battery BAT is shown connected for a negative ground. The hot battery cable is removed from the battery, and a 100-amp shunt SH connected between the hot battery terminal and the battery cable previously removed. The arrow ACC on the shunt SH goes to the accessories. A small voltage drop across the shunt SH is fed to the ammeter

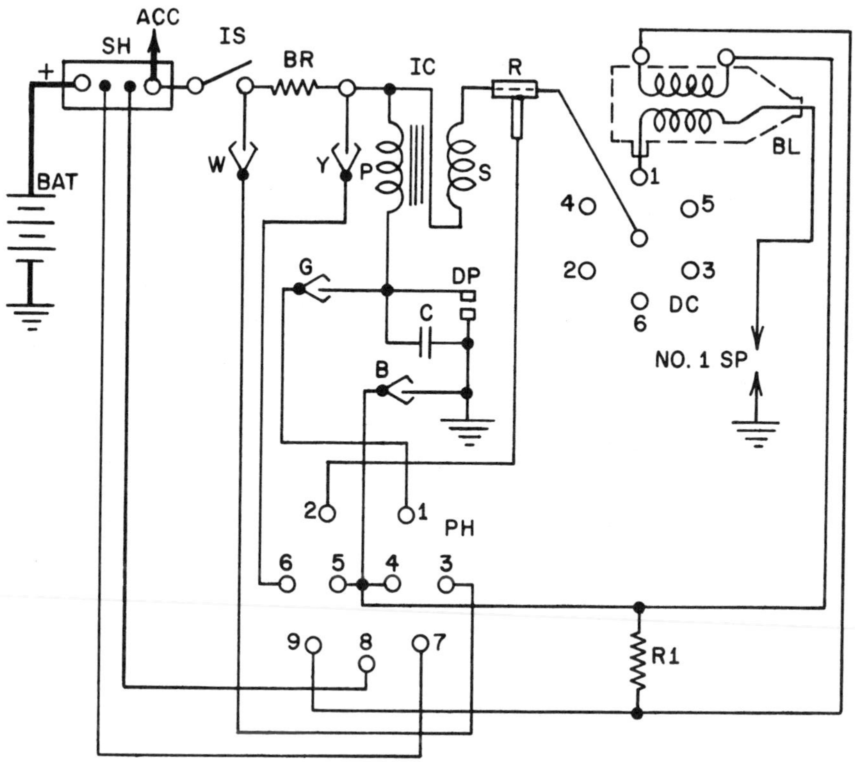

Figure 14-1 Engine harness connections for engine analyzer inputs, typical, Source 1

in the console through cable plug PH, terminal PH7 positive, and PH8 negative.

The ignition switch IS is in the vehicle. Ballast resistor BR is the normal ignition coil 1C ballast resistor. The distributor points DP, capacitor C, and distributor cap DC completes the ignition system.

The white clip W from the harness connects to the hot battery post after the ignition switch IS and before the ballast resistor BR. The white clip W is for the "alternator test" and is connected to the cable harness plug, terminal PH3.

The yellow clip Y is connected to the battery side of coil 1C primary P and goes to the volts/ampere switch in the console through plug PH, terminal PH6.

The green clip G is connected to the distributor point DP side of the ignition coil IC primary P and goes to the primary switch in the console through plug PH, terminal PH1.

The black clip B is connected to the engine ground and goes to the console ground through plug PH, terminals PH4 and PH5.

The red clip R is a specially constructed device consisting of a split cylinder with each half on the jaws of a spring clamp. This clamp wraps around (but has no piercing, or other contact), the insulation of the high-tension cable between the ignition coil IC secondary (center hole) and the center of the distributor cap DC. This clip picks up a signal by capacitive coupling, acting as a very small capacitor. If there is leakage through the high-voltage cable, an arc will jump to the clip and cause the scope pattern to be enlarged and go off scale. This signal goes to the chassis "secondary" switch through cable plug PH, terminal PH2. A much larger capacitor than the value of the capacitive coupling is located in the chassis serving as a capacitor voltage divider with most of the voltage drop occurring across the capacitive coupling at clip R.

The blue BL pick-up head is connected between the No. 1 spark plug and the No. 1 terminal of the distributor cap DC. This head is of molded plastic containing a current transformer with the primary coil in series with the No. 1 plug and distributor. A secondary coil, shunted by resistor R1, is connected to the harness cable plug PH, terminals PH9 and ground terminals PH4, and PH5. Terminal PH9 goes to the "horizontal trigger sweep" circuit in the console, as well as to the "dynamic compression" trigger circuit. The current transformer in this blue lead is like any other current transformer in that the secondary must be loaded to maintain a very low voltage across the secondary. Resistor R1 is a load resistor to provide a total resistance, including the secondary, of 100 ohms.

Similar harnesses are available for permanent installation in vehicles to permit rapid analysis and service on trucks, buses, taxis, and the like by fleet operators.

It is conceivable that some day all vehicles will have a permanently-connected analyzer harness input cable terminating in a plug, or jack, accessible without lifting the hood. Law enforcement officers may even be equipped with portable analyzers to check vehicles suspected of violating pollution and performance control laws.

Functions of Controls and Switches

The functions of the console front panel controls and switches of a typical Source 1, Model 28-01, console-type engine analyzer are given in order below.

Numbers 1 through 10 are all part of the interlocked push button switch. The mechanical linkage of this switch is such that no two push buttons can be depressed at the same time. When any one button is in the depressed position, depressing another button releases the one previously depressed. The other switches and controls operate independently. The following switch and potentiometer number references will be found in various schematic wiring diagrams which follow.

1. *Volt-Amps;* Switch S1A energizes the necessary circuits to allow making battery voltage, cranking voltage, point resistance, charging current and charging voltage checks. The voltmeter is separate from the ammeter (dynamic compression meter) and permits reading volts and amperes simultaneously. The ammeter operates off a shunt.
2. *Alternator Test;* Switch S1B1, is the alternator test push button. When it is depressed, signal and supply voltages are applied to the alternator test circuits. This signal is acted upon and fed to the scope for a visual inspection of the alternator under test.
3. *Timing,* Switch S1C; when the timing push button is depressed, the timing light is caused to flash steadily. The timing advance control of the timing light assembly is not energized; rotation of this control will have no effect; and there will be no visual indication or pattern displayed on the scope screen.
4. *Advance* Switches S1D1 and S1D2; the advance push button is

used to activate the required circuits to measure the ignition advance of the engine under test. The timing light will flash steadily, and rotation of the advance control located on the timing light assembly will cause the pointer of the ignition advance meter to swing up-scale in accordance with the control rotation. There will be no pattern display on the scope screen.

5. *Primary,* Switch S1E; the primary push button couples signals from the primary circuit (across the distributor points) to the scope where it is visible on the screen as a replica of the instantaneous voltages present across the distributor points of the engine under test.

6. *Secondary,* Switch S1F; the secondary push button couples signals from the secondary circuit to the vertical section of the scope. Signals from the primary circuit are also connected, or coupled, to the triggered horizontal sweep section of the scope. These signals are then amplified and applied to the scope and are visible on the screen as a replica of the instanteous voltages present in the secondary circuit of the engine under test.

7. *18 KV,* Switch S1G; when the 18 KV push button is depressed, an analysis of the secondary circuit resistance can be made. Circuits of the horizontal and vertical sections are activated to produce a paraded pattern of all cylinder firing lines. The height of the voltage spike of each cylinder firing is of utmost importance. With reference to the 18 KV scale located on the left side of the scope screen grid, each cylinder spike can be measured and excessive difference in any specific spike noted. The horizontal sweep is triggered by the No. 1 spark plug.

8. *36 KV Dynamic Compression,* Switch S1H; the 36 KV dynamic compression push button is used to measure voltages in excess of 18 KV (coil output, etc.), and to make a dynamic compression check of cylinders in engine under test. The signal is coupled to the horizontal section, vertical section, and the dynamic compression circuit. A paraded pattern will be present on the scope screen and the dynamic compression meter will be sensitive to the rotation of the cylinder selector control (on an engine, but not on a simulator), and will also affect the pattern on the scope screen as it shorts out one cylinder pattern at a time. The horizontal sweep circuit is triggered by the No. 1 spark plug firing.

9. *Carburetor Adjust,* Switch S1J; when the carburetor adjust push button is depressed, the carburetor adjust scale of the ammeter is ener-

gized and is sensitive to the idle speed of the engine under test. There will be no pattern display on the scope screen.

10. *Off*, Switch S1; the off push button releases the last depressed push button of 1 to 9 above. This switch also cuts off all signals from the engine under test, and all supply voltages from the power supply to the chassis. All of the following controls and switches operate independently.

11. *Ground Polarity*, Switch S7; the ground polarity switch is used to select the proper polarity to coincide with the battery ground polarity of the vehicle under test. The ground polarity switch should be in the same position as the battery of the vehicle under test. If the battery of the vehicle positive post is grounded, turn polarity switch to positive; if the negative terminal is grounded, turn the polarity switch to negative.

12. *Volts*, Switch S6; the volts switch is used to select the proper range or scale of the voltmeter. The range selected should be the next one higher than the voltage of the battery under test; 10 for 6-volt system, 20 for 12-volt system, and 40 for 24-volt system.

The "PT-RES" position of the volts switch is used to measure the resistance of the distributor points when they are closed.

13. *PT-RES*, Switch S9; the PT-RES push button is depressed to check the distributor primary circuit (points, leads, etc.), for excessive resistance due to corrosion, pitting, etc. When depressed, the PT-RES push button completes the circuit from the distributor points of the engine under test to the PT-RES scale of the voltmeter. It is important that the breaker points be closed for these measurements.

14. *Pattern Height*, Potentiometer R20; this dial is only used to increase the height of the pattern displayed on the scope screen during the alternator test procedure.

15. *Synchronize*, Potentiometer R19; this knob is used to synchronize the horizontal sweep to hold the pattern stationary on the scope screen only during the alternator test. All other scope pattern tests are automatically synchronized by the triggered horizontal sweep circuits.

16. *Balance*, Potentiometer R28; this control is used to balance the air-fuel ratio meter. The READ, switch S10, button must be depressed to make this adjustment.

17. *Read*, Switch S10; depressing the read push button actuates the air-fuel ratio meter.

18. *Push To Adjust*, Potentiometer R27; this control is used to cali-

brate the air-fuel ratio meter; with the "Read" push button depressed, depress the "Push to Adjust" knob, then rotate it to position the air-fuel ratio meter pointer to the line at the left.

19. *Parade,* Potentiometer R8; this control is used to move the pattern, displayed on the scope screen, horizontally. It will move or parade the cylinder firing line across the scope screen to allow individual inspection and analysis of any specific cylinder firing line, after using the "Expand" control to spread the patterns out to a greater width. It acts as a horizontal position control.

20. *Pattern Selector,* Switch S2; this rotary switch has three positions from left to right, NORMAL, SEPARATED, and PARADED.

21. *Expand,* Potentiometer R5; this control is used to expand a single cylinder firing line to cover the width of the entire scope screen. This allows detailed analysis of each cylinder firing line. After the expand control has been adjusted to the point where a single cylinder firing pattern covers the screen, the parade control can be adjusted to move the cylinder firing patterns in position one by one, in the firing order onto the scope screen. The expand control acts as a width control.

22. *Balance To Line,* Dual Potentiometers R16A and R16B; this control has two functions—to balance the meter during the dynamic compression test using R16B; and to balance the meter when in the CARBURETOR ADJUST position using R16A.

23. *Dynamic Compression No. 1 Cylinder,* Switch S5; this switch is used to effect dynamic compression measurement of cylinder No. 1 only. The theory behind dynamic compression measurements involves the shorting out of each cylinder, in turn, to prevent its firing, and noting the loss of engine speed. Since the initial triggering pulses used in the circuits are derived from the No. 1 cylinder plug, it is obvious that the system would fail to function when the pulse from plug No. 1 is shorted out. The "Dynamic Compression No. 1 Cylinder" switch energizes the circuits to prevent this from happening and allow a compression check on No. 1 cylinder.

24. *Cylinder Selector,* Rheostat R13; this control is used for two functions; to short out one cylinder at a time on the dynamic compression test; and to separate a single cylinder from the superimposed pattern on the SEPARATED position of the Pattern Selector Switch.

25. *Pattern Display, Distant Viewing,* Switch S3; this switch is effective only when the 18 KV push button is depressed. It increases the

brilliancy and width of the plug firing spikes on the scope screen for better visibility at a distance.

26. *RPM*, Switch S8; the rpm selector switch is used to select the proper scale on the tachometer. The three positions of the switch are; 800, 1600 and 10,000. The position should be selected that will keep the pointer reading as close to the middle of the scale as possible.

27. *Panel Lights*, Switch S11; this push button switch is used to turn ON and OFF the console panel lights, and to energize the heater coils of the Clayton cell unit of the air-fuel ratio meter. The panel light switch should be left ON whenever the service lane is in operation whether or not the console is being used.

Pattern Selector Switch

The pattern selector switch usually has three positions; 1. Normal, 2. Separated, and 3. Paraded.

In the NORMAL (1) position, the horizontal sweep is triggered by distributor points pulses to sweep once for each cylinder firing, and the patterns, either primary or secondary, are "superimposed". Each cylinder firing pattern is repeated and all patterns are seen at the same time, one on top of the other.

In the SEPARATED (2) position, the pattern selector switch opens a short across a resistor in the "vertical position" control circuit to raise the patterns on the screen to the upper part of the scope screen. Any single cylinder firing can be separated from the shifted, superimposed group of patterns by obtaining a signal from the plate P of V8B tube of the dynamic compression circuit. The dynamic compression switch tube V9 is deactivated so that the cylinder firings are not shorted out during this operation. This circuit contains the time delay cylinder selector rheostat that shifts the vertical position of the pattern on any preselected cylinder back to its normal position at the bottom of the scope screen. This is accomplished by feeding a signal from the plate P of tube V8B to the input of the scope vertical amplifier of such polarity and magnitude to depress one superimposed pattern at a time to its normal position at the bottom of the scope screen.

The PARADED (3) position of the pattern selector switch connects the No. 1 plug firing to the triggered horizontal sweep circuit so that the scope sweeps only once for all cylinder firings. The vertical input is fed from the center of the coil for secondary patterns, and from the distributor points for primary patterns, to give individual patterns in parade formation across the screen. The patterns can be spread out by the EXPAND control which is a width control; individual cylinder patterns can be selected by the PARADE control which is a horizontal position control.

Vertical Deflection Section of Scope-Type Engine Analyzer

The vertical section of scope-type engine analyzers is actually simpler than a standard oscilloscope vertical deflection section in that there is no concern here about high or extremely low frequency response. The vertical gain control is one potentiometer with no range multiplier switch and it is mounted on the chassis with a screwdriver slot adjustment. The vertical position control, focus, brightness, and spot shape controls are also mounted on the chassis having a screwdriver slot adjustment. There is a fixed resistor in the vertical position control circuit that is normally shorted out by a switch, but is introduced in the circuit by the SEPARATED (2) position of the pattern selector switch to move the pattern to the top of the scope screen.

Figure 14-2 shows the vertical amplifier section of a typical Source 1, Model 28-01, engine analyzer where VI is the vertical input signal to the amplifier which passes through capacitor C5 (0.01 mfd) and splits to go to the observer scope reference B REF B and also to the vertical calibration potentiometer R1 (3 megohms screwdriver adjustment on chassis), through resistor R38 (2.2 megohms) and R39 (330 Kohms) to ground. The wiper of R1 goes to R41 (10 Kohms) and to the grid of tube V1A (12AU7). The tube V1B (12AU7) grid is biased from the vertical position potentiometer R2 (50 Kohms) which is a screw slot adjustment on the chassis. Resistor R45 (22 Kohms) is in series with the bias voltage and R2, and is normally shorted out by a section of the pattern selector switch S2, but is open only during the SEPARATED position of the pattern selector switch to cause the vertical position of the pattern to be raised higher on the scope screen.

The plates of the two tube sections V1A and V1B go to the vertical deflection plates VP of the cathode ray tube (8CP31). The plate of V1A, REF A also goes to the brightness amplifier input shown in Figure 14-3 for distant viewing. Other components and values are; resistors in ohms R44 (220 K), R40 (75 K), R43 (47 K), R46 (220 K), R42 (330 K), R47 (560 K), capacitor C12 (20 mfd).

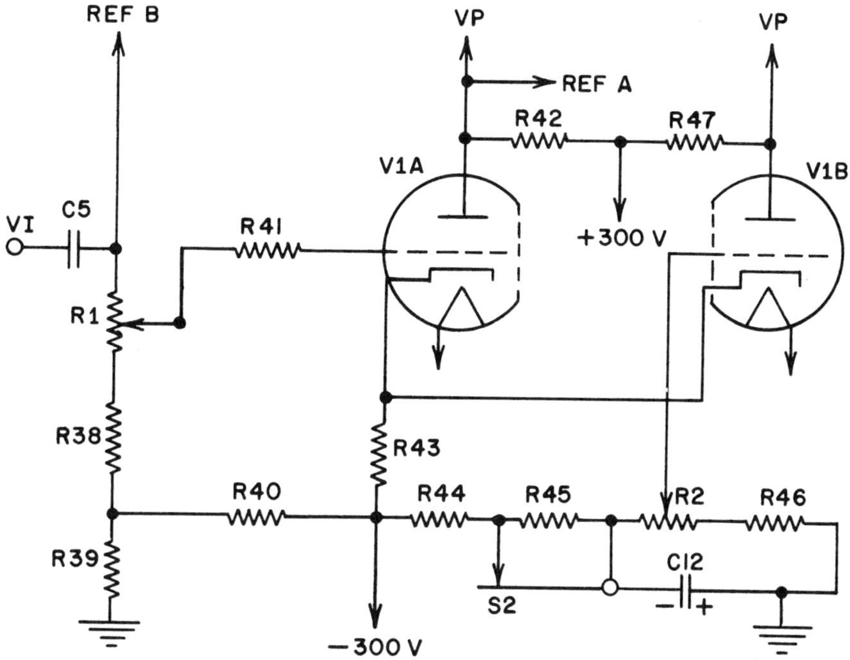

Figure 14-2 Vertical section of scope-type engine analyzer, typical, Source 1, Model 28-01

Brightness Amplifier for Distant Viewing

There is a PATTERN DISPLAY push button on the Source 1, Model 28-01, engine analyzer for the purpose of increasing the width and brightness of the plug firing spikes of the pattern display on the scope screen. This allows better visibility at a distance.

Figure 14-3 shows, schematically, the brightness amplifier where the input REF A is from one of the CRT vertical plates as shown in Figure 14-2 at the plate of tube V1A, and goes through capacitor C6 (180 mmfd) to the grid of tube V2A. The circuit is only energized when the 18 KV push button is depressed, and by depressing the PATTERN DISPLAY push button S3 which applies plate voltage to tube V2A (12AU7) through resistor R50 (15 Kohms).

The output of the amplifier tube V2A feeds pulses from, not only the plate through capacitors C9 (0.01 mfd) and C30 (500 mmfd), to the spot shape SS control and to grid 9 of the CRT to widen the spot shape of

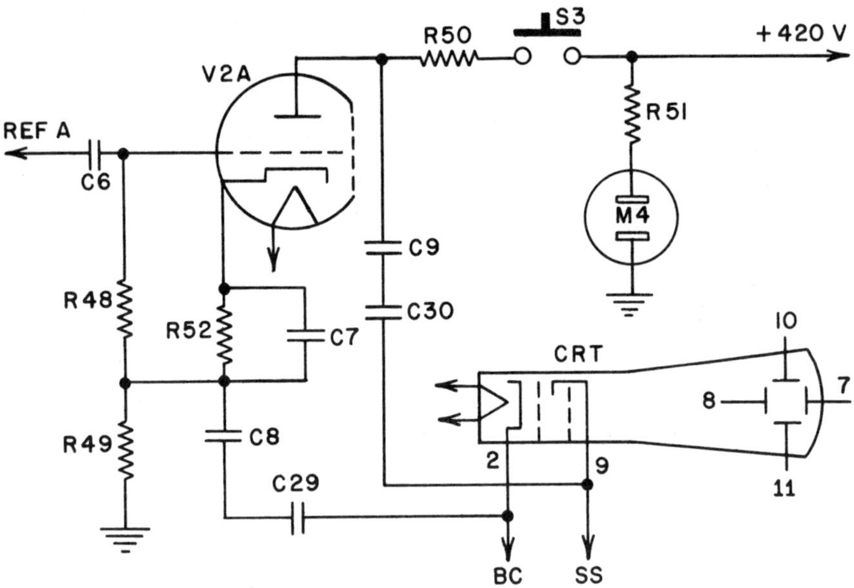

Figure 14-3 Brightness amplifier of vertical section shown in Figure 14-2, Source 1, Model 28-01

the pattern, but also from the cathode circuit of V2A through capacitor C8 (0.01 mfd) and C29 (500 mmfd), to the brightness control BC and to the cathode 2 of the CRT to brighten the pattern.

Other values are; resistors in ohms R48 (2.2 meg), R49 (15 K), R 52 (3.3 K), R51 (100 K), capacitors C7 (0.001 mfd), and neon lamp M4, the distant viewing, or pattern display pilot light.

Vertical Section Amplifier for Observer Scope

The Source 1, Model 28-01, engine analyzer console has provision for an observer scope for remote observations of the same patterns appearing on the console screen. An additional vertical amplifier and filter is required to supply the observer scope vertical input.

Figure 14-4 shows this amplifier schematically, where the input REF B is from the same vertical input source as the console vertical section shown in Figure 14-2. This signal goes through vertical output preamplifier tube V2B (12AU7A) and transistor Q1 (2N3392), with the output from the emitter of transistor Q1 going to the vertical output VO RED plug that is connected to the observer scope vertical input VI RED. The transistor Q2 (2N3392) functions as a vertical output filter.

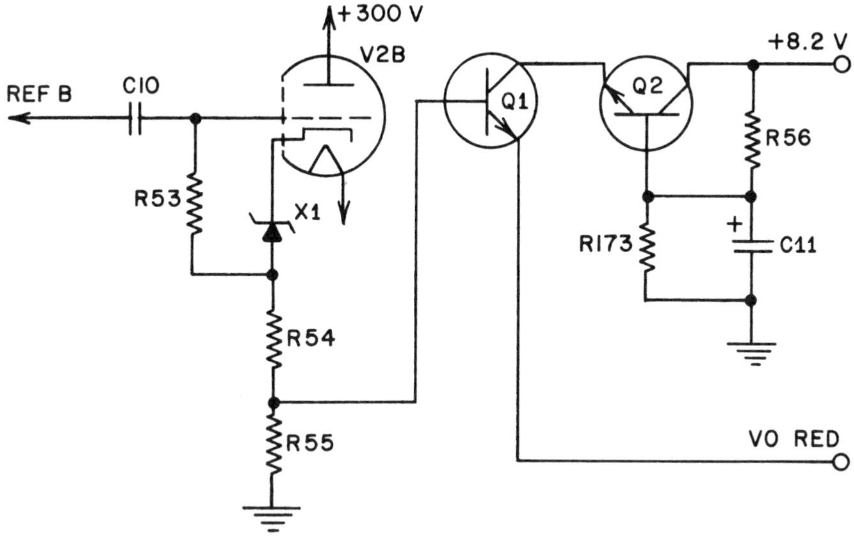

Figure 14-4 Vertical section amplifier for observer scope, Source 1, Model 28-01

Other component values are; resistors in ohms R53 (4.7 meg), R54 (18 K), R55 (510), R56 (5.6 K), R173 (47 K); capacitors in mfd C10 (0.1), C11 (22), zener diode X1 (10 V).

Triggered Horizontal Sweep Circuit

The first stage in the triggered horizontal sweep circuit of a typical scope analyzer Source 1, Model 28-01, is shown in Figure 14-5. This stage is called the TRIGGER GENERATOR and serves to take the reduced voltage from the No. 1 spark plug entering at terminal PH9 where it goes through tubes V3A (½ 12AX7) and V3B (½ 12AX7) and out at reference REF A to the trigger stage shown in Figure 14-6. In going through this stage in Figure 14-5, the negative pulses are clipped and the damped wave, or hash, is eliminated, resulting in clean positive spikes leaving the stage at REF A. The progress of the signal from the No. 1 spark plug is to terminal PH9 through R59 (10 Kohms) shunted by C13 (0.01 mfd), to the grid 2 of vacuum tube V3A (½ 12AX7). The output of V3A is from the plate 1 through R62 (220 Kohms) to the grid 7 of V3B. The output of V3B (½ 12AX7) is from plate 6 to REF A through C15 (0.005 mfd). The output REF A goes to the input of the triggered sweep stage input shown in Figure 14-6.

Other component values are; resistors in ohms R58 (270 K), R60 (2.7 K), R61 (62 K), R63 (220 K), R64 (270 K); capacitors C14 (0.001 mfd), C15 (0.005 mfd).

The second stage in the triggered horizontal sweep is called the SWEEP TRIGGER STAGE and is shown schematically in Figure 14-6.

The purpose of this stage is to take the clean positive pulse signals from stage 1 shown in Figure 14-5, and create a sawtooth output wave form for the horizontal sweep which is fed to the horizontal amplifier Figure 14-7, and to the horizontal plates of the cathode ray tube CRT.

The path of the signal is from terminal REF A through R66 (1 megohm) and to the grid 1 of thyratron tube V4 (2D21). When the positive

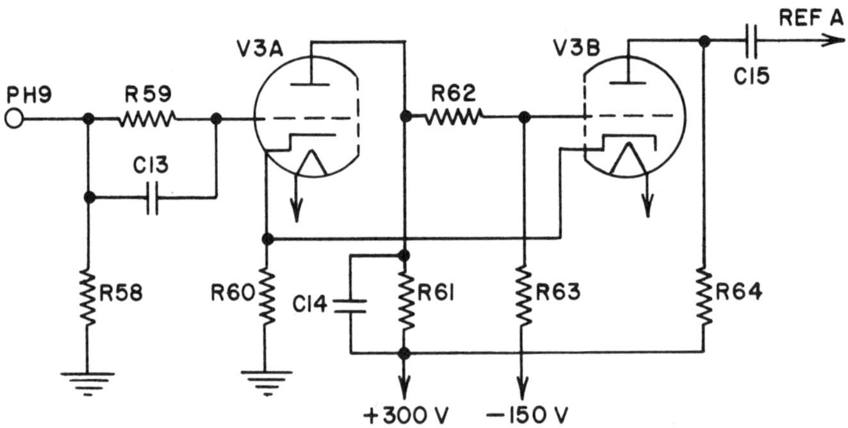

Figure 14-5 Horizontal sweep trigger generator, Source 1, Model 28-01

input pulse overcomes the negative bias, the tube V4 fires, oscillating at triggered frequency, and generates a sawtooth wave across capacitor C16 (sawtooth forming capacitor), which is fed to the plate 9 of tube V5A (½ 6BA8) and to the grid 2 of V5B (½ 6BA8). The output of V5B is in the cathode circuit across R5 (500 Kohms) expand potentiometer and is then fed to the horizontal amplifier shown in Figure 14-7, through REF B.

Other components serve to shape, and place in proper phase position, the sawtooth pattern. The values are; resistors in ohms R65 (1 meg), R67 (4.7), R68 (470 K), R69 (1 meg), R3 (3 K horizontal linearity control for right-side of screen adjustment, by screw slot adjustment on chassis), R4

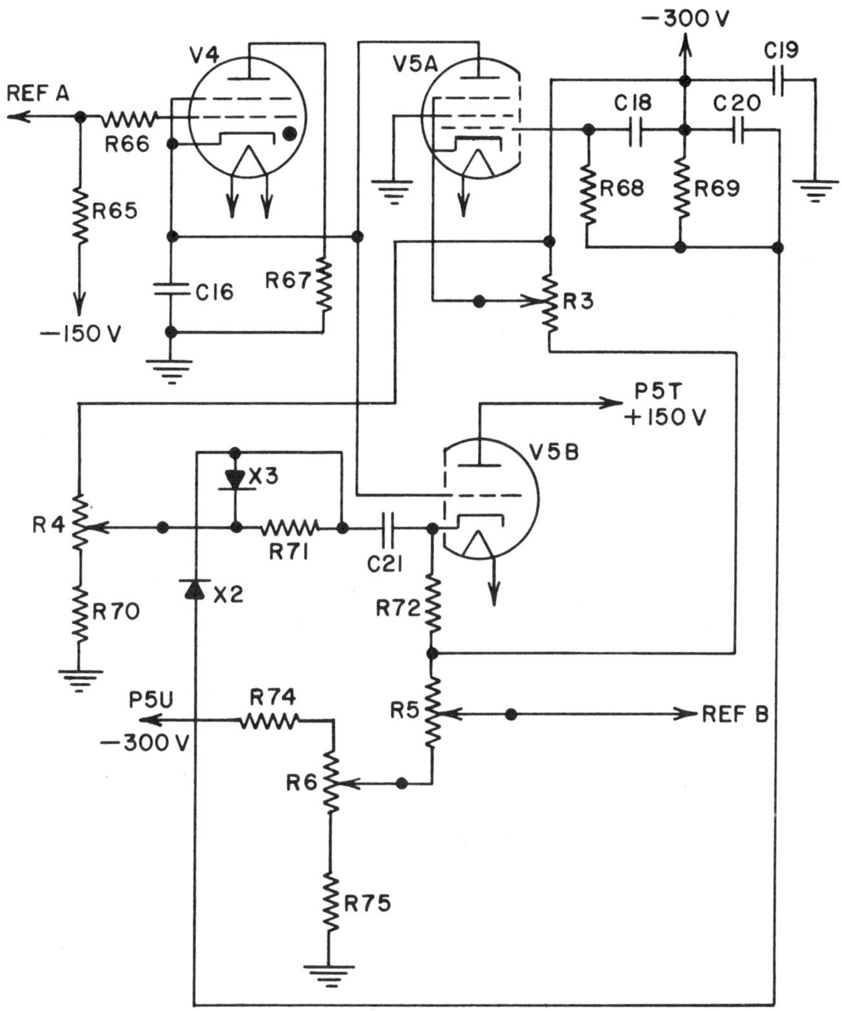

Figure 14-6 Horizontal sweep trigger stage, Source 1, Model 28-01

(100 K horizontal bias to stabilize the pattern by screw slot adjustment on the chassis), R70 (100 K), R71 (6.2 meg), R72 (82 K), R5 (500 K expand control by knob on panel for adjusting width of pattern), R74 (82 K), R6 (50 K left edge clamping adjustment by screw slot on the chassis to hold the pattern locked in at the left edge of the screen regardless of the position of the expand control. This left edge clamping is ac-

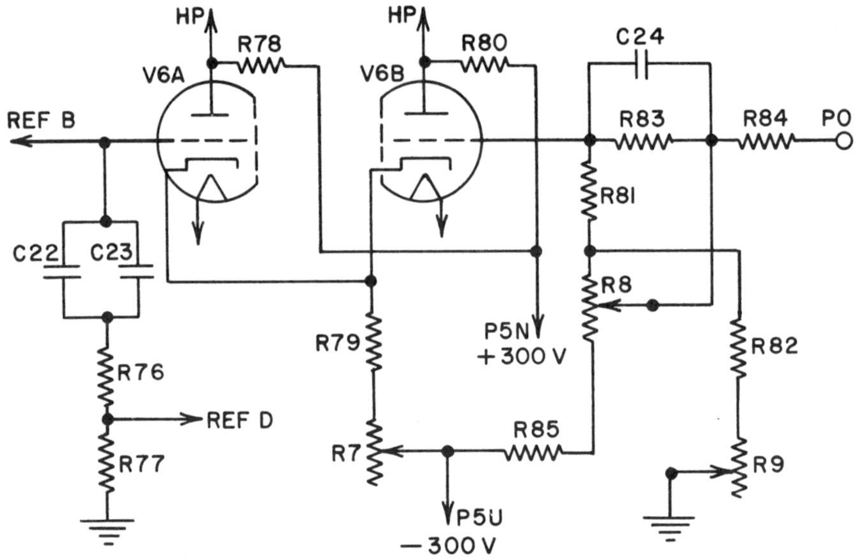

Figure 14-7 Horizontal sweep amplifier of scope analyzer, Source 1, Model 28-01

complished by the filter clamp circuit consisting of capacitors C18 and C20, resistors R68 and R69, and the clamping diode X3 to form an average voltage regulator with V5A. A constant height sawtooth is maintained in the cathode follower circuit of V5B regardless of the pulse trigger rate. The expand control R5 can then expand the pattern only to the right of the screen.), R75 (62 K); capacitors in mfd C16 (0.056), C18 (3.0), C19 (0.47), C20 (0.22), C21 (1.0); diodes X2 and X3 (PA3).

The third and last stage is shown in Figure 14-7 comprising the horizontal amplifier that drives the horizontal plates of the CRT.

In Figure 14-7, the signal from the previous stage, Figure 14-6, enters at REF B and goes to the grid 2 of V6A (½ 12AX7) driving the two tube sections V6A and V6B in a push-pull arrangement with the output of the plate of V6A 180° out of phase with the plate of V6B. The two outputs at HP go to the horizontal plates of the CRT. Other component values are; resistors in ohms R76 (2.2 meg), R77 (220 K), R78 (470 K), R79 (150 K), R80 (560 K), R81 (47 K), R82 (27 K), R83 (10 K), R84 (100 K), R85 (30 K); capacitors C22, C23 and C24 (0.05 mfd).

REF D is a connection to the horizontal output amplifier for the observer scope. Parade output PO is the connection to the parade output (black) for the observer scope.

Potentiometer R7 (100 Kohms), is the horizontal drive control for adjusting linearity on the left side of the scope screen and is a screw slot adjustment located on the chassis; potentiometer R8 (50 Kohms) is the parade control by knob on the console panel; and potentiometer R9 (50 Kohms), is the horizontal position control adjusted by screw slot on the chassis.

Horizontal Output Amplifier for Observer Scope

The Source 1, Model 28-01, engine analyzer has an additional horizontal amplifier and filter circuit to supply horizontal output signals from the console to the observer scope module.

Figure 14-8 shows this amplifier schematically, where the input signal is derived from REF D Figure 14-7. This signal is amplified by transistors Q3 (2N3392) and Q4 (2N3392). The emitter output of Q4 goes to the horizontal output HO (white) plug which is connected to the input jack of the observer scope. Transistor Q5 (2N3392), and associated components, functions as an output filter.

Other component values are; resistors in ohms R86 (220 K), R87 (18 K), R88 (3.3 K), R89 (30), R90 (5.6 K), R172 (47 K); capacitors C25 and C26 (22.0).

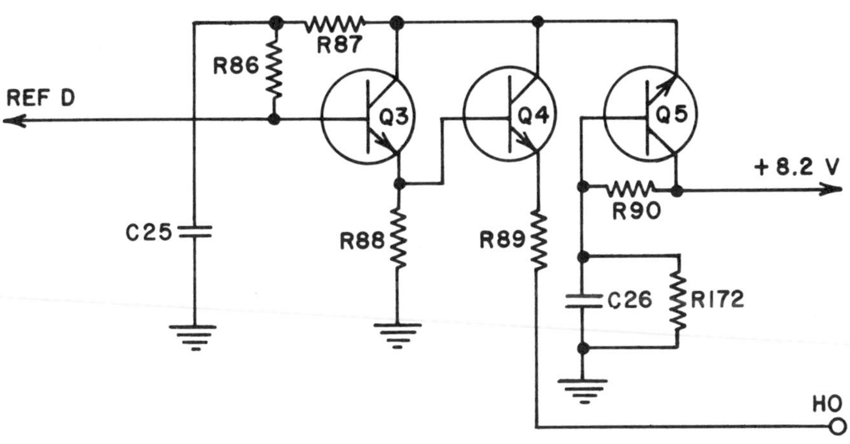

Figure 14-8 Horizontal output amplifier for observer scope, Source 1, Model 28-01

Alternator Test Section Circuit

The alternator test section of the Model 28-01 scope-type analyzer is used to examine the wave shape of the engine alternator output. Any deviation in the normally uniform rectified three-phase ripple on the scope screen indicates certain defects in the alternator.

Since these ripples bear no frequency relationship to the pulses from the No. 1 spark plug, the first stage of the horizontal sweep trigger generator (Figure 14-5), is eliminated for this test. Instead, a separate horizontal oscillator, synchronized with the alternator output pulses, supplies the synchronized horizontal sweep. This is fed to the horizontal trigger stage (Figure 14-6), at input terminal REF A. The alternator test section has a voltage preamplifier with its own pattern height control on the front panel. The output of this amplifier is fed directly to the vertical amplifier section of the scope.

Figure 14-9 shows the schematic wiring diagram of the alternator test section of the scope analyzer. The alternator output is connected to terminal PH3 where the rectified three-phase ripple signal goes to the grid

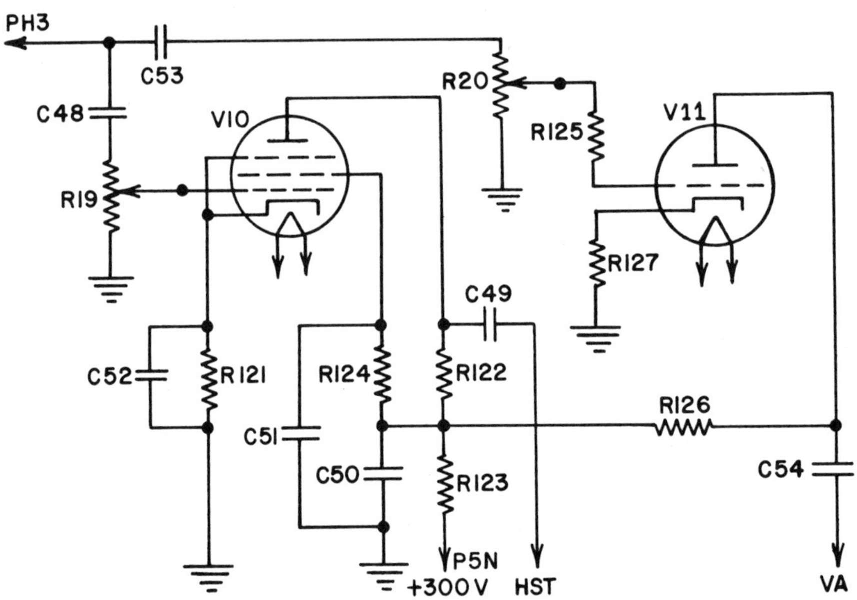

Figure 14-9 Alternator test section of scope analyzer, Source 1, Model 28-01

circuits of tube V10 (6AU6), and tube V11 (6C4). Signal voltage is applied to V10 across synchronizing control R19 (500 Kohms, located on front panel); through capacitor C48 (0.01 mfd); and to tube V11 (6C4), across the pattern height control R20 (500 Kohms, located on console front panel), through capacitor C53 (0.01 mfd).

The output of V10 is from plate 5 through capacitor C49 (0.01 mfd) to the horizontal sweep trigger stage HST, input reference A REF A in Figure 14-6.

The output of V11 is from its plate, through capacitor C54 (0.005 mfd) to the input of the scope vertical amplifier VA.

Other component values are; resistors in ohms R121 (1 K), R122 (220 K), R123 (27 K), R124 (470 K), R125 (1 K), R126 (100 K), R127 (4.7 K); capacitors in mfd C50 (1.0), C51 (0.10), C52 (100).

In the absence of an alternator for making this test, use a battery charger across the test battery, and connect the alternator clip (white) to the battery charger clip. A 60-Hz ripple will then appear on the scope screen after being synchronized. The synchronizing control R19 is capable of covering 60-Hz current.

A small ac transformer can also be used for this purpose having a secondary voltage of about 12 V, in which case the full sine wave will appear on the screen. A calibrated transformer for this purpose is available from Source 1 for testing their analyzers. A tube tester can provide 12 V from appropriate tube socket holes for this purpose. When ac is used for this test, it is better to use a half-wave diode, connected for positive output, in series with the ac source to simulate closely the alternator test patterns.

Dynamic Compression Meters and Circuits

The theory behind dynamic compression, or power check, measurements involves the shorting out of the ignition to each engine cylinder, in turn, and noting the loss of engine total power output as indicated by loss of rpm shown on a special expanded-range tachometer. This tachometer shows small changes in rpm that would be hard to read on a standard tachometer.

The dynamic compression test is accomplished electronically, but can be done by shorting each spark plug with a screwdriver. To detect the reduction of speed as each plug is shorted out requires a tachometer sensitive enough to show these small changes. Such a meter is called a dynamic compression meter.

The dynamic compression meter is usually incorporated in the console of a scope-type engine analyzer, although they are available as separate units. There are advantages in the combination instrument as some of the circuits have components in common. It is also helpful to observe the scope pattern as each cylinder firing is shorted out to determine the dynamic compression reading for each cylinder in the firing order and identify it.

The purpose of the dynamic compression meter readings is to determine if all cylinders are contributing equally to the total power output, as well as to determine and identify any cylinders weak in compression. This system is more reliable than the method of using a pressure gauge in each cylinder plug hole under cranking conditions, as it measures the relative compression under running conditions.

The dynamic compression meter circuit section of a typical scope console Source 1, Model 28-01, is shown in three illustrations, each showing a different function.

Figure 14-10 shows the input circuit from the pick-up head connected to spark plug No. 1. It functions as a CYLINDER-MARKER GENERATOR-TRIGGER stage, and includes the dynamic compression meter and carburetor adjustment meter.

Figure 14-11 shows the second circuit in line which functions as a PULSE GENERATOR comprising a cylinder selector and pulse width control.

Figure 14-12 shows the last circuit in line and functions as a DYNAMIC COMPRESSION SWITCH to conduct current around the engine distributor points to short out the firing of any selected cylinder in the firing order, as observed on the scope screen. This circuit also includes a plug No. 1 trigger stage which functions only during the shorting out of the No. 1 cylinder, to provide normal trigger sweep functions when the No. 1 plug pulses are missing.

In Figure 14-10, the input terminal PH9 is connected to the pick-up head, which has a built-in current transformer, and is connected to the No. 1 spark plug. The pulse signal goes through terminal PH9, capacitor C36 (0.05 mfd), C35 (0.05 mfd), branches off at HZ to the horizontal trigger sweep C39 (0.05 mfd), resistor R106 (1 megohm) to the grid of cylinder-marker generator-trigger thyratron tube V7 (2D21). Output of V7 is from the plate (pin 6), through resistor R110 (2.2 megohms), to the grid of tube V8A (½ 12AX7), Figure 14-11 at reference A REF A.

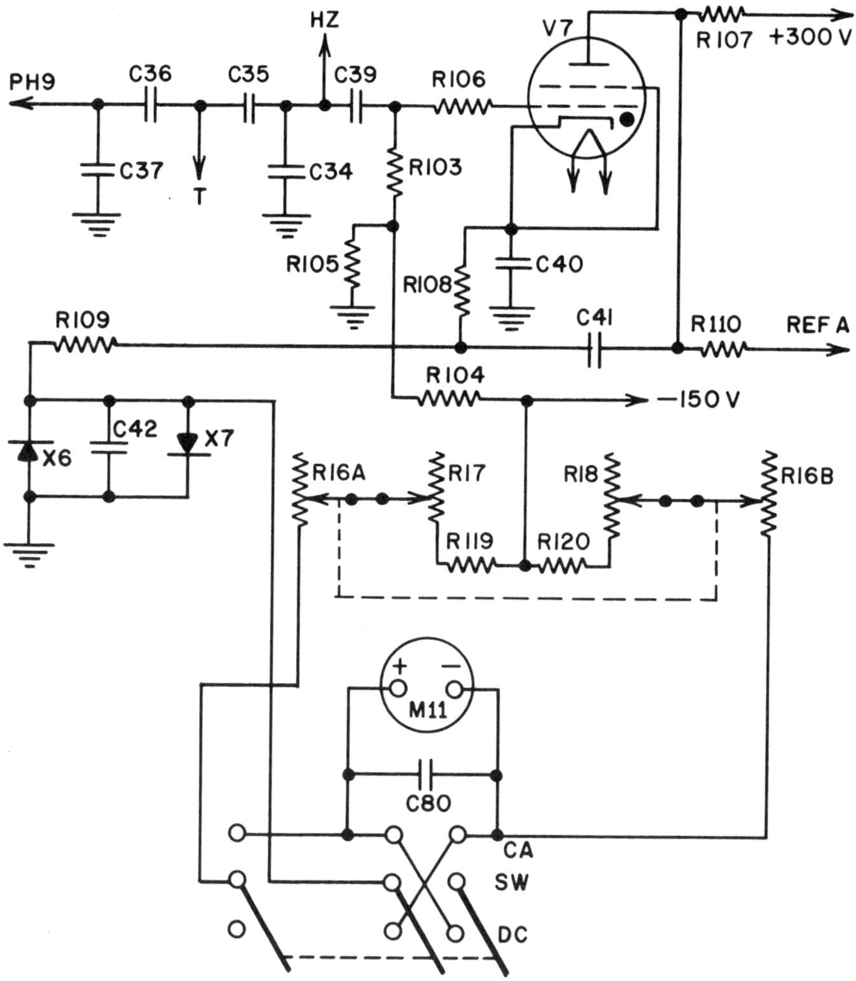

Figure 14-10 "Cylinder marker generator trigger," dynamic compression and carburetor meter circuit, Source 1, Model 28-01

In Figure 14-10 the dynamic compression meter M11 also has two other scales that function for amperes and carburetor adjustment. Meter M11 is connected in the cathode return-to-ground circuit and serves as an accurate tachometer for small changes in rpm. Balance-to-line rheostat R16B (10 Kohms) is adjusted by a panel knob after warm-up to set the meter M11 at zero on the dynamic compression scale. Dynamic compression zero adjustment control R18 (50 Kohms) is a screw slot adjustment located in the chassis.

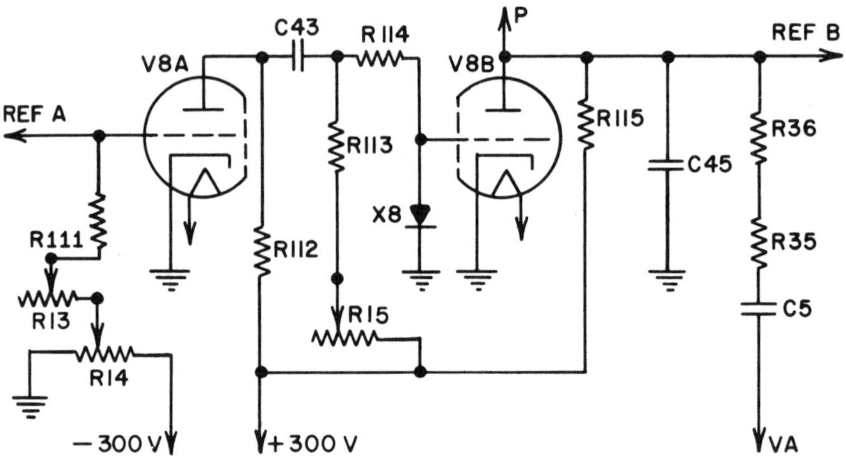

Figure 14-11 Dynamic compression meter "pulse generator," Source 1, Model 28-01

Other component values are; capacitors in mfd C37 (0.005), C34 (0.005), C40 (0.005), C80 and C42 (1000), C41 (2.0); resistors in ohms R103 (1 meg), R104 (220 K), R105 (22 K), R107 (30 K), R108 (30), R109 (1 K), R120 (15 K); diodes X6 and X7 (PA3). Terminal T is only connected during the cylinder No. 1 dynamic compression test, when a stepped-up pulse is fed back to reinforce the weak pulse signal from the No. 1 spark plug. This keeps the trigger and sweep circuits functioning normally.

Meter M11 reads up-scale on the dynamic compression test for a decrease in engine rpm to indicate a higher number (the numbers are arbitrary and mean nothing in absolute values, but are relative), for a cylinder that has a higher loss in rpm when shorted out.

Meter M11 reads up-scale (zeroed at mid-scale) on the carburetor adjustment scale for an increase in rpm.

The DYNAMIC COMPRESSION push button switch SW on the console connects the meter M11 as shown in Figure 14-10 in the DC position. When the CARBURETOR ADJUST push button switch is depressed, it pops out the dynamic compression DC button, since they are interlocked. The switch SW is a simplified connection simulating the two push button switches. When SW is in the upper CARB. ADJ. CA position, the meter polarity is reversed, and three resistors are inserted in the circuit; R119 (15 Kohms), R17 (30 Kohms, carburetor adjustment control mounted on the chassis having a screw slot adjustment), and R16A (10 Kohms, balance-to-line control is ganged to operate in unison with R16B by one knob on the console panel marked "balance to line").

The ammeter scale of meter M11 is used when it is shunted, by switching means, by the ammeter shunt located on the console engine harness. The pushbutton switch is labeled Volts-Amps. The volts are read on a separate meter.

Figure 14-11 shows the pulse generator of the dynamic compression meter. The signal enters at reference A REF A, from the cylinder-marker generator-trigger shown in Figure 14-10 tube V7 output, to the grid of vacuum tube V8A (½ 12AX7). The output of V8A is from the plate through capacitor C43 (0.02 mfd), resistor R114 (22 Kohms), to the grid of tube V8B (½ 12AX7). The output of V8B is from the plate to reference B REF B which connects to reference B REF B in Figure 14-12

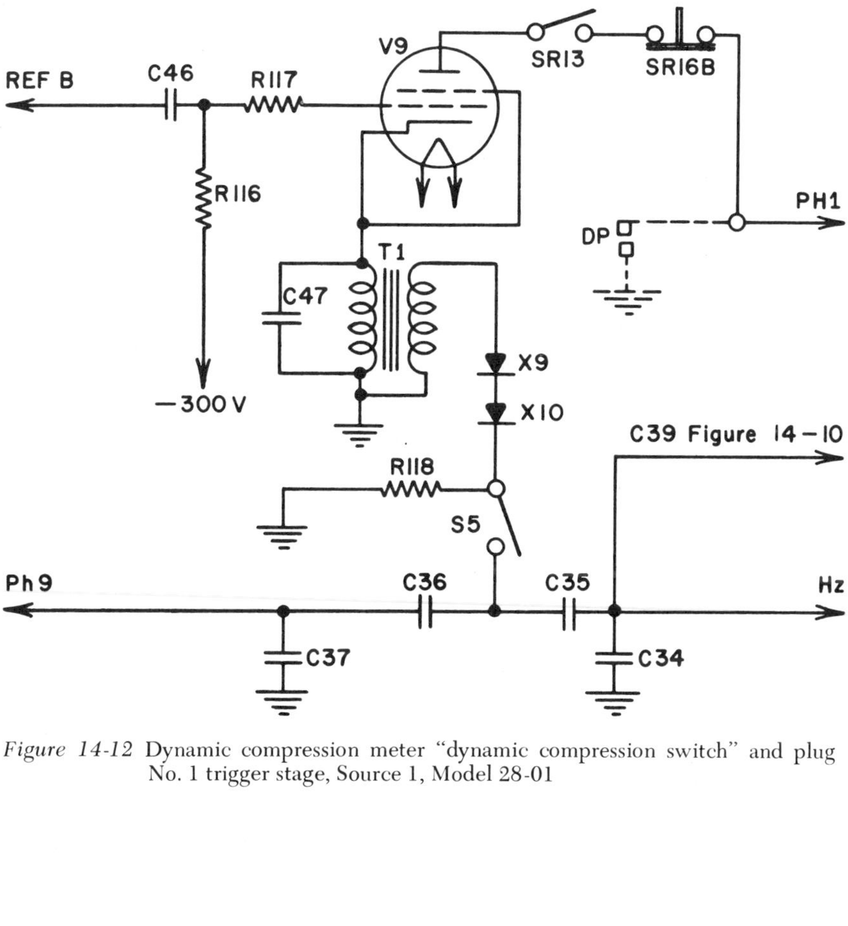

Figure 14-12 Dynamic compression meter "dynamic compression switch" and plug No. 1 trigger stage, Source 1, Model 28-01

input. The plate output of V8B is also connected to the scope vertical amplifier VA through resistors R35 (6.2 megohm), R36 (12 megohm), and capacitor C5 (0.01 mfd). The CYLINDER SELECTOR rheostat R13 (4 megohm) is operated by a knob from the control panel, whereas DC BIAS potentiometer R14 (500 Kohms) is a screw slot adjusted calibrator located in the chassis. The purpose of these two controls is to introduce a time delay for shorting out of each cylinder firing in order. The shorting out circuit is shown in Figure 14-12.

The PULSE DURATION rheostat R15 (1 megohm) is located in the chassis. This is adjusted in conjunction with the CYLINDER SELECTOR control R13 to adjust the width of the shorting out pulse wave so that it shorts out one complete cylinder firing spike without affecting the two adjacent cylinder firing spikes, while observing the patterns on the scope screen.

Figure 14-12 shows the dynamic compression switch and the plug No. 1 trigger stage. The input to this stage is from reference B REF B Figure 14-11 to REF B, and to the grid of the thyratron tube V9 (2D21), through capacitor C46 (0.47 mfd), and resistor R117 (270 Kohms). The output of V9 goes through switch SR13 (a part of cylinder selector rheostat R13 Figure 14-11 on–off), through normally closed push-to-balance switch SR16B (a part of balance-to-line rheostat R16B Figure 14-10), and to terminal PH1; which is the cable clamp connection to the distributor points DP of the engine under test.

When V9 is triggered into conduction, it places a low enough resistance across the engine distributor points DP to keep the spark plug from firing. It does not, however, kill the high-voltage output of the ignition coil completely.

The cylinder selector control R13 in Figure 14-11 is turned to introduce a delay so that tube V9 conducts during the time any one cylinder is scheduled to fire. This prevents that cylinder from firing and it shows up on the scope screen by reducing that particular voltage spike.

When it comes to shorting out cylinder No. 1, it is obvious that the dynamic compression test system would break down, because this circuit is triggered into conduction by pulses from spark plug No. 1.

To avert this situation, a circuit is employed in the cathode of V9 so that when plug No. 1 pulse is shorted out, a pulse at plug No. 1 frequency rate is developed in the cathode of V9. This pulse is stepped up by trigger transformer T1. This stepped-up voltage is fed through rectifiers X9 and X10, No. 1 cylinder switch S5, and back to the input of the dynamic conpression trigger stage V7, Figure 14-10, at a point T between C35 and C36. This rectified positive pulse is also fed to the horizontal sweep HZ to keep all trigger and sweep circuits operating normally.

Timing Advance Meters

1. *Timing Advance Meters—General.* The distributor on internal-combustion engines is adjustable to set the firing of the No. 1 cylinder on top dead center TDC, or at any specified angle of retard or advance. For slow or idling speeds, the spark is timed to fire near TDC, but as the engine speeds up it is necessary that the spark timing be advanced, that is, occur before TDC.

On early engines the spark was adjusted manually to suit the speed and driving conditions. All modern engines have automatic spark timing adjustment. After the initial timing is made according to specifications, the distributor automatically adjusts itself by at least two means. A governor in the distributor automatically advances the spark timing as the speed increases. A diaphragm operated by the vacuum in the intake manifold is linked to the distributor to advance the spark on high vacuum and retard it on low vacuum. There is a lower vacuum in the manifold when the throttle is opened, and at lower engine speeds. So when the accelerator opens the throttle valve for speeding up, the spark retards slightly to prevent knocking. As the speed increases, both the vacuum increases, and the governor operates to advance the spark.

To test the spark advance in degrees at various speeds properly, the distributor can be tested in a distributor test stand. It can be determined if the advance mechanisms are working by using a timing light on the existing timing marks without removing the distributor from the engine.

Timing advance meters are available as a separate unit or combined in a scope-type engine analyzer, providing means of accurately measuring the timing advance without removing the distributor from the engine.

These all work on the same general principle, using an electric meter calibrated in degrees-timing-advance, and a timing light that has a built-in time delay circuit.

The procedure is; first, direct the timing light of the timing advance meter on the engine timing mark at a specified idle speed. Then adjust the distributor for the specified timing degrees, usually TDC. If it is not on TDC, make a note of the degrees advance or retard. Next, increase the engine speed to 2000 rpm (or specified rpm). Direct the timing light on the indicator and adjust the timing advance control knob on the timing light until the timing mark appears at TDC, or the angle found in the first step. The degree reading as shown on the timing advance meter should correspond to the specifications. If not, service and repair of the distributor is indicated.

The timing advance knob introduces a delay circuit that causes the timing light to flash late by the degrees indicated on the timing advance meter. This delay circuit is introduced in the electronic circuit and consists essentially of a rheostat and capacitor in series. The timing advance knob operates the rheostat.

This resistance-capacitor (R-C) delay circuit is shown in Figure

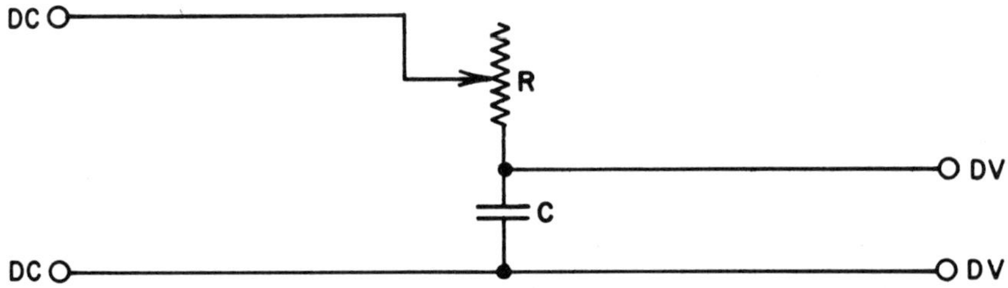

Figure 14-13 Resistor-capacitor (R-C) delay circuit

14-13 where dc source DC is applied across the series connected rheostat R and capacitor C. When the DC voltage (steady or pulsating) is applied, the rheostat limits the charging current to capacitor C. This delays the build-up of voltage at the delay voltage points DV. For a constant dc voltage DC, if rheostat R has a resistance of 1 megohm and capacitor C has a value of 1.0 mfd, then it will take approximately 1 second for the delay voltage DV to reach 63% of the value of dc voltage DC. This is practically linear, so a change of values of either rheostat R or capacitor C will make a proportional change in the time delay. For example, if capacitor C is 0.5 mfd and rheostat R is 1 megohm, the time delay would be one-half second. Also, if rheostat R is 0.5 megohm and capacitor C is 1 mfd the time delay would be one-half second. For pulsating dc voltage DC the time delay will be proportional to the average applied voltage.

This delay circuit is applied to the timing light electronic circuit and the rheostat constitutes the TIMING ADVANCE control knob.

This time delay circuit is also used in dynamic compression test circuits functioning as the cylinder selector control.

2. *Calibrating Timing Advance Meters.* Commercially available distributor test stands can be used to check and calibrate timing advance meters, or the ignition simulator shown in Figure 3-1 can be used.

To use the simulator shown in Figure 3-1, connect the timing light of the timing advance meter to spark plug SP No. 1 and ground G. With the advance control on the timing light at zero (no delay), focus the timing light on the revolving disc wheel W (running at low speed). Adjust the manual spark advance SA (and/or increase the speed of the simulator to operate the advance mechanism), until the No. 1 mark on wheel W lines up with the 20° mark on dial DL. This shows an advance on the distributor of 20°, which is an advance of 40° on the crankshaft.

Turn the advance control on the timing light until the No. 1 cylinder mark on wheel W is returned to zero mark on dial DL. The timing advance meter should read 40° indicating the angle of advance on the crankshaft.

The simulator dial DL is operated by the distributor which turns at one-half the speed of the crankshaft. Therefore the simulator dial DL will show one-half the angle of the spark advance on the crankshaft damper or flywheel.

Adjust the timing advance calibrating control R21, Figure 14-14, until the meter reads the correct 40° of advance for 20° on dial DL.

3. *Typical Timing Advance Meter Circuit, Source 1.* The timing advance meter section of Source 1, scope-type engine analyzer, Model 28-01, is shown in Figure 14-14.

The timing advance circuit consists of two thyratron tubes V12 and V13 and associated components, timing light assembly TL, and degrees timing advance meter AM.

Signal pulses from engine plug No. 1, from the voltage divider cable pick-up, is applied to the signal input terminal PH9. This pulse is applied to the grid of V12 through a capacitor network including C55. V12 will fire and conduct at the same frequency rate as the plug No. 1.

The advance control (600 Kohm rheostat ADV) located in the timing light assembly TL (enclosed in dotted lines), is connected in the grid-cathode circuit of V13 through timing light plug terminals TL8 and TL9.

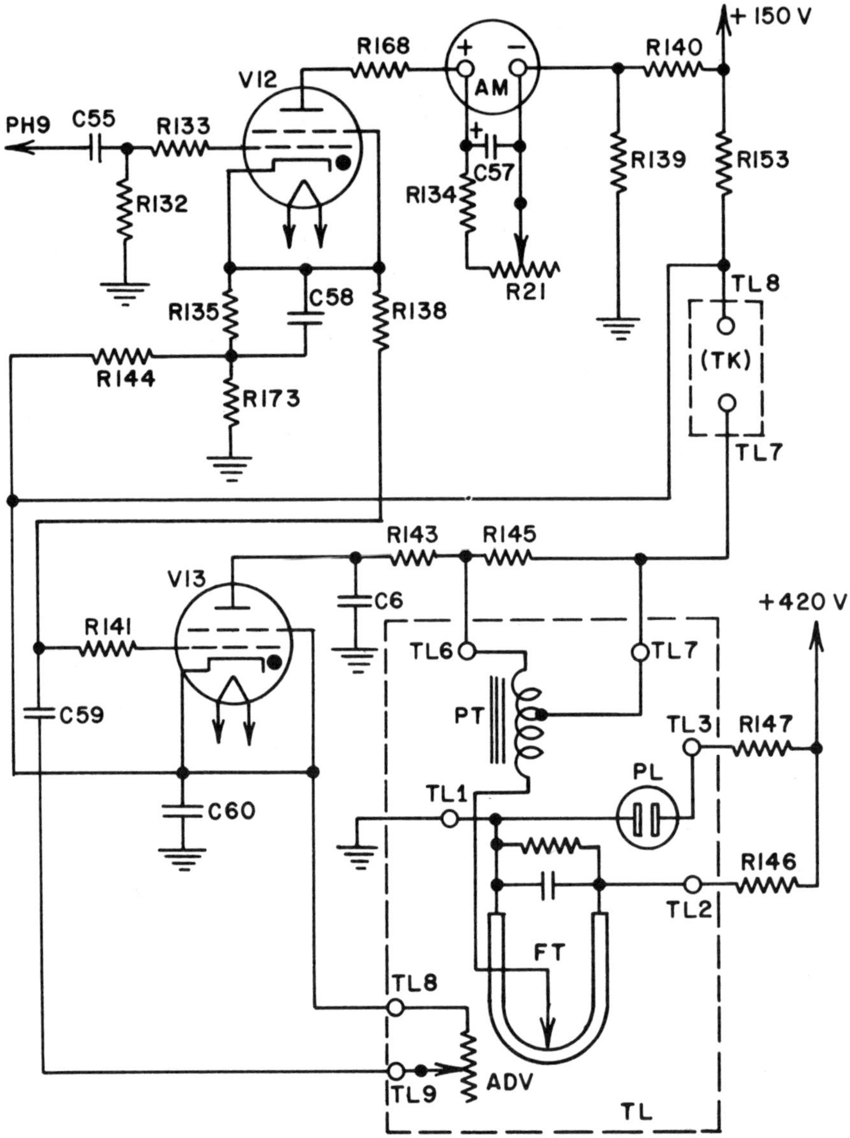

Figure 14-14 Timing advance meter section of scope analyzer, Source 1, Model 28-01

Primary winding of the pulse transformer PT of the timing light is connected in the plate circuit of V13 through timing light plug terminals TL6 and TL7. When the advance control ADV is turned to its OFF (no resistance) position, V13 will conduct instantaneously with V12, causing the pulse transformer PT to flash the flash tube FT in accordance with the plug No. 1 firing pulses as in any timing light. However, when the advance control ADV is turned away from the OFF position, resistance is added to the grid-cathode circuit of V13. Due to the time delay introduced by the resistance of the advance control ADV and capacitor C59, V13 will now conduct a time interval later than V12 depending on the amount of resistance added by rheostat advance control ADV. Pulses at the plate of V13 are applied to primary winding of pulse transformer PT, and timing light flash tube FT will flash at a time interval later than the firing of plug No. 1. A measure of advance timing of plug No. 1 firing pulse is accomplished as follows;

With the advance control ADV of the timing light turned to its OFF position, and the timing light focused on the timing mark provided on the engine, an acceleration of the engine speed to 2000 rpm operates the distributor advance mechanism to cause the timing light mark to move away (advance) from the stationary timing pointer, against the direction of rotation of the engine crankshaft. Rotation of the advance control ADV delays the conduction time of V13. This delays the pulse to the primary winding of pulse transformer PT, which, in turn, delays the flash of the timing light flash tube FT to cause the timing marks to move in the same direction of rotation as the engine crankshaft. The delay time provided by rotation of the advance control ADV necessary to cause alignment of the timing mark with the stationary pointer on the engine is measured in the plate circuit of V12 by advance meter AM. V12 is held in conduction during the delay time, or the time that V13 is cut off.

The average current developed in the plate circuit of V12 represents the advance timing measurement and is indicated in degrees on the timing advance meter AM.

The timing advance meter AM is calibrated by rheostat R21. Meter AM should read less than 3° when advance control ADV is in the OFF or no resistance position.

The tachometer TK is shown in block form and uses some of these same components.

The values of other components are; capacitors in mfd C6 (0.005), C37 (0.05), C55 (0.01), C57 (100), C58 (0.005), C59 (0.05), and C60 (0.001); resistors in ohms R21 (5 K), R132 (47 K), R133 (270 K), R134 (4.7 K), R135 (15 K), R138 (560 K), R139 (33 K), R140 (10 K), R141 (470 K), R143 (15), R144 (2.7 K), R145 (15), R146 (5 K), R147 (150 K), R153 (27 K), R168 (1111 in slave control) R173 (4.7 K); tubes V12 and V13 (2D21 thyratrons).

Tachometer Section of Scope Console

Figure 14-15 shows the tachometer section of a typical scope-type analyzer. This circuit uses many components that are common to the timing advance meter section described in Figure 14-14. The component values are the same. Additional component values used only in the tachometer section are given below. This circuit differs from the timing advance meter section (Figure 14-14), only in that C56 and R142 are switched in the circuit by switch S1D1, and the advance meter AM is removed from the circuit.

The signal pulse from the plug No. 1 is fed through terminal PH9, capacitors C55 and C56, switch S1D1, and resistor R142 directly to the grid of tube V13. The plate circuit feeds pulses to a conventional capacitor-resistor frequency-conscious tachometer circuit. Since the pulses come from the No. 1 plug, the tachometer will read the same for any number of cylinders. This is because the crankshaft makes two revolutions between the firing of the No. 1 cylinder (4-cycle) regardless of the number of cylinders the engine has. Switch S8 is shown in the 800 rpm position, the second position in the direction of the arrow is 1600 rpm, and the third position is the 10,000 rpm position.

To calibrate the tachometer (with the plug No. 1 pulse input); first, calibrate at 1500 rpm on the 1600 rpm scale by adjusting R23; second, calibrate at 1500 rpm on the 10,000 rpm scale by adjusting R24. Disconnect the pulse input signal and adjust the coarse, R26, and fine, R25, tachometer zero adjust controls until tachometer pointer is at zero. Next, reconnect the input pulse signal and check the 1500 rpm on the 10,000 rpm scale. If it does not read correctly, repeat above calibration procedure until it is correct.

Calibrate 700 rpm on the 800 rpm scale by adjusting R22.

Recheck all above calibration points and repeat until all scales are accurate. If unable to achieve this, there is probably a defective component in the circuit.

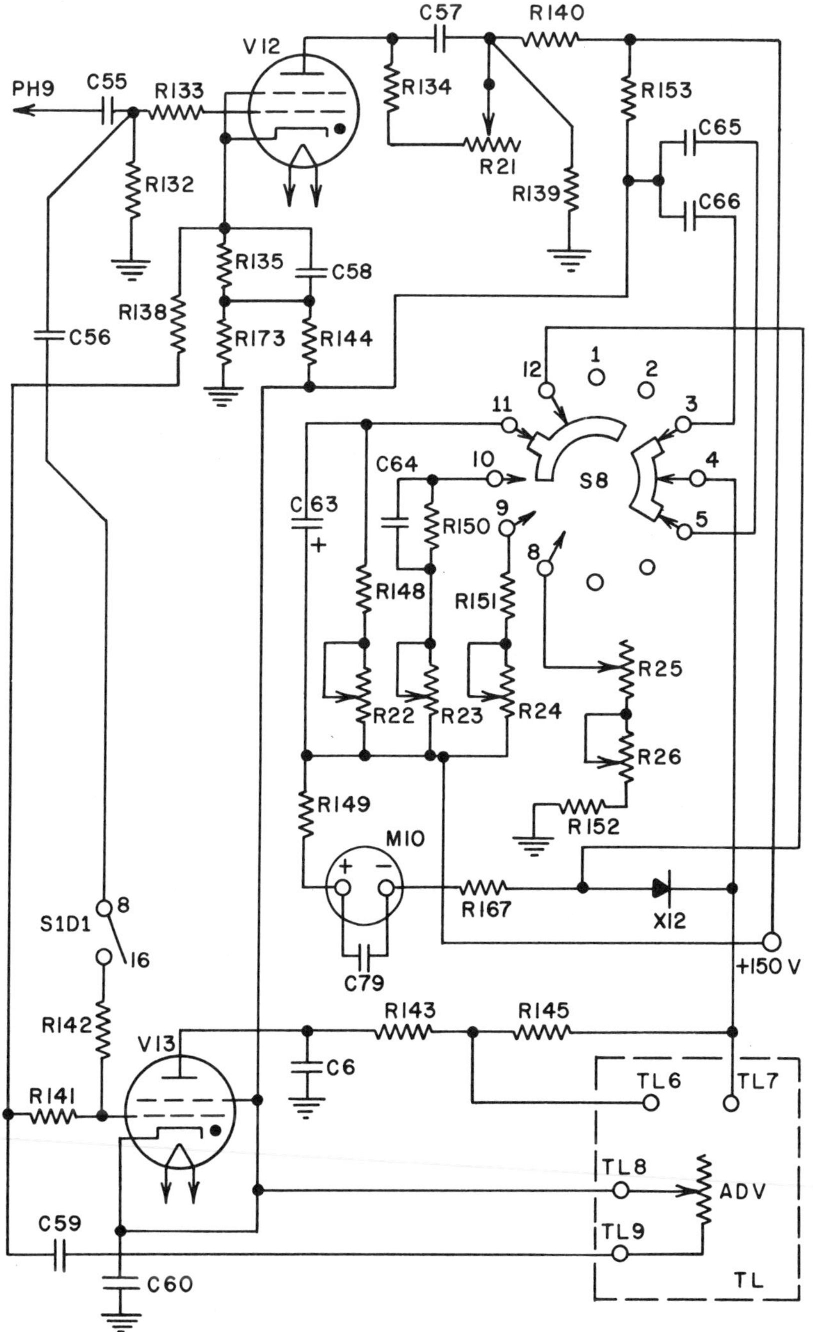

Figure 14-15 Tachometer section of console, typical, Source 1, Model 28-01

The component values are as follows (for other values see description accompanying Figure 14-14); condensers in mfd C56 (0.01), C63 (100), C64 (500), C65 (0.68), C66 (0.22), C79 (500); resistors in ohms R22 (3 K), R23 (2 K), R24 (500), R25 (50 K), R26 (250 K), R142 (560 K), R148 (220), R149 (180), R150 (220), R151 (22), R152 (47 K), R167 (1111 in slave control); meter M10 tachometer; diode X12 (PA3).

Air-Fuel Ratio Meter Section

Figure 14-16 shows the schematic of the air-fuel ratio meter used in the engine analyzer Source 1, Model 28-01. Meter M14 is the air-fuel ratio AFR meter, switch S10 is air-fuel ratio READ push button switch DPDT to activate the meter M14 to read the ratio after calibration. Potentiometer R28 (2 ohms) is the BALANCE adjustment to balance the AFR meter. The READ button, switch S10, must be depressed to make this adjustment. The push-to-adjust potentiometer R27 (50 ohms) is used to calibrate the AFR meter. With the read push button depressed, depress the push-to-adjust knob, then rotate it to position AFR meter pointer to the line at the left.

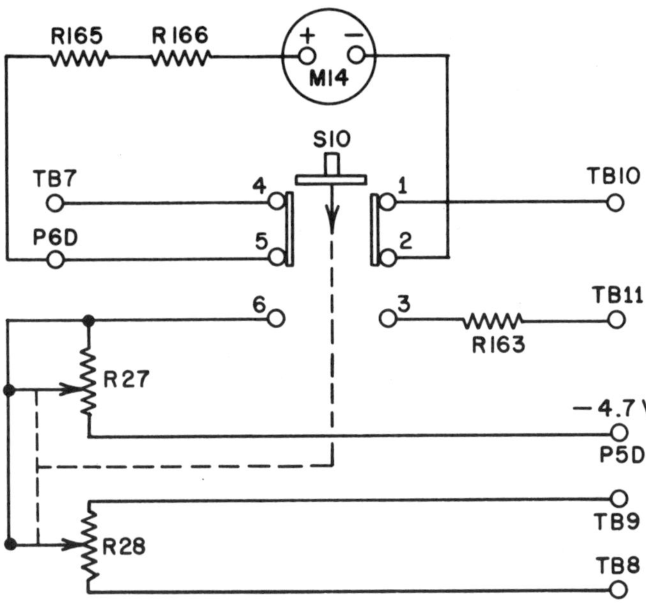

Figure 14-16 Air-fuel ratio meter, typical of Source 1, Model 28-01, scope console

Other component values are; resistors in ohms R163 (4K), R165 (26.1), and R166 (8.2).

The terminals TB7, TB8, TB9, TB10, and TB11 connect to the Clayton cell in the exhaust sampler unit. This Clayton cell is heated by elements from 120 V ac source. The terminals TB7 to TB11 connect to a two-compartment cell unit consisting of an electrical bridge that changes resistance value depending on the heat conductivity of the exhaust gases.

High-Voltage Power Supply for Scope Analyzer

The power supplies for a scope-type engine analyzer usually consist of two independent power supplies.

The usual cathode-ray tube power supply is used for high voltage, with a bleeder network supplying adjustable voltages for spot shape, focus, and brightness. The controls are usually inside the chassis and may be near the CRT mounting bracket. The CRT most commonly used is the 8-inch size, however other sizes are used such as 5, 7, 9, 12, up to 21 inches, and possibly larger.

A second power supply furnishes filament voltages and regulated dc voltages for the vacuum tubes, thyratrons, and transistors. Voltage regulators in the high plate voltages use the gaseous types such as OA2, OA3, OB2, OB3, OC3, and OD3. Whereas, in the lower voltages used for bias, the zener diode is used either alone or in conjunction with a power transistor to regulate currents higher than the zener diode rating.

Figure 14-17 shows schematically a high voltage CRT power supply typical of Source 1, Model 28-01, engine analyzer. The CRT V14 (8CP31) has two vertical deflection plates VP fed from the output of the vertical section shown in Figure 14-2. The two horizontal deflection plates HP are fed from the output of the horizontal section shown in Figure 14-7. Power to the other terminals is supplied from 115 V ac lines L1 and L2, through transformer T2, and out at secondary through high-voltage rectifiers X4 and X5, and across bleeder network consisting of series-connected resistors (in ohms) R94 (2.7 meg), R95, R96, R97, R98, R99, and R100 (3 meg), R10 (3 meg spot shape), R11 (2 meg focus), R101 and R102 (2 meg), and R12 (500 K brightness). Capacitors C28 and C32 (0.15 mfd) are high voltage filter capacitors.

Other component values are; resistors in ohms R91 (100 K), R92 (100 K), R93 (330 K); capacitors C27 (0.68 mfd), C31 (47 mmfd). Refer-

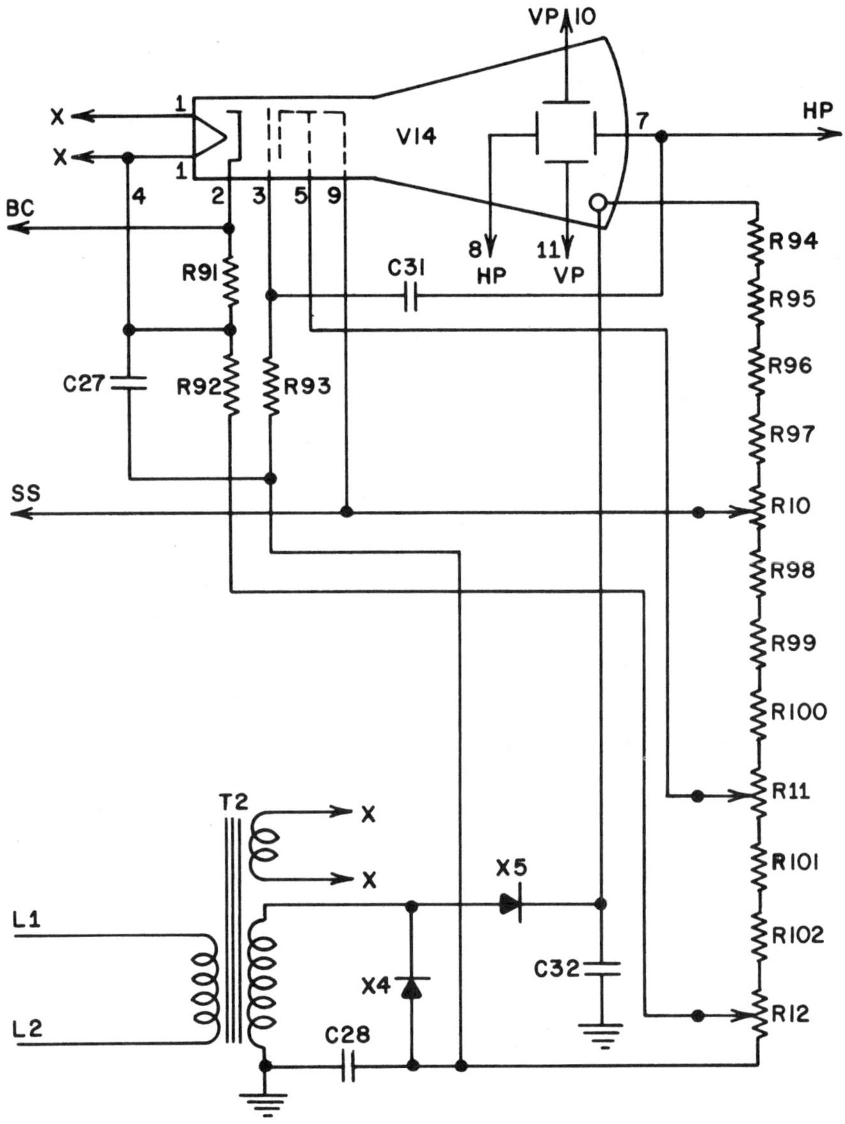

Figure 14-17 High-voltage CRT power supply, typical of Source 1, Model 28-01, engine analyzer

ence BC for brightness control and SS for spot shape is connected to the brightness amplifier shown in Figure 14-3.

Main Power Supply

Figure 14-18 shows schematically the main power supply typical of Source 1, Model 28-01, engine analyzer where there are four separate power supply sources from three transformers.

The two high-voltage supplies shown in Figure 14-18a from transformer T3 provide positive 420 V unregulated, positive 300 and 150 V regulated by voltage regulator tubes V15 (OA2) and V16 (OA2). Transformer T3 also provides negative 300 and 150 V regulated by voltage regulator tubes V17 and V18 (OA2). A third power supply shown in Figure 14-18b is from transformer T4 to provide 4.7 V regulated by zener diode X23 (1NF8.T10) and rectifier X24 (1N1519), through full-wave bridge rectifier X19, X20, X21, and X22 (all 1N4822). A fourth power supply shown in Figure 14-18c is from transformer T5 providing regulated 8.2 V source regulated by zener diode X26 (SZ8.2) and transistor Q6 (2N554).

Other component values are; resistors in ohms R154 (30), R155 (3 K), R156 (360), R157 (3 K), R158 (360), R159 (50), R160 (22), R161 (75), R162 (30); capacitors in mfd C69 (40), C70 (40), C71 (0.05), C72 (40), C73 (40), C74 (0.05), C75, C76, C77, and C78 (1000); rectifiers X13 to X18 and X25 (1N4822). The heater supply secondaries on transformer T3 are; X-X supplies V1, V2, V12, and V13, Y-Y to tubes V3, V4, V5, and V6, Z-G to tubes V7, V8, V9, V10, and V11. These tube numbers will be found in various schematics shown previously. The primary leads L1 and L2 of transformers T3, T4, and T5 are connected to 115 V ac by switching means.

Scope Patterns, General

Much has been written, and is still being written on the subject of analyzing scope patterns to diagnose engine performance characteristics. Any patterns other than normal are, therefore, beyond the scope of this book. To diagnose patterns requires specialized knowledge. (In the medical profession, a trained specialist reads and makes diagnoses of electrocardiograms (EKGs), which are similar to scope patterns.)

To simplify the use of scope engine analyzers, some have a roll chart, in addition to the operating manual patterns, showing various patterns

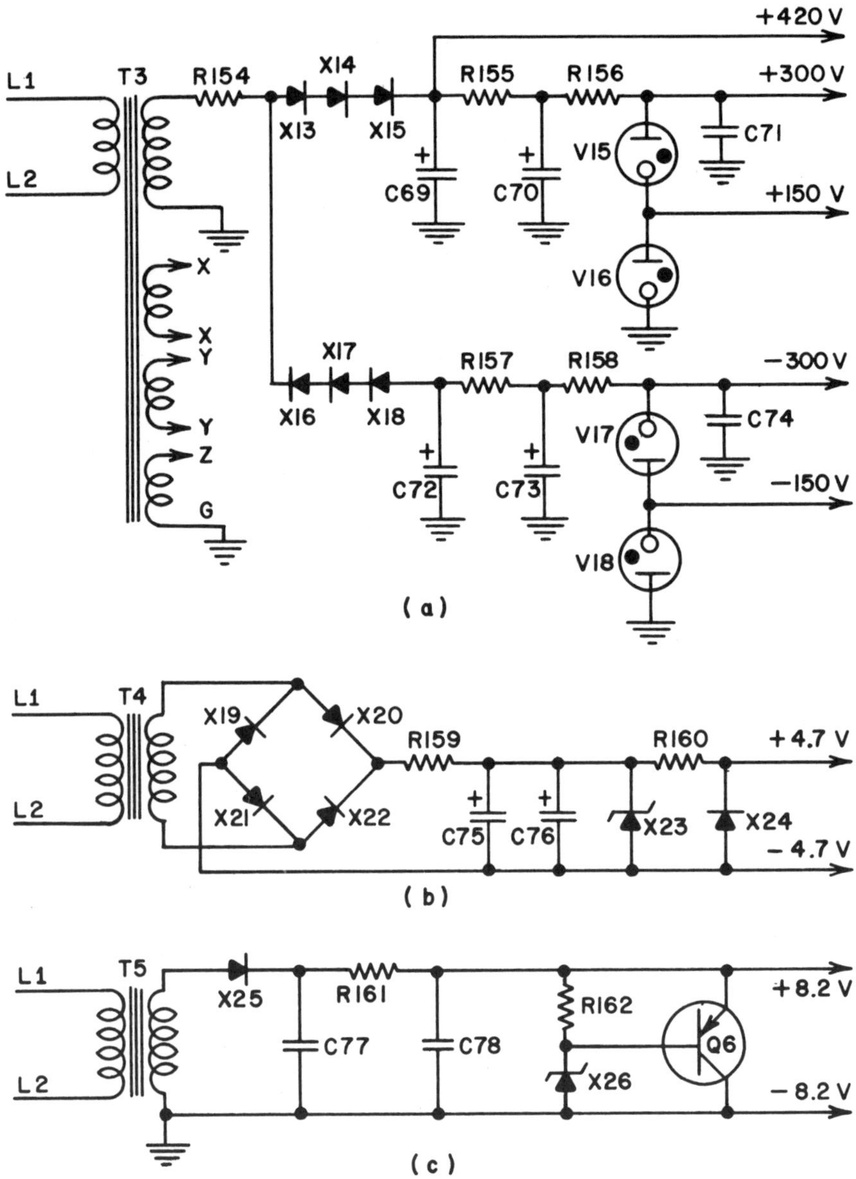

Figure 14-18 Main power supply, typical of Source 1, Model 28-01, engine analyzer

to compare with observed patterns that do not look normal. It then tells what to look for to correct the abnormal patern. Recommended service procedures are then followed until the normal pattern appears.

An abnormal scope pattern may also be caused by defects in, or improper operation of, the scope. This is not very likely to be confusing because any malfunction of the scope would be easily identified by one experienced in the use of the scope.

Scope Patterns of Primary Circuit Superimposed

With the pattern selector switch in the NORMAL position, depress the PRIMARY push button. Adjust the PARADE control to position the left end of the pattern at the left edge of the degree scale. Adjust the EXPAND control to position the right end of the pattern to the zero line of the degree scale. The complete pattern should be precisely within the zero to 90° area of the scale. Figure 14-19 shows a normal superimposed primary pattern for conventional ignition systems, with all cylinders displayed in patterns superimposed one on top of the other; where A is the distributor points opening, A–B is the plug firing oscillations resulting from the spark jumping the plug gaps; B–C is the coil and capacitor oscillations resulting from unused energy after the plug stops firing; point C is where the distributor points close and indicates on the scale the dwell angle in degrees; C–D is the time the points remain closed, and point D is where the distributor points open again and the scope spot of light jumps back to point A to repeat the cycle.

The point C, where the points close, should show a sharply defined right angle pattern with distributor points known to be in good condition. If the point C is not a right angle, but instead is either rounded off or has an upward spike, then adjust the primary trimmer capacitor C1 (5–80 mmfd) which is located in the primary pick-up lead inside the chassis.

Scope Patterns of Secondary Circuit Superimposed

With the pattern selector switch in the NORMAL position, depress the SECONDARY push button. Figure 14-20 shows a normal secondary pattern for conventional ignition systems with all cylinder patterns superimposed, one on top of the other. Point A is the points opening point; A–B is the firing action of the plugs; B–C is the coil and capacitor oscillations; C is the points closing point; and point D is the points

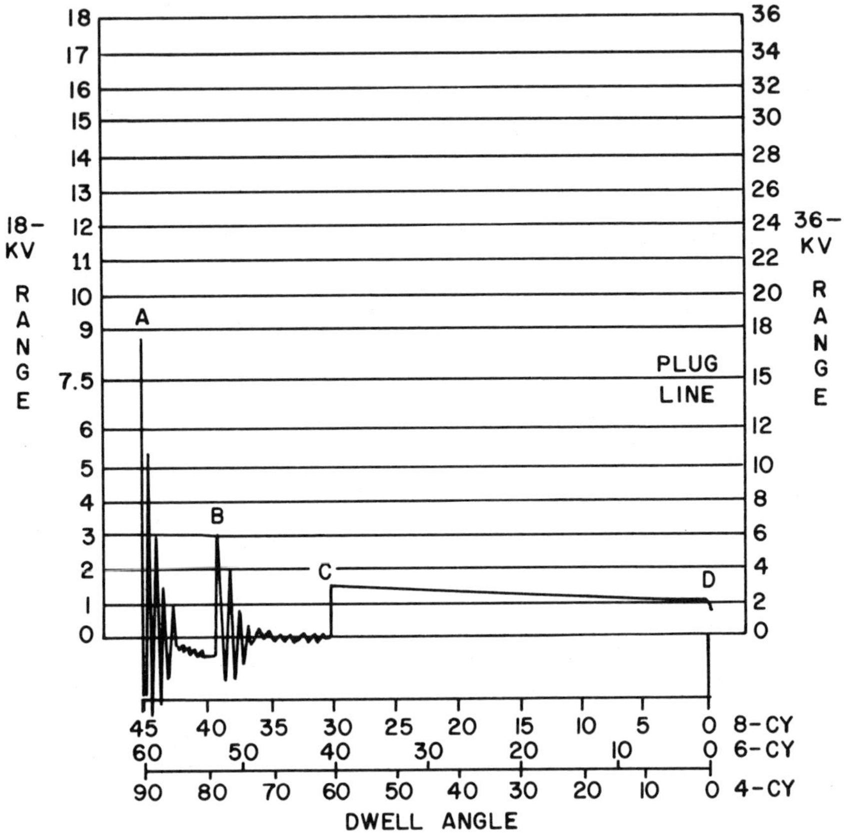

Figure 14-19 Normal superimposed primary pattern display on scope screen

opening point where the next plug fires a voltage spike. The horizontal sweep is triggered to cause the scope spot of light to jump back to point A where the rest of the secondary pattern is shown.

Scope Patterns Separated, Primary or Secondary

With the pattern selector switch in the SEPARATED position, depressing either the PRIMARY or SECONDARY push button will shift the superimposed pattern higher on the scope screen as shown in Figure 14-21. When viewing superimposed patterns in which a fault is found, we can determine what is wrong in general; but not in detail or which cylinder is affected. Figure 14-21 shows a primary superimposed pattern with a fault appearing in the area between points B and C where an upward sloping line is cutting the coil-capacitor oscillations.

Turning the CYLINDER SELECTOR control slowly will cause the faulty cylinder to drop down from the other superimposed patterns as shown in Figure 14-22. Here the details of the fault can be examined in greater detail.

To determine which cylinder is faulty (without moving the cylinder selector control) switch the PATTERN SELECTOR switch to PARADED. The result is shown in Figure 14-23. The faulty cylinder is dropped down

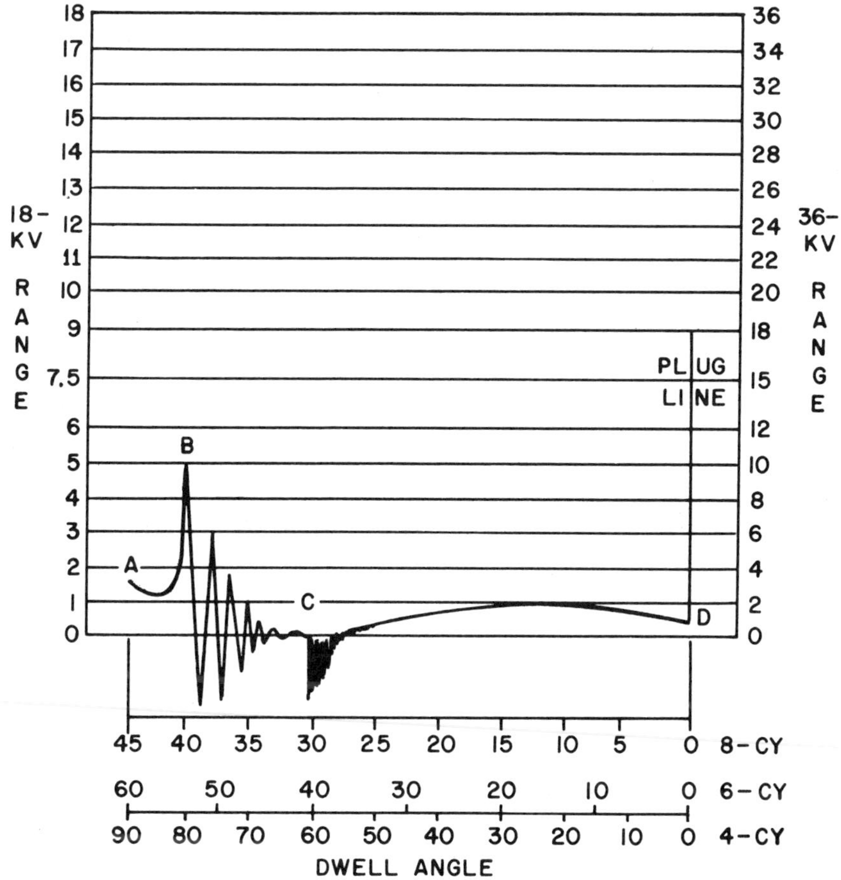

Figure 14-20 Normal superimposed secondary pattern display on scope screen

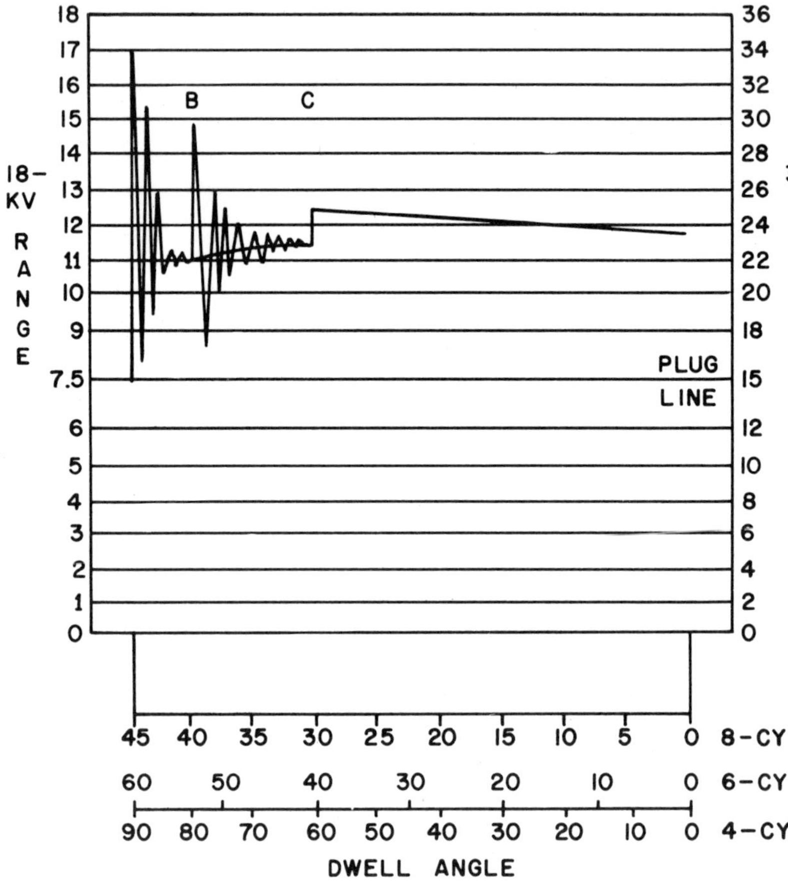

Figure 14-21 Superimposed pattern with pattern selector switch in SEPARATED position

and identified as cylinder No. 8 in the firing order of numbers 1, 2, 7, 8, 4, 5, 6, 3 shown above the patterns.

Paraded Patterns, Primary or Secondary

With the pattern selector switch in the PARADED position, depressing either the PRIMARY or SECONDARY push button will result in each cylinder pattern displayed side by side across the scope screen in correct firing order as counted from the left.

Figure 14-24 shows the secondary paraded pattern, but the primary pattern would look like the normal superimposed primary pattern except each cylinder firing would be displayed separately, just as the secondary pattern is shown in Figure 14-24.

It should be noted in Figure 14-24 that the spike for the No. 1 cylinder is the one at the extreme right on the screen whereas the balance of the No. 1 cylinder firing pattern is at the extreme left of the screen. This is because the No. 1 cylinder plug is used to trigger the horizontal sweep circuits.

The paraded pattern (either primary or secondary), can be expanded to show only one individual cylinder firing covering the full width of the

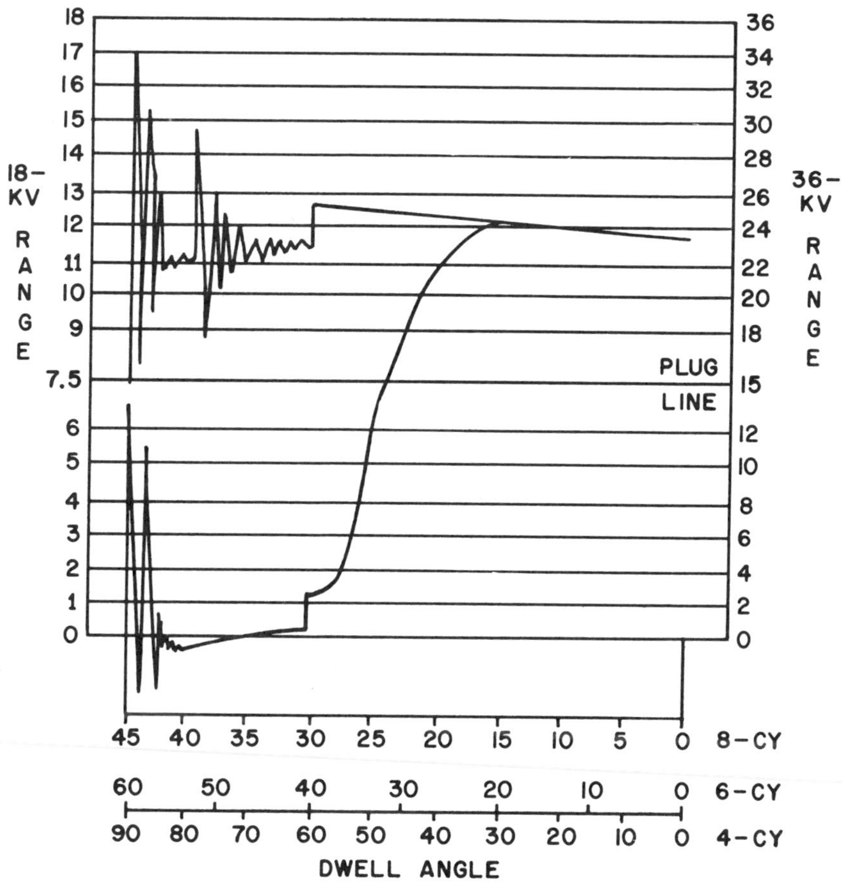

Figure 14-22 Superimposed pattern with pattern selector switch in SEPARATED position and "cylinder selector" control turned

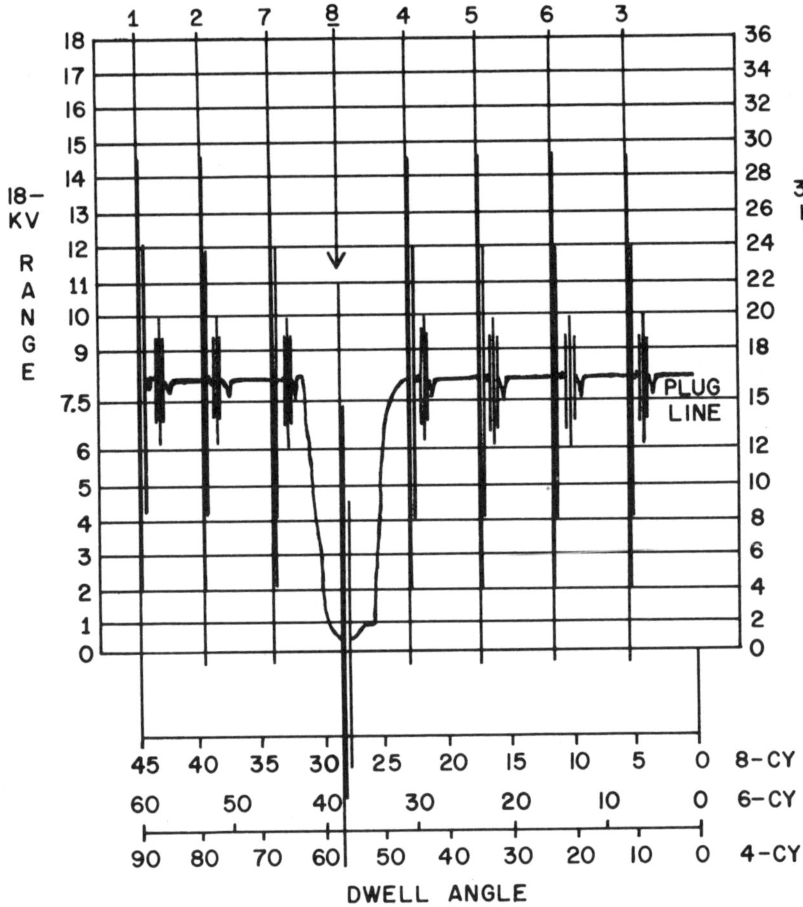

Figure 14-23 Paraded pattern with cylinder No. 8 separated

scope screen. This is accomplished by turning the EXPAND control. Any one cylinder can be selected by parading the patterns across the screen using the PARADE control.

Alternator Test Pattern

The alternator test consists of viewing the three-phase rectified output voltage pattern on the scope screen. A normal pattern is shown in Figure 14-25 where each rectified ac ripple from the alternator above the battery voltage is uniform and in a straight line. Any deviation from this normal pattern indicates a fault in the alternator.

The procedure for alternator test is as follows.

1. Depress the ALTERNATOR TEST push button and operate engine at 1000 rpm.
2. Adjust PATTERN HEIGHT control to full clockwise rotation.
3. Adjust SYNCHRONIZE control until a single pattern is stationary on the screen.

This circuit can be tested, in the absence of an alternator-equipped vehicle, by using an ignition simulator. Connect a high-rate battery charger across the simulator battery and proceed as above. The 60-Hz

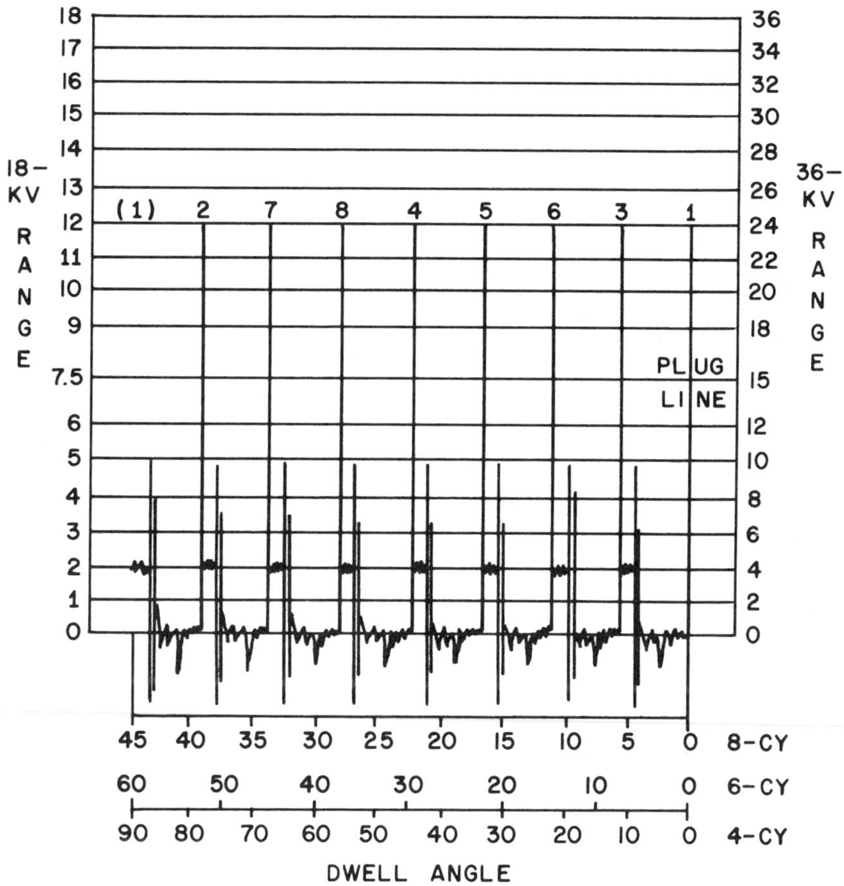

Figure 14-24 Paraded secondary pattern displayed on scope screen

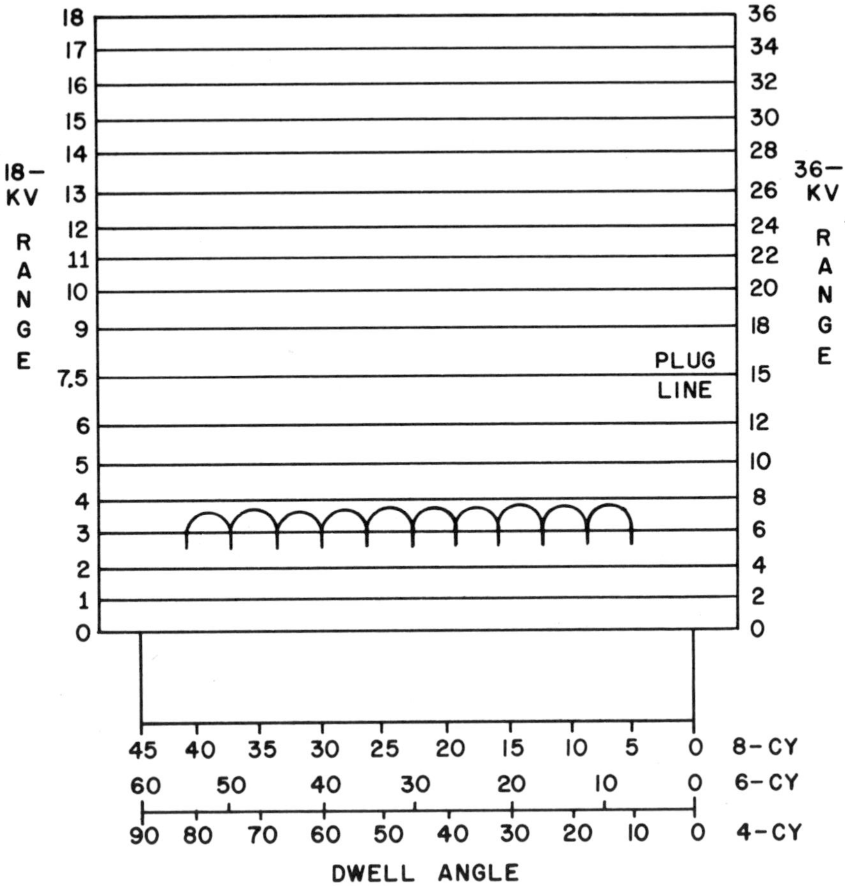

Figure 14-25 Alternator test pattern on scope screen

ripple of the charger can be synchronized with this circuit. An alternate method is to use a small 12 V ac transformer fed through a small half-wave positive output rectifier. This 12 V ac can be readily obtained from the appropriate pins of the tube sockets of any tube tester. Connect this test voltage to the alternator test leads. Adjust the HEIGHT control to keep the pattern within the scope screen.

Observer Scope, General

Some scope-type engine analyzers have provision for plugging into the main console, one or more observer (or slave) scopes that can be viewed remotely for purposes of instruction, supervision, and the like. They usually have their own power supply but obtain the signals from the output of the console chassis. They operate on the same general principles as the main console.

The Source 1, Model 28-01, console has provision for connecting an observer scope to the main console for remote viewing. It has a CRT, power supply, horizontal and vertical amplifiers, but gets the signals from the main console. When the observer scope is switched to certain functions, a "slave control" is provided to maintain a constant impedance in the circuits affected.

The main console has additional amplifiers in the horizontal and vertical sections to provide needed signal strength to the observer scope, particularly over long transmission lines. The horizontal, vertical, and paraded signals are the only three connections between the main console and the observer scope.

The scope tube is the same as used in the main console (8CP31) and the high-voltage power supply to the scope tube is the same except the brightness amplifier is not connected to the terminals BC or SS.

The main power supply is simpler in the observer scope than it is in the main console. The power supply is the same as shown in Figure 14-18a less the unregulated positive 420 V source.

1. *Vertical Section, Observer Scope.* The vertical amplifier section of the observer scope is shown schematically in Figure 14-26 where the input signal enters at V1 RED terminal from the V0 RED terminal of the main console. This signal is fed through tube V1 (6AV6) into the push-pull output stage V2A (12AU7) and V2B (12AU7), the plates of which are connected to the vertical deflection plates VP of the scope tube (8CP31).

Potentiometer R1 (50 ohms) is a gain or height control and R2 (50 Kohms) is a vertical position control.

Other component values are; resistors in ohms R17 (10 k), R18 (470 K), R19 (10 K), R24 (330 K), R25 (220 K), R26 (330 K); capacitors in mfd C7 (20), C9 (0.1), C10 (0.05).

2. *Horizontal Section, Observer Scope.* The horizontal amplifier section of the observer scope is shown schematically in Figure 14-27 where the horizontal input HI WHITE goes through amplifier tubes V3 (6AU6) and V4 (6C4), then through bias transistor Q1 (2N2895 or

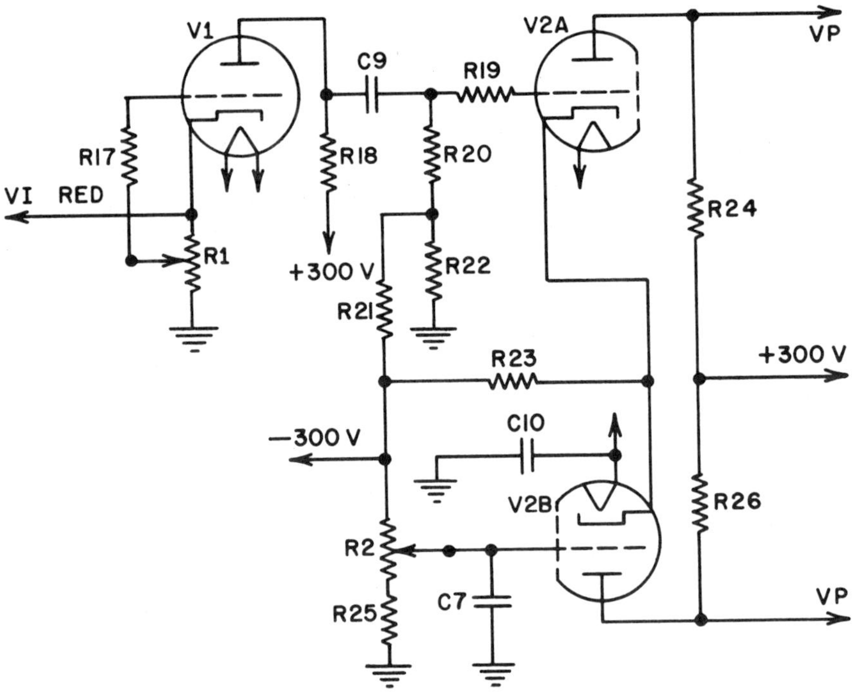

Figure 14-26 Observer scope vertical amplifier, Source 1, Model 28-01

2N2102), and through the push-pull output tube V5A and V5B (12AX7), to the horizontal deflection plates HP of the scope CRT (8CP31).

Potentiometer values are; R3 (100 Kohms is the horizontal gain or width control), R5 (50 Kohms is the left edge adjust control), R6 (500 Kohms is the parade control), R7 (50 Kohms is the parade clamp control), R8 (100 Kohms is the horizontal linearity control), and R4 (5 Kohms is the bias control).

The parade input terminal PI BLACK is connected to the parade output terminal PO of the console scope horizontal amplifier.

Other component values are; resistors in ohms R27 (4.7 meg), R28 (47 K), R29 (100 K), R30 (51), R31 (390 K), R32 (4.7 meg), R33 (22 K), R34 (4.7 meg), R35 (82 K), R36 (62 K), R37 (100 K), R38 (470 K), R39 (75 K), R40 (33 K), R41 (150 K), R42 (560 K); capacitors in mfd C11 (0.22), C12 (0.05), C13 (500 mmfd), C14 (0.1), C15 (0.05), C16 (0.05), C17 (2), C18 (2), C8 (20); diode X11 (500 ma 600 PIV silicon limiter).

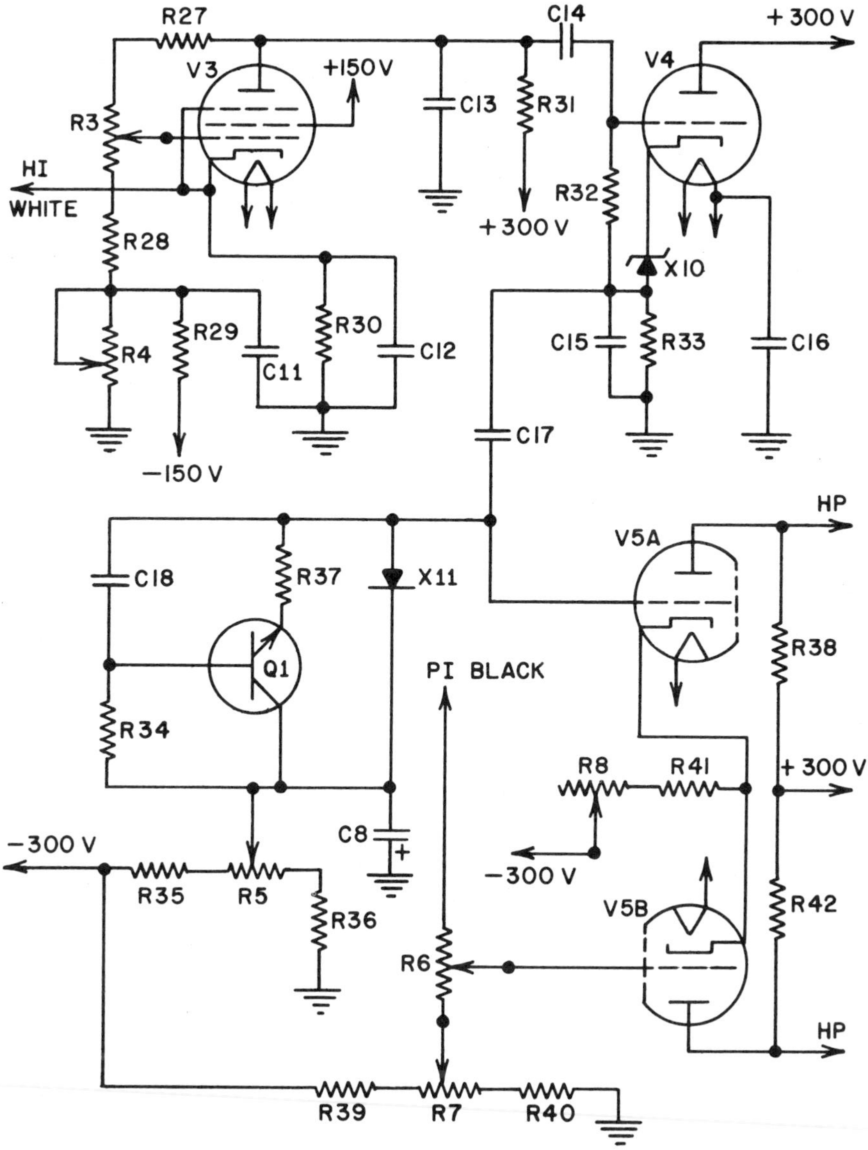

Figure 14-27 Observer scope horizontal amplifier, Source 1, Model 28-01

Observer Scope Slave Control

The main console can be connected to the observer scope as well as to a large variety of external meters or modules.

A "slave control" is plugged in between the main console and the modules. The purpose is to maintain a constant impedance, allow the meters to read the same, and maintain the same accuracy when the slave control push buttons are depressed.

When the slave control push buttons are depressed, the switch removes a resistor from the circuit having the same value of resistance as the newly inserted meter or module. The like meters are operated in series. When no meter is in series, a resistor is substituted to maintain the same resistance by the slave control.

The slave control is shown schematically in Figure 14-28 for four modules.

In Figure 14-28a the VOLTS module push button switch S12 is shown

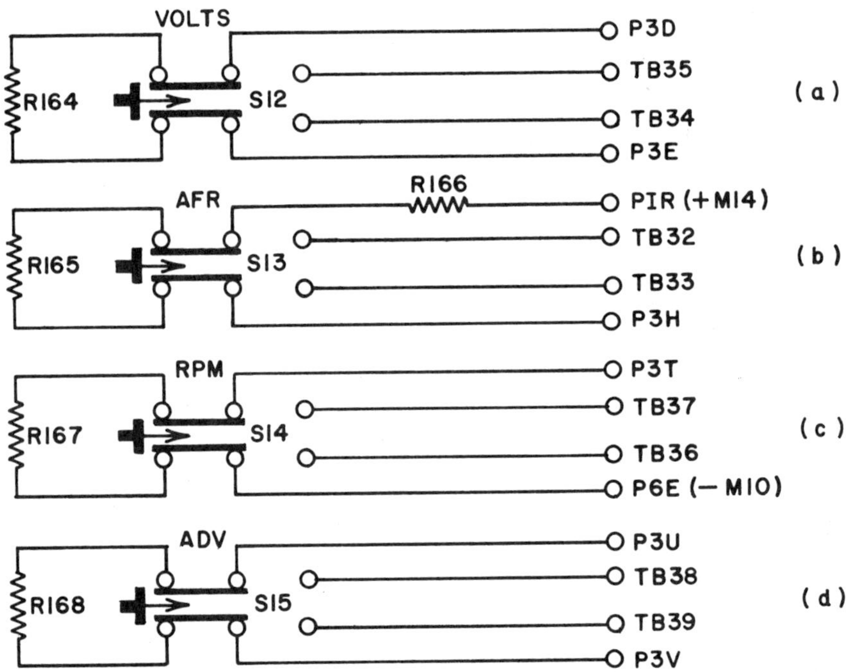

Figure 14-28 Observer scope slave control, Source 1, Model 28-01

in the normal position when the external module is deactivated. The resistor R164 (1111 ohms) is in series with the console voltmeter through terminals P3D and P3E. When switch S12 is depressed, the resistor R164 is removed from the meter circuit and replaced with the module voltmeter having an equal resistance to R164 (1111 ohms). This places the console meter in series with the module meter, maintaining the same impedance and readings. The module meter is connected to a master terminal board, terminals TB34 and TB35.

Figure 14-28b shows the air-fuel ratio AFR meter module operated by switch S13. Resistor R165 (26.1 ohms) is switched in and out of the circuit, but R166 (8.2 ohms) is common to both circuits.

Figure 14-28c shows the tachometer rpm module control where switch S14 switches resistor R167 (1111 ohms) in and out of the circuit.

Figure 14-28d shows the ignition advance (timing advance) ADV module control where switch S15 switches R168 (1111 ohms) in and out of the circuit.

All of the TB connections go to a master terminal board to which the modules, observer scope, main scope console and other equipment are connected. All plug P connections go to the console. The P references and the resistor designations are the same as other schematics of the console shown and described under appropriate headings elsewhere.

Automotive Diagnostic Center

A complete automotive diagnostic center or service lane comprises all available test equipment arranged in a certain order to render fast, efficient analysis of any conceivable malfunction of vehicle performance.

The master console may be mounted at eye level on an overhead trolley. The various modules are designed to acquaint the customer with electronic testing of a motor vehicle.

Connections of the master console, observer scope, observer modules and all associated electrical equipment is shown in block form in Figure 14-29 where the master terminal board TB interconnects all related electrical units. The reference to any TB numbers in the schematics previously described refers to a corresponding number on this board.

The observer modules having direct connection to the console through the terminal board TB reflect the readings of their counterparts on the console.

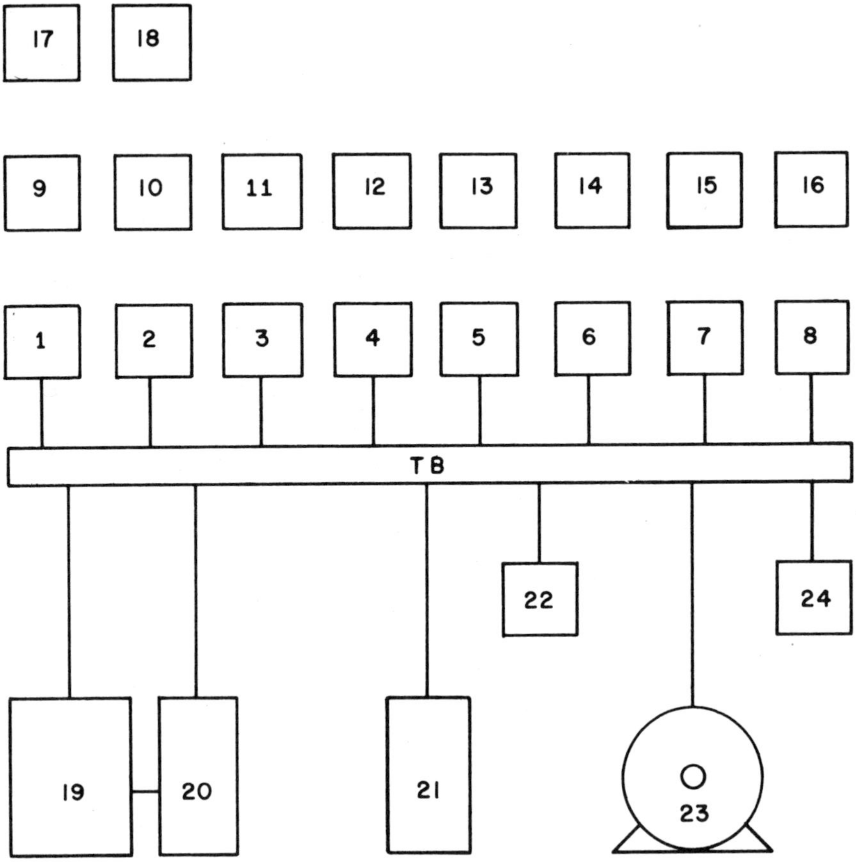

Figure 14-29 Automotive diagnostic center, console and modules

Some of the available modules and accessories are shown in block diagram form in Figure 14-29. The numbered blocks are identified and described as follows.

1. Start and charge voltmeter Source 1, Model 28-32, connects to TB34 and TB35 (positive lead to TB34, and negative lead to TB35).
2. Distributor advance meter, Model 28-41, connects positive lead to TB38 and negative lead to TB39.
3. Dwell meter, Model 28-42, connects positive to TB36 and negative to TB37 (same as engine rpm tachometer).
4. Engine rpm (tachometer), Model 28-34, connects positive to TB36 and negative to TB37.

5. Ignition observer scope module, Model 28-31, reproduces patterns displayed on console scope during various test procedures. Connected to console through terminal board TB40 to 45 inclusive.

6. Air-fuel ratio meter AFR module, Model 28-40, connects positive to TB33 and negative to TB32.

7. Road speed meter (mph) module, Model 28-35, connects positive to TB13 and negative to TB30.

A road speed meter calibrated in miles per hour would, of necessity, have to be fed a current from a tach generator driven by the rear wheel tire to give accurate readings for all wheel and tire sizes. This tach generator is usually incorporated in the dynamometer used for the road performance meter (hp) tests.

In the absence of a road speed meter, there are other ways of measuring the road speed for such purposes as adjusting the speed for wheel balancing, checking the speedometer, and related tests requiring a knowledge of the road speed in miles per hour. A mechanical tachometer applied to the spinning rear axle will read the rpm. A stroboscopic type of tachometer will read the rear wheel rpm when adjusted to make the wheel appear to be standing still. With the rpm on one wheel known (with the other wheel blocked), refer to Table 12-1 to convert rpm into miles per hour. These tables allow for one wheel turning twice as fast due to differential action while the other is blocked.

8. Road performance meter (hp) module, Model 28-39, connects positive to TB15 and negative to TB31.

A road performance meter functions to read the actual road hp of the vehicle under test. A dynamometer is driven by the rear wheels of the vehicle, with a remote control pendant for adjusting the load provided. The dynamometer usually has a tach generator incorporated to show the actual road speed in mph on a separate road speed meter.

A dynamometer is any device that measures the load or hp of an engine. They may be all electric, in which case a generator driven by the engine is loaded with resistors that are adjustable, or the field rheostat is used to make the adjustment. The electrical power output is calibrated to read the driving hp, or the generator operates a torque arm and scale.

Dynamometers may be all mechanical, in which case a friction load such as a brake drum is applied and a torque arm connected to a scale to show the foot-pounds, or hp output.

Dynamometers may also use some form of hydraulic pump to load the engine, in which case the pump output back pressure is adjusted to vary the load. The output is measured by a pressure gauge and/or a torque arm and scale. This output reading is either calibrated, or calculated to read the hp output.

The expression BRAKE HORSEPOWER is defined as the net horsepower output of an engine, turbine and the like, as measured by a band brake or a dynamometer.

ROAD HORSEPOWER is the actual horsepower delivered to the road through the rear, or driving wheels, which is somewhat less than the hp delivered by the engine due to losses in the transmission, drive shaft, differential and tires.

9. Introduction module, Model 28-33, no connection to terminal board TB.

10. Preparation and visual inspection module, Model 28-45 or 28-48, no connection to terminal board TB.

11. Brake effort module, Model 28-43, no connection to terminal board TB, but hydraulically connected to the vehicle brake system.

12. Vacuum test module, Model 28-49, no connection to terminal board TB, but pneumatically connected to vacuum gauge on console and tubing to service lane.

13. Wheel alignment observer module, no connection to terminal board TB, but is connected by conduit to wheel alignment pit.

14. Shock absorber observer module, Model 28-53, no connection to terminal board TB.

15. Wheel balance observer module, Model 28-54, no connection to terminal board TB.

16. Headlight observer module, Model 28-55, no connection to TB.

17. Under vehicle observer module, Model 28-38, no connection to TB.

18. Front end suspension observer module, Model 28-52, no connection to TB.

19. Master console engine analyzer scope, Model 28-01, connected to terminal board TB and to slave control 20.

20. Slave control for use with observer scope and connected to console 19 and to TB.

21. Clayton exhaust sampler assembly connected to air-fuel ratio AFR meter through TB.

22. Road performance dynamometer junction box connected to TB.

23. Road performance test dynamometer for loading vehicle to determine horsepower output at the driving wheels.
24. Portable load control pendant to adjust the dynamometer load.

Check Out Procedure, Calibration and Adjustments of Source 1, Model 28-01, Engine Analyzer

Although the following procedure is specifically prepared for Source 1, Model 28-01, engine analyzer, the same principles apply to any model or brand. There may be differences in the designations of the various controls and functions between manufacturers. If they serve the same purpose, these same procedures can usually be followed.

Usually the various calibrating and adjusting controls are labeled on the chassis or on a tube layout sheet pasted to the inside of the cabinet.

1. *VOLTS meter calibration and check.*
   a. Connect black and yellow insulated clip of the engine harness to an adjustable dc power supply, such as a battery and potentiometer, observing correct polarity.
   b. Connect a standard, accurate dc voltmeter to the black and yellow clip.
   c. "Zero" the console voltmeter M9 by adjusting the meter zero screwdriver adjustment located in the middle portion of the meter-cover directly below the words "volts" on the meter face.
   d. Turn polarity switch to its "negative" position.
   e. Rotate "volts" switch to the "10" position.
   f. Adjust dc power supply to 7 V on the standard dc voltmeter.
   g. Depress VOLT-AMPS push button. Compare readings of voltmeter on console to that of the standard voltmeter (after readjusting to exactly 7 V). Both meters should read 7 V ± one-half a division of console voltmeter 0–10 V scale.
   h. Turn "volts" switch to "20" position.
   i. Adjust dc power supply to 14 V.
   j. Compare readings of the two meters. The console voltmeter should read 14 V within ± one-half a division on the console voltmeter 0–20 V scale.
   k. Turn "volts" switch to "40" position.
   l. Adjust dc power supply to 24 V.

m. Compare the two meters. They should be the same within ± one-half a division on the console 0–40 V scale.
n. This particular model has no calibrating resistors.

NOTE: If a separate calibrating rheostat is provided for each range, the meter can be easily adjusted and calibrated against an accurate dc voltmeter of at least ½% accuracy. When there are no calibrating adjustments provided, it is necessary to analyze the degree of error on each scale. If all scales are out of tolerance in the same direction and at the same *percentage,* it usually indicates an aging (losing magnetism) meter. The meter should be replaced or provided with a correction chart. New blocks can also be drawn on the meter face for charge, "ignition," and "cranking" at the corrected position. If the meter reads too high at the same percentage, then one resistor across the meter, or a resistance in series with the meter, will correct the error on all ranges. Some analyzers have a built-in calibration system using zener regulated voltages as a reference.

2. *Points resistance check.*
   a. Connect all leads from the engine harness to the simulator as shown in Figure 14-1.
   b. Depress VOLT-AMPS push button on console front panel.
   c. Rotate "volts" switch to "PT-RES" position.
   d. Close the points on the simulator by revolving the drive shaft by hand until the points close (low meter reading). When the points are open, the meter will read full scale. (On a Source 1, Model 27-29, the meter will read the battery volts on the 0–40 volt scale).
   e. Depress the PT-RES push button on console front panel. Keep depressed while viewing the volts meter. Pointer of volts meter should read in the "PT-RES" area of meter scale corresponding to the battery voltage.
   f. Normally the "PT-RES" area is calibrated to read a maximum of 50 millivolts MV voltage drop for a 6-V system, 100 MV for 12-V, and 200 MV for a 24-V system. Some engine manufacturers may specify a higher voltage drop (on double breaker points, for ex-

ample), such as 250 MV for a 12-V system. These millivolt readings are read on the 0–10-V scale and multiplied by 100. For example, number 2 indicates 200 MV.

3. *Ammeter and shunt check.*
   a. Adjust console ammeter needle to zero by screwdriver adjustment on lower front of meter cover. Connect the shunt lead (screw terminal) of the engine harness to the positive of a 12-V battery. Connect a battery charger, having an adjustable rate, negative clip to the battery negative, and positive clip to a standard accurate ammeter positive terminal. The ammeter negative terminal is connected to the spade terminal of the shunt (which is connected to the battery positive).
   b. Set polarity switch to NEGATIVE (negative ground).
   c. Adjust battery charger to approximately 20 amp and compare meter readings (read standard ammeter, ignore ammeter on charger). They should agree within ± one division on the console "amperes" meter.
   d. Repeat at 40 and 60 amp. The meters should agree within ± one division on the console "amperes" meter.
   e. Reduce battery charger output to 10 amp or less.
   f. Turn the polarity switch to POSITIVE.
   g. The console meter should read negative (to the left of zero) the same amps as read on the standard ammeter ± one division.
   h. Return polarity switch to NEGATIVE.

4. *Horizontal sweep calibration, primary pattern.*
   a. Connect engine harness to simulator as shown in Figure 14-1.
   b. Set the pattern selector switch to the NORMAL position.
   c. Turn the simulator on and adjust for an engine speed of 1500 rpm as read on the tachometer.
   d. Depress the PRIMARY push button switch.
   e. Alternately adjust the PARADE control on the front panel of the console, and the HORIZONTAL POSITION control R9, (Figure 14-7), on the console chassis, until a normal primary pattern appears on the scope screen.
   f. If the pattern is not normal, turn the following console chassis controls to a point approximately ¾ of a turn clockwise and

readjust as needed: Horizontal linearity R3, Figure 14-6; Horizontal drive R7, Figure 14-7; Horizontal Bias R4, Figure 14-6; Brightness R12, Figure 14-17; Vertical position R2, Figure 14-2; and Focus R11, Figure 14-17.

g. Turn VERTICAL CALIBRATION control R1, Figure 14-2, on chassis, fully clockwise.

NOTE: If primary pattern is unstable, adjust the horizontal bias control R4, Figure 14-6, on chassis until pattern stabilizes.

h. Adjust spot shape R10, Figure 14-17, focus R11, Figure 14-17, and brightness R12, Figure 14-17, controls for clearest, sharpest scope trace possible.
i. Rotate PARADE control to its full clockwise position. In this position the white line on the parade control knob will be in line with the "dot" on the parade arrow located immediately above the knob.
j. Rotate the EXPAND control to its full counterclockwise position. In this position the white line on the expand control knob will be in line with the "dot" on the expand arrow located immediately above the knob.
k. Adjust the HORIZONTAL POSITION control R9, Figure 14-7, until the left end, or edge of the primary pattern is on the 45, 60, 90° mark of the horizontal scale of the scope screen.
l. Adjust the HORIZONTAL BIAS control R4, Figure 14-6, until the overall length of the primary pattern extends from the 45° mark on the left side to the 10° mark on the right side (using the 0–45° scale on the screen).

5. *Left edge clamp adjustment.*
   a. Complete steps (a) through (l) of 4. *Horizontal sweep calibration, primary pattern* above.
   b. Keep pattern on scope screen and simulator running at 1500 rpm.
   c. To check left edge clamp adjustment, slowly turn the EXPAND control on console front panel clockwise. The left edge of the pattern should remain fixed at the 45, 60, 90° point on the scope screen scale. If it does not remain fixed, adjust left edge clamp as follows;

(1). Simultaneously, rotate the EXPAND control on the console front panel, and the left edge adjust control R6, Figure 14-6, on chassis. The expand control should be rotated fully clockwise and fully counterclockwise through the entire rotation. As the expand control is being turned through its complete rotation, keep the left edge of the pattern on the 45, 60, 90° mark by adjusting the LEFT EDGE adjust control. For best results, rotate these controls very slowly—a very small amount at a time. Should the pattern disappear or go off screen at any time during these adjustments, return to position with the HORIZONTAL POSITION control R9, Figure 14-7, on the chassis. Adjust the left edge control and horizontal position control simultaneously until the left edge of the pattern remains stationary (locks in) at the 45, 60, 90° point when the expand control is turned to extreme clockwise position.

(2). After the left edge is clamped at the 45, 60, 90° mark of the screen, return the EXPAND control to its full counterclockwise position. Set the PARADE control to its full clockwise position. The left edge of the pattern should be at the 45° mark and the right edge should be at the 10° mark of the 0–45° scale. When the expand control is rotated clockwise from its full counterclockwise position, the left edge of the pattern should remain fixed and the right edge should expand outward to the right.

(3). After completing the left edge clamp adjustment, adjust the HORIZONTAL POSITION control R9, Figure 14-7, to place the left edge of the pattern to the middle of the scope screen scale. Adjust it to be at 30° on the 0–60° scale. Adjust the PARADE control on the console front panel to position the left edge of the pattern at the 45, 60, 90° mark.

6. *Primary Trimmer adjustment.*

a. Follow steps (a) through (l) of 4. *Horizontal sweep calibration, primary pattern* to obtain the primary pattern on the scope screen.

b. Adjust the primary trimmer capacitor C1 (not shown, but is in the primary lead inside the chassis), so that the dwell line at point C in Figure 14-19 forms a right angle as shown. If there is an upward spike, or the corner is rounded off, the trimmer is out of adjustment.

7. *Secondary pattern, superimposed.*
   a. Connect engine harness to simulator as shown in Figure 14-1.
   b. Turn PATTERN SELECTOR switch located on console front panel to NORMAL.
   c. Turn simulator on, and adjust for an engine speed of 1500 rpm on the tachometer.
   d. Depress the SECONDARY push button switch on the console front panel.
   e. Set the left edge of the secondary pattern to the 45, 60, 90° mark by adjusting the PARADE control on the console front panel. Adjust the right edge of the pattern to zero at the right end of the scope screen with EXPAND control. The normal pattern should appear.

8. *18KV pattern calibration, paraded.*
   a. Follow steps (a) through (c) of 7. *Secondary pattern, superimposed* above.
   b. Depress 18 KV push button on console front panel. All cylinder firing patterns will appear on the scope screen in their normal firing order.
   c. Adjust the VERTICAL POSITION control R2, Figure 14-2, until the pattern dwell line is on the horizontal zero line on the scope screen scale.
   d. Using the PARADE and EXPAND controls located on console front panel, adjust the pattern until it extends from 0 to 45° on the 0–45° scale. Observe the spacing between each cylinder plug spike, using the dwell angle scale on the screen. If the pattern shows unequal spacing of the plug firing lines on the right portion of the scope screen, adjust the HORIZONTAL LINEARITY control R3, Figure 14-6, to equalize the spacing. If unequal spacing is evident on the left portion of the scope screen, adjust the HORIZONTAL DRIVE control R7, Figure 14-7, for equal spacing of that portion of the scope screen.
   e. Adjust the EXPAND control slowly until one complete cylinder firing line is spaced between 0–45° marks on the scope screen 0–45° scale.

f. Slowly rotate PARADE control to parade each cylinder firing pattern across the scope screen. Each cylinder should center between the 0–45° marks and fill the space entirely. If the cylinder firing patterns change brightness or tend to lock up as they are paraded across the screen, touch up adjustments of HORIZONTAL LINEARITY control R3, Figure 14-6, and HORIZONTAL DRIVE control R7, Figure 14-7. Use caution when making these adjustments, and maintain equal spacing of cylinder plug firing lines. Use the dwell angle calibration marks as a guide.

9. *Cam angle or dwell calibration and check.*
   a. Connect engine harness as shown in Figure 14-1.
   b. Connect a standard dwell meter, known to be accurate, to the distributor and ground on the simulator.
   c. Adjust simulator for a reading of 1500 rpm on the console tachometer.
   d. Depress the PRIMARY push button switch on console front panel and maintain the pattern between zero and the 45, 60, 90° marks on the scope screen.
   e. Compare the reading on the standard dwell meter to the *average* of the cam angle as shown on the scope screen. The readings should be the same. If not, correct the console scope readings very carefully by adjusting the HORIZONTAL DRIVE control R7, Figure 14-7, on the console chassis. If it is also necessary to adjust the HORIZONTAL LINEARITY control R3, Figure 14-6, and the HORIZONTAL DRIVE control R7 to correct the dwell reading, steps (a) through (f) of No. 8, *18 KV pattern calibration, paraded,* will have to be repeated.

10. *18 KV and 36 KV vertical height calibration.*
   a. Connect engine harness to simulator as shown in Figure 14-1 except for the red clip R lead. Remove red clip R lead from simulator coil lead and connect it to the distributor, or primary connection, on the simulator.
   b. Turn simulator on and adjust speed for a reading of 1500 rpm on the console tachometer.
   c. Depress the 18 KV push button on the console front panel.
   d. Adjust the 18 KV vertical spike trimmer C3 (not shown, but is located in series with the secondary pick-up lead inside the chas-

sis), so that the left end of the dwell line is horizontal and straight with no upward spike showing.

e. Disconnect the red clip R lead from the simulator primary connection and reconnect it to the coil secondary lead of the simulator. Note the height of the plug firing spikes with reference to the 18 KV scale on the left side of the scope screen.
f. Depress the 36 KV push button switch on the console front panel.
g. Adjust the 36 KV vertical spike trimmer C4 (not shown, but is located in series with the secondary pick-up lead inside the chassis and in series with 18 KV vertical spike trimmer C3), until the height of the plug firing spikes is one-half the value noted in step (e), with reference to the 36 KV scale located on the right side of the scope screen.

11. *Tachometer calibration and check out procedure.*

> NOTE: The console tachometer is a 320 microampere full-scale movement with 120 microampere suppressed zero. Therefore, the meter zero adjust (on lower portion of meter case), will not move the pointer because it is not engaged with the pointer movement.

a. Connect a standard tachometer of known accuracy to the simulator to monitor engine speed.
b. Adjust the speed of the simulator to 1500 rpm as indicated on the standard tachometer.
c. Turn RPM switch on the console to "1600".
d. Depress 18 KV push button switch.
e. Adjust 1600 rpm adjust control R23, Figure 14-15, on console chassis until console tachometer reads 1500 rpm (or the exact rpm indicated on the standard tachometer).
f. Turn RPM switch to "10,000".
g. Adjust 10,000 rpm adjust control R24, Figure 14-15, until console tachometer reads 1500 rpm on the 10,000 rpm scale.
h. Disconnect the engine harness from the console panel (plug PH).
i. Adjust the COARSE TACH zero adjust control R26, Figure 14-15, and FINE TACH zero adjust control R25, Figure 14-15, on the console chassis until the tachometer pointer is at zero. The meter ad-

justment controls are extremely sensitive, therefore, care should be exercised when adjusting the coarse and fine tach zero adjust controls.

j. Reconnect the engine harness to the console. Tachometer should read 1500 rpm with the RPM switch in the "10,000" position. If the tachometer does not read correctly in step j, repeat steps f, g, h, i, and j.

   It will probably be necessary to repeat this procedure several times to get proper balance between tachometer zero and 1500 rpm on the 10,000 rpm scale.

k. Adjust simulator speed to 700 rpm and adjust console tachometer 800 rpm adjust control R22, Figure 14-15, to read 700 rpm on console tachometer.

l. Recheck rpm calibration on all three (800, 1600, 10,000) ranges.

12. *Dynamic compression calibration.*

a. Connect engine harness to simulator as shown in Figure 14-1, and adjust speed to 1500 rpm on console tachometer.

b. Depress 36 KV DYNAMIC COMPRESSION push button switch.

c. Turn "balance to line" control on console front panel until the white line of balance-to-line knob is in a vertical position.

d. Adjust the "dc zero" adjust control R18, Figure 14-10, until the pointer of the dynamic compression meter rests on the balance-to-line zero.

e. Rotate the cylinder selector control to its OFF position.

f. Adjust parade control and expand control to position the pattern between zero and 45° marks on the 0–45° scale.

g. Turn the "dc bias" control R14, Figure 14-11, until the downward or cylinder marker pulse is positioned between the end of No. 1 cylinder firing line and the beginning of No. 2 cylinder firing line.

h. Rotate the "cylinder selector" control slowly to the right. Note the downward or cylinder marker pulse "short out" each plug firing spike in turn, as it moves across the scope screen.

i. Adjust parade and expand controls on console front panel to center the pattern between zero and 45° marks on the 0–45° scale. The pattern should fill the entire space between these 0–45 marks.

j. Rotate the cylinder selector control to place the cylinder marker pulse between any two spikes.
k. Adjust "pulse duration" control R15, Figure 14-11, on the console chassis to spread the cylinder marker pulse full-width between the two spikes, being careful not to short out either one of the adjacent spikes.
l. (1). Starting from the cylinder selector OFF position, slowly rotate the cylinder selector control clockwise. Each plug firing line should short out as the cylinder marker pulse moves through it.
(2). To short out the last plug on the right (No. 1 spark plug), rotate the cylinder selector control to its maximum clockwise position. Turn the No. 1 cylinder switch (dynamic compression), on the console front panel to ON. Slowly rotate the cylinder selector control counterclockwise until the No. 1 spike is shorted out.
(3). Return the No. 1 cylinder switch, and cylinder selector control to their OFF position.

13. *Timing light check*
a. Connect engine harness to simulator as shown in Figure 14-1.
b. Turn simulator on and adjust speed to 1500 rpm.
c. Depress TIMING push button switch.
d. The timing light should fire steadily. No pattern should appear on the scope screen. Rotation of timing advance control located on the timing light should not advance the pointer on the advance meter.

14. *Ignition advance check.*
a. Repeat steps (a) and (b) of No. 13, *Timing light check* above.
b. Depress ADVANCE push button switch.
c. Timing light should flash steadily, and when the control knob of timing light is turned clockwise, the pointer on the advance meter should swing up-scale.

15. *Advance calibration.*
a. Connect console engine harness, shown in Figure 14-1, to ignition simulator, shown in Figure 3-1, as follows.

(1). Blue clip to No. 1 plug.
(2). Green clip to distributor points.
(3). Black clip to ground.
(4). Red clip to coil H.T. lead.
(5). Yellow clip to positive side of coil.

b. Operate simulator at 1500 rpm as observed on console tachometer.

c. Depress ADVANCE push button switch on console front panel. Timing light should fire steadily. No pattern should be visible on the scope screen.

d. Turn advance control knob of timing light to its fullest counterclockwise position. The console ignition advance meter should not read over 3° on the 60° scale.

e. Referring to the ignition simulator shown in Figure 3-1, direct the timing light on wheel W and dial DL. Adjust the spark advance SA of the simulator until the cylinder No. 1 mark on wheel W is in line with zero on dial DL. Cylinder No. 2 mark should be in line with 60° on dial DL. Turn advance control knob of timing light to its full clockwise position. The No. 2 cylinder mark on wheel W should now have moved from 60° to approximately 40° on dial DL. This indicates a difference of 60 minus 40, or 20° time delay on the simulator distributor. Since the distributor turns at one-half the speed of the crankshaft, this 20° represents a 40° timing light delay, or crankshaft ignition time delay as read on the console ignition advance meter. The advance control introduces a 40° delay in the firing of the timing light. Adjust the advance control on the timing light until dial DL is exactly 40° (20° distributor time delay).

f. Adjust the advance adjust control R21, Figure 14-14, on the console chassis so that the pointer on the console ignition advance meter reads exactly 40° (or twice the degree difference on dial DL), on the crank 60° scale.

16. *Carburetor adjust calibration.*

a. Connect console harness to simulator as shown in Figure 14-1.

b. Set BALANCE TO LINE control on front panel to full counterclockwise position.

c. Turn simulator on and adjust for engine speed of 1200 rpm.

d. Adjust the CARB. ADJ. control R17, Figure 14-10, on console chassis until pointer of amperes meter is set on the zero of carburetor adjustment balance to line scale.

17. *Primary pattern, separated and paraded.*
   a. Connect harness to simulator as shown in Figure 14-1.
   b. Turn simulator on and adjust to 1500 rpm.
   c. Turn PATTERN SELECTOR switch on front panel to NORMAL.
   d. Depress PRIMARY push button switch on front panel.
   e. A normal primary pattern should appear on scope screen.
   f. Turn pattern selector switch to SEPARATED.
   g. Rotate the cylinder selector control to the OFF position. Primary pattern should appear normal, but moved to the upper portion of the scope screen.
   h. Turn the cylinder selector control slowly clockwise. As control is slowly turned, cylinder firing lines will "drop out" of the superimposed primary pattern and move to the lower portion of the scope screen. Each cylinder firing line will "drop down" in their normal firing order. This separation will allow individual inspection and analysis of a specific cylinder firing line.
   i. Adjust cylinder selector control until a single cylinder firing line is dropped down and centered as shown in Figure 14-22.
   j. Turn pattern selector switch to PARADED.
   k. All cylinder firing lines in firing order will appear on the upper portion of the scope screen. The specific firing line selected in step (i) will be pulled down, allowing identification of that particular cylinder. (See Figure 14-23).
   l. Adjust EXPAND control so that one cylinder firing is expanded from zero to 45° on the 0–45° scale.
   m. Slowly rotate PARADE control to parade each cylinder firing line across the scope screen.
   n. Return pattern selector switch to NORMAL position.

18. *Alternator test check.*
   a. Connect engine harness to simulator as shown in Figure 14-1, and connect a battery charger across the battery BAT. Adjust the charger to a high rate to simulate the normal alternator ripple.
   b. Turn simulator on and adjust speed to 1000 rpm.

c. Adjust synchronize control and pattern height control on console front panel to obtain a rectified sine wave as shown in Figure 14-25.

19 *Vacuum gauge check.*

a. Connect the hose of console harness to a vacuum supply pump, and a standard vacuum gauge of known accuracy.

b. Compare readings of console vacuum gauge to those obtained on the standard gauge at 5, 10, 15, 20, and 25 in. of vacuum. Readings should be the same within a tolerance of ± ½ in. of vacuum at each check point.

20. *Distant viewing check.*

a. Connect engine harness to simulator as shown in Figure 14-1 and adjust to a speed of 1500 rpm.

b. Depress 18 KV push button switch. Note width of trace on plug firing lines.

c. Depress PATTERN DISPLAY (distant viewing) push button. Note increase width of plug spike trace. No change indicates defective component, as there are no adjustments.

14-5 Typical Console Scope-Type Engine Analyzer, Model 1000

Front Panel Controls and Functions

Scope-type engine analyzers typical of Source 15 differ from Source 1 previously described, in panel arrangements, instrumentation, and functions. One typical model, Model 1000, has the following described front panel arrangements and functions.

1. *Scope.* The instantaneous operating characteristics of either the primary or secondary circuits in the ignition system are visually displayed on the scope screen in the form of patterns and waves.

The scope screen is calibrated with four different measuring scales. The vertical scale at the left edge of the screen is used to measure ignition pulses in the range of magnitude from 0 to 18 KV, such as spark plug voltage requirements. The vertical scale at the right edge of the screen is used to measure ignition pulses in the higher range of magnitude from 0 to 36 KV, such as ignition coil voltage output. The two horizontal scales at the bottom of the screen are calibrated in degrees and are used to measure the distributor dwell or cam angle. The cam angle measurement is a measure of time during which the ignition breaker points remain closed in proportion to the total cycle between breaker point opening points. The upper scale is used to measure the cam angle of 8-cylinder engines. The lower scale is used to measure the cam angle of 6-cylinder engines. The 8-lobe scale reading, multiplied by 2 is used to measure the cam angle of 4-cylinder engines.

2. *Engine Speed Meter.* The attainment of specific engine speeds is required when making most of the analyzer tests. A means of measuring this engine speed in revolutions per minute, rpm, is provided by the engine speed meter.

The meter is divided into three separate scales which are linearly calibrated in rpm to cover the following speed ranges; 200–1200 rpm, 1000–6000 rpm, and 2000–12,000 rpm. Only the scale in use, as selected by the engine speed scale selector switch, becomes illuminated.

> NOTE; When volts-adv.-power scale selector is turned to the POWER CHECK position, the 200–1200 rpm scale on the engine speed meter becomes illuminated regardless of the existing position of the engine speed scale selector. Since the power check compression test is made at a recommended engine speed of 800–1000 rpm, both the measuring circuit and the scale on which the engine speed should be measured are automatically selected.

3. *Volts-Advance-Power Meter.* This multi-function meter is used to measure the results obtained from three different engine analysis tests. Only the scale in use as selected by the VOLTS-ADV.-POWER scale selector switch, becomes illuminated.

(a). VOLTS SCALE. The volts scale at the top of the meter is used to make voltage analysis tests in the primary circuit of the ignition system. It is also used to test the proper operation of the high-voltage power supply through voltage dropping resistors in the analyzer. Two ranges of voltage can be measured on the 0–18 V and 0–1.8 V scales. The 0–1.8 V range is used to measure small voltage losses which may have become excessive due to increased electrical resistance in the wiring, or components associated with the primary ignition and other battery voltage circuits. Examples of circuit points at which the voltage loss measurements may be taken are as follows; distributor points, ignition switch, starter switch and solenoid, battery cables and terminals, and the like. The 0–18 V range is used to measure higher voltages in the primary circuit such as; battery voltage, voltage regulator output, voltage drop across ballast resistor or primary resistance wire, coil primary and battery voltage under starting conditions, and the like.

(b) INDIVIDUAL CYLINDER OUTPUT SCALE. This middle scale of the meter is used to indicate the results obtained in rpm loss from the electronic compression tests made when shorting out one cylinder at a time. Variations in cylinder output greater than 50% (as shown by the rpm reading change on the linearly calibrated 0–150 rpm scale) indicates that cylinder performance is poor and that the engine is in need of repairs.

(c). CRANKSHAFT DEGREES SCALE. The crankshaft degrees scale at the bottom of the meter is linearly calibrated from 0 to 60° to measure the results obtained from ignition timing tests of the distributor spark advance mechanisms. The proper value of distributor advance for a particular domestic car is indicated on the Dyna-Tune specifications roll chart which is built into the analyzer panel.

4. *Dyna-Grams Roll Chart.* This roll chart enables the operator to compare the test findings observed on the scope screen conveniently with the normal and unsatisfactory patterns shown on the roll chart. The brief, step-by-step diagnosis procedures graphically illustrated on the roll chart eliminate constant referrals to the operation instruction manual.

5. *Dyna-Tune Specification Roll Chart.* This roll chart lists engine tune-up specifications of all late model, domestic-manufactured cars. This provides the operator with convenient and ready reference information needed for the proper diagnosis and adjustment of many of the engine tests.

6. *Power Switch.* This 3-position switch controls the application of power to the analyzer. The middle position of the switch is OFF and removes all power to the analyzer. The right-hand position of the switch applies power to both roll chart lights and to the lumiline light on the front panel. The left-hand position of the switch applies power to the lights and also to the analyzer power supply circuit.

7. *Test Selector Switch.* The test selector switch consists of five

wafer sections (front and rear), and has nine positions. It is used to select the circuits that enable the analyzer to perform the following engine diagnostic tests.

(a). PRIMARY CIRCUITS.

Position 1. Superimposed cylinders (cranking speed).
Position 2. Superimposed cylinders.
Position 3. Any single cylinder.

(b). SECONDARY CIRCUITS.

Position 4. Superimposed cylinders.
Position 5. Any single cylinder.
Position 6. Parade cylinders 18 KV.
Position 7. Parade cylinders 36 KV.

(c) ACCESSORIES.

Position 8. Accessory.
Position 9. Timing light.

Test Position 8 selects analyzer circuitry that permits additional tests to be incorporated into the instrument with future developments of accessory items. The connections are provided through a jack J8 shown under the heading *Ignition Simulators.*

8. *Engine Speed Scale Selector Switch.* This switch selects the proper measuring circuit and speed-range scale for indicating the engine speed on the engine speed meter. The speed range selected automatically illuminates the proper scale on the engine speed meter. See NOTE under topic heading No. 2, Engine Speed Meter, above.

9. *Volts-Adv.-Power Scale Selector Switch.* This switch selects the proper measuring circuits and scale to indicated readings on the volts-advance-power meter for the three following test measurements; voltage, advance ignition timing, and power check. The measuring circuit selected automatically illuminates the proper scale on the volts-advance-power meter. See NOTE under topic heading No. 2, Engine Speed Meter, above.

10. *Car Ground Switch.* This switch is set to correspond to the polarity of the grounded battery post in the primary circuit of the engine ignition system. In this manner, the correct polarity of engine test signals is automatically selected to produce proper diagnosis indications.

11. *Rpm Meter Adjust.* This control adjusts the amplitude of triggering pulses being fed into the analyzer from the plug No. 1 lead so that a steady and unwavering reading may be obtained on the engine

speed meter. The adjustment of this control will also affect the stability of the scope patterns in the test selector positions 3, 5, 6, 7, and 8. Aligning the pointer on the control knob within the region marked NORMAL will usually obtain the desired stability requirements.

12. *Hold Control.* This control adjusts the amplitude of the trigger pulses being fed into the analyzer from the primary lead so that stable and proper presentation of scope patterns may be observed in test selector positions 1, 2, and 4. Proper scope pattern stability is usually obtained with pointer on control knob aligned within region marked normal.

13. *Engine Selector.* This control is set to correspond to the number of cylinders used in the engine under test. The adjustment of this control varies the width of the marker pulse which may be used to identify or mark (pulse selector turned to mark only, or power check positions), the individual cylinder pulses as they are paraded in their sequential firing order on the scope screen in test selector positions 6 or 7. Clockwise rotation of this control narrows the marker pulse width; counterclockwise rotation broadens the pulse width. In this manner, the trace time represented by the pulse width can be varied to correspond to a proportionate length of time occupied by the pulse from each individual cylinder.

14. *Spread Control.* This control adjusts the horizontal sweep width of the scope trace. With the control knob turned to AUTO (extreme counterclockwise position), the amplitude of the generated sawtooth sweep voltage is automatically adjusted so that the scope trace extends between the vertical 18 KV and 36 KV scales on the scope screen. The "auto" position of this control is used to observe scope patterns in the test selector positions 1, 2, 4, 6, and 7.

Turning this control clockwise from "auto" position allows the operator to vary the amplitude of the generated sawtooth voltage. In this manner, the scope pattern represented from any individual cylinder in test selector positions 3 and 5 (cylinder selector turned clockwise from OFF position), may be expanded. It can be expanded to fill the width of the screen to give a detailed inspection of each individual cylinder test pattern.

15. *Cylinder Selector.* Clockwise rotation of this control from its OFF position increasingly delays the triggering of the horizontal sweep generator so that, in effect, the sweep trace may be initiated at varying

times to synchronize with a pulse from any one of the individual cylinders. In addition, this clockwise rotation of the cylinder selector increases the delay in the triggering of the marker pulse generator so that the marker pulse may be made to synchronize with a pulse from any one of the individual cylinders. Therefore, the test pattern from any individual cylinder may be separately selected for closer observation (test selector positions 3 and 5). Also, any individual cylinder may be separately selected for power check tests (test selector position 6).

16. *Pulse Selector Switch.* This switch is used to select the operation of the marker pulse.

The OFF position of this switch disables the marker and power check circuitry. The MARK ONLY position of this switch applies the marker pulse on the scope screen trace so that each cylinder may be separately selected by the cylinder selector (test selector in position 6 or 7). POWER CHECK positions 2–8 of this switch enable the cylinder selector to mark or select any one of the cylinder pulses from No. 2 through No. 8 for the power check electronic compression test (test selector in position 6).

The power check No. 1 position is used to make the electronic compression check of cylinder No. 1 (test selector in position 6). This switch position performs the same circuit functions as power check positions 2–8, except that, in addition, the now weakened pulse from cylinder No. 1 is stepped up by a transformer and coupled back into the analyzer plug No. 1 triggering circuit. This keeps analyzer circuits triggered into operation during the time cylinder No. 1 is being power checked (shorted out).

17. *Balance To Line.* This control is adjusted to obtain a zero rpm reading on the individual cylinder output scale of the VOLTS-ADVANCE-POWER meter prior to making power check measurements.

18. *Oscilloscope crt adjustments.* The scope controls are located behind a small door at the bottom of the front panel similar to some TV sets.

(a). BRIGHTNESS. This control adjusts brightness of the scope tube pattern.

(b). FOCUS. This control adjusts the clarity or sharpness of the scope trace pattern.

(c). HOR. ADJ. LEFT EDGE. The horizontal adjust left edge control lengthens or shortens the left edge of the scope pattern.

(d). HOR. ADJ. RIGHT EDGE (AUTO). The horizontal adjust right edge (auto) control lengthens or shortens the right edge of the scope pattern (spread control should be pre-positioned in "auto" position).

(e). VERT. CENT. The vertical centering control positions the scope pattern vertically. The base line of the scope trace should be positioned vertically between the two horizontal cam angle scales at the bottom of the screen.

(f). H.V. CHECK. The high voltage check switch does not actively affect the scope pattern. It is provided as a convenience to the operator for checking the operation of the high voltage CRT power supply. Loss of high voltage will disable the scope trace.

With the VOLTS-ADVANCE-POWER scale selector in the 0–18 V position, the depressing of this push button switch should give a reading of approximately 14–15.4 V on the volts scale of the volts-advance-power meter as an indication that the high voltage power supply is functioning normally.

(g). OTHER CONTROLS. All other calibrating adjustments, intended only for properly trained and equipped technicians, are located in a row behind the removable lumiline light and protective shade. This light and shade can be removed from the front panel to expose the controls.

Typical Features and Circuits of Engine Analyzer, Source 15, Model 1000.

The scope-type engine analyzers typical of Source 15 operate on the same general principles as those previously explained for analyzers typical of Source 1 and others. However, the following differences are worthy of description.

1. *The High Voltage Supply* to the CRT is obtained in the same way as that of some TV sets. The schematic is shown in Figure 14–30 with the component reference numbers as found in Source 15, Model 1000. These reference numbers are usually printed adjacent to the component on the chassis. An RF (radio frequency) oscillator V10 (6AQ5) generates a 300 kc (kilocycle) high voltage that is rectified by two tubes, V4 and V5 (1X2), to provide a positive 3500 V and a negative 2700 V. It has a high-voltage regulator V7C (1/3 6EZ8) to control the output of the oscillator. Tube V5 provides the source voltage of negative 2700 V, for controlling the brightness and focus of the scope trace. Tube V4 provides the source voltage of positive 3500 V to the anode of the CRT (8CP31).

The high voltage adjust control R41 (1 megohm on front panel behind

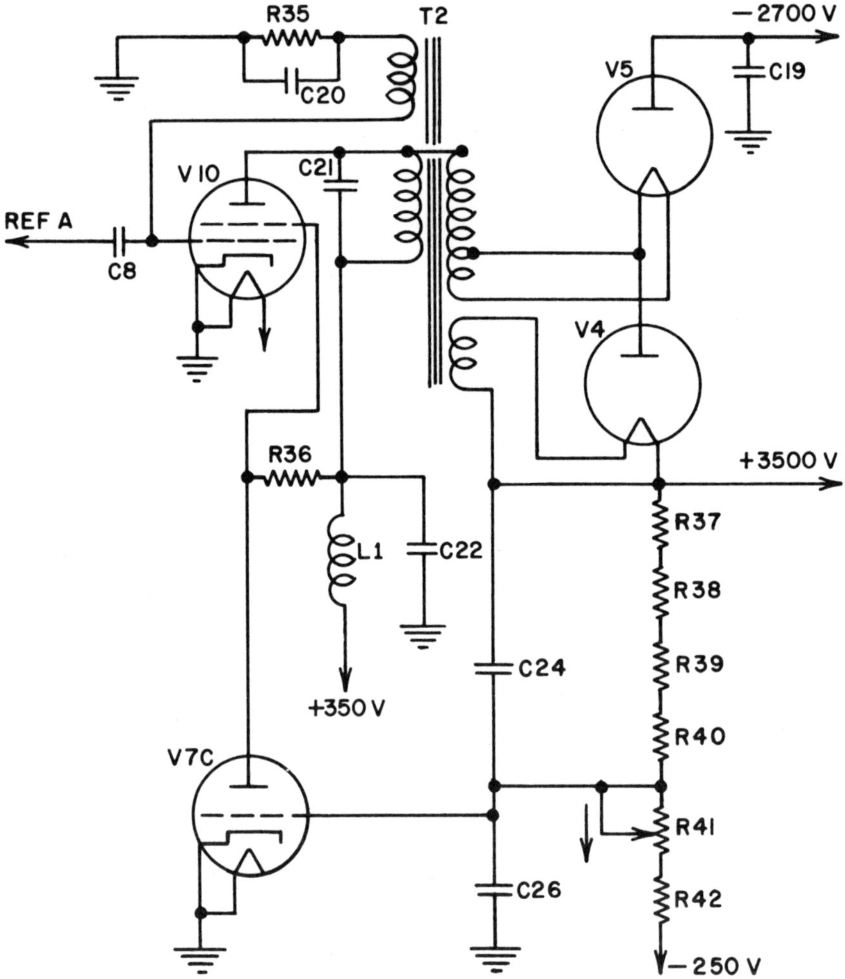

Figure 14-30 High-voltage power supply, typical, Source 15, Model 1000

light), adjusts the bias of the triode section of V7C (1/3 6EZ8) to control the operation of the RF oscillator V10 (6AQ5). This, in turn, effectively regulates the high-voltage output from high-voltage rectifiers V4 and V5.

Reference A REF A connects to the grid of the automatic sweep limit tube V7A (6EZ8) in Figure 14-31.

Other values are; capacitors in mfd, C8 (50 mmfd), C19 (4.7 K mmfd 4 KV), C20 (0.001), C21 (0.001 1 KV), C22 (0.05), C24 (3.3 K mmfd 5

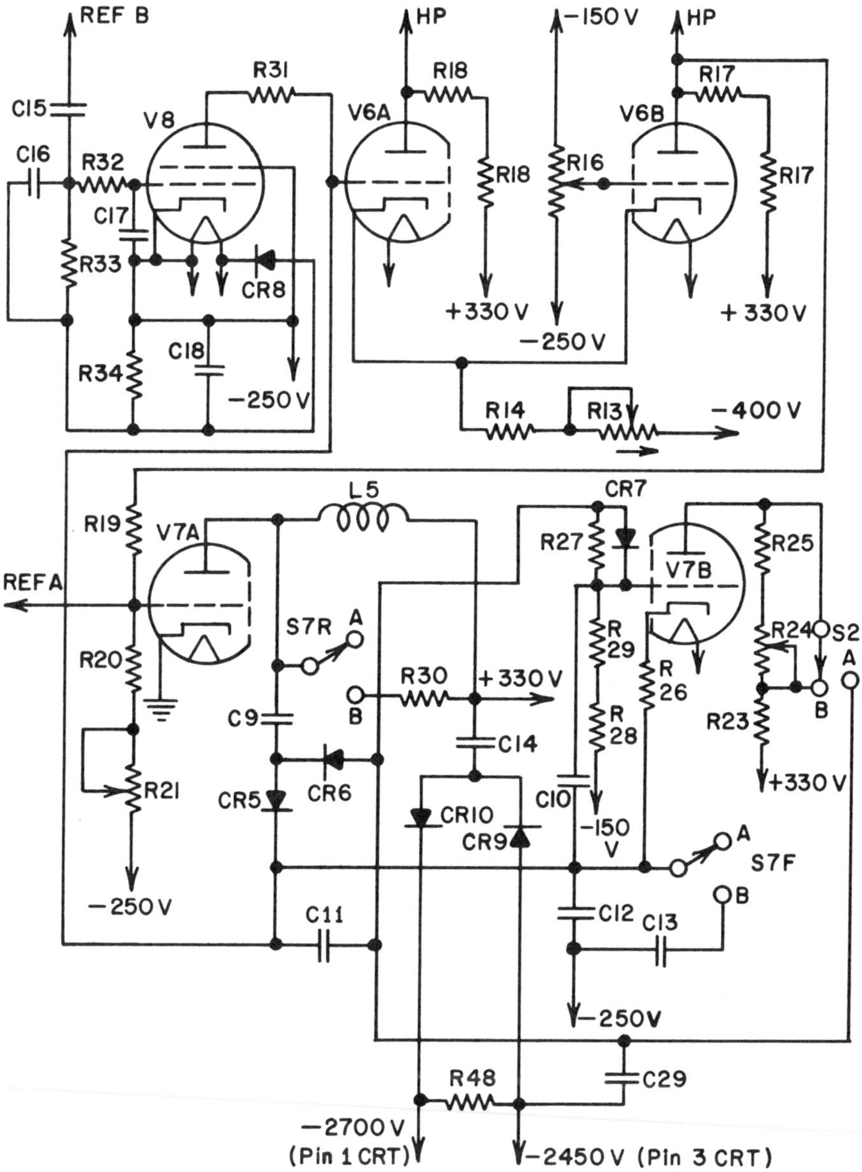

Figure 14-31 Automatic right-edge sweep limit control circuit, typical, Source 15, Model 1000

KV), C26 (0.15); resistors in ohms, R35 (47 K), R36 (56 K), R37, R38, R39, and R40 (6.8 meg), R41 (1 meg high voltage adjustment), R42 (1.5 meg); choke L1 (4 mh); transformer T2 (high-voltage oscillator).

2. *High Voltage Check.* To check for proper high voltage from the power supply, a dc voltage is tapped off the bleeder circuit and fed to the dc voltmeter 0–18 V range through multiplier resistors. A voltage reading of 14–15.4 V indicates proper high voltage circuit operation. This voltage can be adjusted by control R41 (1 megohm) Figure 14-30.

3. *Automatic Right-Edge Sweep Limit Control Circuit.* The horizontal sweep, sweep width, and automatic right-edge sweep limit circuit is shown schematically in Figure 14-31. This circuit generates and controls the sawtooth deflection voltage applied to the horizontal deflection plates of the CRT, enabling the display of test patterns on the scope screen. It consists of the following tubes and their associated circuitry; horizontal sweep generator V8 (5969), automatic sweep limit V7A (6EZ8), horizontal sweep width control V7B (6EZ8), and horizontal amplifiers V6A and V6B (12AU7).

Under static conditions with no trigger pulses applied to reference B REF B, the sweep generator V8 is biased to cutoff by a negative 7 V bias voltage applied to its control grid. The dc bias supply circuit incorporated in the filament circuit of V8 consists of CR8, C18 (20 mfd), and R34 (10 Kohms). When a trigger pulse is applied to the grid of V8 through capacitor C15 (0.001 mfd), V8 is driven into conduction to generate the sawtooth sweep voltage. In test selector positions 1, 6, 7, and 8 (position B of switch S7F), the sawtooth voltage is developed across parallel connected capacitors C12 (0.2 mfd) and C13 (1 mfd). In test selector positions 2, 3, 4, and 5 (position A of switch S7F), the sawtooth wave is developed across C12 only. The amplitude of the sawtooth voltage (corresponding to width of scope pattern), is controlled by V7A and V7B (6EZ8). Switch S7R is in position A for test selector switch positions 1, 3, 5, 6, 7, and 9. It is in position B for test selector positions 2, 4, and 8.

With the SPREAD control switch S2 in the AUTO position B as shown, it shorts out R24 (2 megohm spread control located on the front panel). The amplified sawtooth wave developed at the plate of V6B is coupled, through R19 (1.2 megohm), to the grid of V7A along with a 300 kc RF signal obtained from the control grid circuit of V10 (see REF A Figure 14–30). Tube V7A is driven into conduction by the amplitude of the generated sawtooth wave. With V7A conducting, the 300 kc signal is amplified and then rectified by diodes CR5 and CR6 to provide a negative dc bias control voltage. This bias control voltage is coupled to the grid of V7B to control the current conduction through V7B. Therefore, in summation, the amplitude of the generated sawtooth wave controls the current flow through V7B. This, in turn, determines the amplitude of the generated sawtooth wave sweep voltage to provide a fixed scope pattern width that extends just between the vertical KV scales on the right and left sides of the scope screen.

When the spread control R24 (2 megohm), is turned clockwise out of AUTO position switch S2 (from contact B to A), it removes the short from across R24 and R25 and places a short across tube section V7B. Under these conditions, tube V7B is eliminated from the sweep width control circuit. Varying the spread control R24 now adjusts the amplitude of the sawtooth wave to provide a scope pattern width that can be expanded beyond the automatic limits.

Test selector switch positions 1, 2, and 4 enables pulses which are developed in the primary circuit of the engine ignition system to be coupled to the analyzer through the primary lead. These pulses trigger the horizontal sweep circuit into operation at the pulse frequency rate which is determined by the engine rpm and number of engine cylinders.

Test selector switch positions 3, 5, 6, 7, and 8 enable pulses which are developed at plug No. 1 in the secondary of the engine ignition system to be coupled to the analyzer through plug No. 1 lead to trigger the horizontal sweep circuit into operation at the plug No. 1 pulse frequency rate. This rate is less than the primary rate (at the same rpm) in direct proportion to the number of cylinders. For example, in an 8-cylinder engine the plug No. 1 pulse rate will be ⅛ that of the primary pulse rate at the same engine rpm.

Other values of components are; resistors in ohms, R13 (25 K potentiometer, horizontal amplifier adjust), R14 (27 K), R16 (1 meg potentiometer, horizontal adjust left-edge), R17 (2) (47 K), R18 (2) (47 K), R20 (560 K), R21 (500 K pot., horizontal adjust right-edge, auto), R23 (180 K), R26 (3.3 K), R27 (10 meg), R28 and R29 (22 meg), R30 (10 meg), R31 (47), R32 (150 K), R33 (56 K), R48 (470 K); capacitors in mfd C9 (22 mmfd), C10 (0.5), C11 (0.005), C14 (50 mmfd 3KV), C16 (0.001), C17 (100 mmfd); choke L5 (2.4 mh); diodes CR5, CR6, and CR7 (silicon 1N441B 200 PIV 10 amp surge), CR8 (selenium 1N1635 30 PIV 3 ma average), CR9 and CR10 (Silicon 1N1695 400 PIV 50 amp surge).

4. *Marking* or identifying individual cylinders with test selector in positions 6 and 7.

A marker pulse generator is triggered by the same time-delay circuit that selects individual cylinders to short out during the power check or dynamic compression test.

With the pulse selector in the MARK ONLY position, the marker pulse is coupled to the CRT vertical deflection plates where it appears on the scope pattern. Thus, as the cylinder selector is turned clockwise, the triggering of the marker pulse generated is increasingly delayed so that the marker pulse may be moved from left to right on the scope screen to intersect the pulse from each individual plug circuit or cylinder. When the marker pulse is positioned so that it intersects (marks) a pulse from the particular cylinder, the marker pulse is actually occurring at the same time as the cylinder pulse which is being marked.

After the pulse from an individual plug circuit, or cylinder, has been identified or marked, the operating characteristics of the marked cylinder may be more closely inspected. This is done by expanding the individual pulse to fill the width of the scope screen by switching the test selector to position 3 (primary patterns), or to position 5 (secondary patterns).

The time-delay circuitry of the cylinder selector control is again incorporated, but this time the delayed pulses trigger the horizontal sweep circuit into operation. This is done at a time that synchronizes with the marked cylinder pulse. In this way, any individual, marked cylinder pulse is automatically expanded for detailed analysis.

With the test selector in position 3, the pulses applied to the vertical deflection plates of the CRT are derived from the primary circuit of the ignition system through the primary lead. Thus, the expanded scope pattern is representative of any one selected cylinder pulse which is developed in the primary circuit of the ignition system.

With the test selector in position 5, the pulses applied to the vertical deflection plates of the CRT are derived from the secondary circuit of the ignition system through the secondary lead. Thus, the expanded scope pattern is representative of any one selected cylinder pulse which is developed in the secondary circuit of the ignition system.

5. *Cathode Ray Tube (CRT) Deflection System.* Although most scope-type engine analyzers use CRT tubes in sizes ranging from 5 to 9

in. (using electrostatic deflection), Source 15 has at least one known model, 40-250, that uses a 21-inch black and white (B&W) picture tube. They use an external magnetic yoke deflection system as in TV circuits. At the present time the cost of a 21-inch B&W CRT is far less than the cost of an 8-inch scope tube.

As the state of the art progresses, and costs are reduced, it will be highly possible and practicable to use color TV picture tubes in engine analyzers. The possibilities are unlimited. For example, it would be possible to identify the different patterns, or parts of patterns, by the color or hue of the scope traces.

6. *High Voltage Pickup Heads.* The high voltage pickup heads of Source 15, Source 1, and others, are similar in function but they do appear in different shapes and colors.

Figure 14-32a shows the green No. 1 spark plug head which has a small current transformer T molded into the pickup head.

Figure 14-32b is a torpedo-shaped high-voltage pickup head containing a small oil filled high voltage capacitor C. These may be green for the No. 1 spark plug, or yellow for the coil secondary pickup head. A larger capacitor in the chassis, in series with C, forms a voltage divider network to reduce the voltage entering the chassis and going to the triggering circuits. Most of the voltage drop is across C.

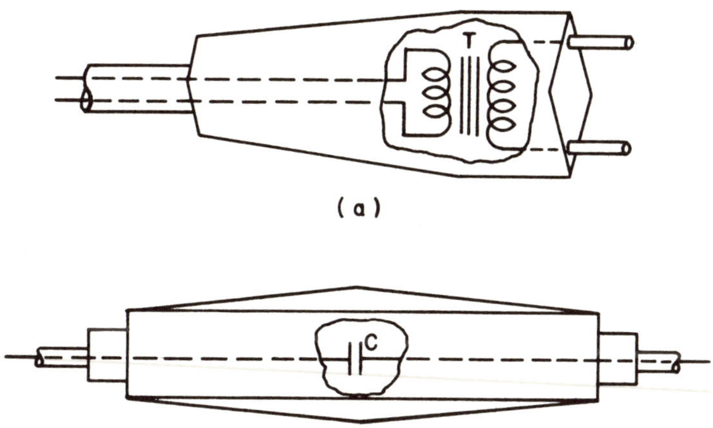

(a)

(b)

Figure 14-32 Internal connections of high-voltage pickup heads, typical, Source 15

Not shown here is the clamp-around type of capacitive coupled secondary coil pickup (R), and the blue current transformer head BL as shown in Figure 14-1.

14-6 General Trends in Design of Scope-Type Engine Analyzers

The general trend of all brands of scope-type engine analyzers is in step with the trends of other electronic equipment industries. Larger CRTs are being used. Battery powered portable units are available.

The main trend is toward all solid state electronics with or without plug-in printed circuit boards (PCB). The PCBs contain a complete functional component, such as; horizontal amplifier, vertical amplifier, sweep, spark advance, and so forth. The trend is also toward the use of more integrated circuits (IC).

The plug-in PCB permits rapid service on-the-job. A quick diagnosis will usually isolate the trouble to a particular circuit. Simply by changing boards, the fault may be corrected. The defective board can then be repaired in the shop or returned to the factory for repair, credit, or exchange.

For example, a typical engine analyzer, Source 15, Model 40-200, will have the PC boards listed below. They contain all components except the front panel-mounted controls and switches, under panel adjustment board, main power supply, and CRT. The PC boards are identified as follows.

A. Dwell angle meter board.
B. Power check and voltmeter.
C. Number 1 counter, provides synchronization between ignition pulses and scope circuits, primary and secondary.
D. Number 2 counter, the same as *C*, above.
E. Spark advance.
F. Input, reset, and synchronizing.
G. Power supply, positive 10- and 20-V dc.
H. Power supply, negative 10- and 20-V dc.
I. Ohms and combustion analyzer.
J. Horizontal amplifier.
K. Vertical amplifier.

L. Raster.
M. Sweep and vertical pre-amplifier.
N. Horizontal sweep and inverter.
O. Expanded tachometer, ammeter, and blanking.
P. Tachometer and timing mark locater.

14-7 Analog Engine Analyzer

The analog engine analyzer differs from the fully metered-type of engine analyzer in that most or all of the indicating meters are replaced by other forms of indicating devices such as annunciators and lights. These indicate a circuit or engine component as either GOOD or FAIL. A scope is usually a part of the analog engine analyzer.

The analog engine analyzer has advantages in high-speed production lines, busy service lanes, safety inspection check points, and other situations requiring a minimum of time training, and experience on the part of the operator.

The analog engine analyzer uses many circuits common to analog computers dealing directly with physical and electrical quantities such as voltage, current, time, etc. Since these circuits are rather lengthy, involving a multiplicity of solid state devices such as transistors and integrated circuit boards, it is beyond the scope of this book to break down and explain the details of each function of a complete analog analyzer. However, an effort will be made to explain the general principles of operation where they differ from other analyzers.

The signals are obtained for the analog in the same manner as any other analyzer. The signals, however, are translated and calibrated to operate relays and annunciators that indicate to the operator whether the test is normal or abnormal. Abnormal could be above normal or below a specified tolerance limit of normal.

The many relays involved may be mounted on plug-in boards and operated from a 28 V dc power supply, to turn indicating lights on and off. The relays are arranged in interlocking pairs so that no two GOOD and/or FAIL lamps can be lit simultaneously.

The annunciators are like relays except they do not actuate contacts. Instead, they drop a flag in a window opening—the flag has a message to be read by the operator.

The sequence of tests is advanced by "stepping" relays, either from the console panel or by remote control.

Each lamp circuit relay and annunciator is fed by a "flip-flop" circuit, arranged by cross-coupling, so that only a GOOD or FAIL lamp may be turned on at any one time. The "flip-flops" are controlled by three gated inputs; R (reset), S (set), and C (clear).

A negative going pulse (positive 28 V to ground), or ground on the set (S) line will latch the output high (positive 28 V) and turn that particular lamp on.

A negative going pulse (positive 28 V to ground) or ground on the reset (R) line will latch the output low (ground) and turn that particular lamp off.

A negative going pulse (positive 28 V to ground) or ground on the clear (C) line will turn all lamps off. All clear (C) lines are tied together and controlled by the TELL-A-CLEAR button on the console panel. This is used as a master CLEAR to turn all good/fail lamps off.

The TELL-A-TEST latch board has seventeen relays, sixteen of which are used to control the test nomenclature lamps on all of the Tell-A-Test front panel adjacent to the good/fail lamps. These are the test number and nomenclature lamps such as; timing at idle, timing advance, and so forth. Each of the sixteen test nomenclatures has two lamps. When lit they indicate the particular portion of the engine under test. Thus each relay drives two test nomenclature lamps.

14-8 Computerized Diagnostic Console

The latest development in the field of engine analyzers requires very little time to make a complete check on engine performance. An electronic computer compares the operation through built-in transducers to preset specifications and then prints out a complete report. The transducers are built into some experimental automobiles; therefore, it is necessary only to plug the tester into a jack located under the hood for a full diagnosis, completed in a few minutes. The scope tube is eliminated.

14-9 Thyratron Tube Ratings and Characteristics

Standard vacuum tubes and transistors of the type found in radio, TV, and electronics are used in automotive tune-up and test equipment. Since information on these is so readily available, it is thought superfluous to include the subject here. However, the thyratron tube is less understood outside the field of industrial electronics and, therefore, this versatile tube should be given more exposure.

Many thyratron tubes are used in the circuits of scope-type engine analyzers, wheel balancers, as well as other automotive tune-up and test equipment. These are usually type 2D21 or 2050. They each have a heater voltage of 6.3 V at a current of 0.6 amp and are of the shielded grid-type. They each have an anode current rating of 100 ma average. The 2D21 will handle a peak anode current of 500 ma whereas the 2050 is rated at a peak anode current of 1 amp.

Figure 14-33 shows the tube base connections and symbols for the thyratron type 2D21 where plate P (anode) is terminal 6, cathode C is terminal 2, heater H goes to terminals 3 and 4, control grid G is terminal 1, and screen grid SG goes to terminals 5 and 7. Type 2050 tube terminals are shown in circles.

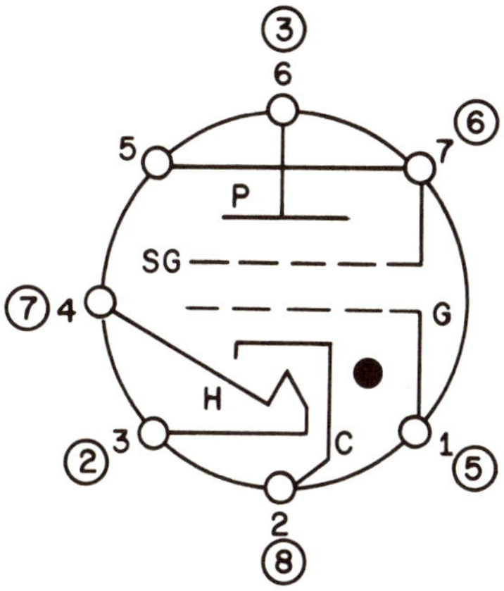

Figure 14-33 Thyratron tube base connections and symbols

The thyratron tube is gas filled, as indicated by the black dot in the symbol, and does not behave like a vacuum tube. In a vacuum tube, a change in the grid voltage causes a corresponding change in the flow of current between the plate and the cathode. Whereas, in a thyratron tube, the grid at a critical bias voltage acts as a trigger to start the flow of current from the plate to the cathode. Thereafter, the grid loses control and a change of bias (within limits) will not affect the plate-to-cathode current, or stop the tube from conducting from plate to cathode. The voltage drop across the tube plate to cathode is approximately 10 V, regardless of the current, within limits.

The tube can be stopped from conducting by removing the plate voltage, or by reducing the voltage substantially below the voltage drop of

10 V. Of course, if the plate of the tube is supplied by alternating current, then the tube stops conducting and is restored to normal as each half-cycle goes through zero voltage. Although the grid bias has no control of the current normally, it can be made to stop conducting if a negative bias is applied in the order of about 150 V. A positive pulse of at least 150 V applied to the grid will offset this negative bias and cause the tube to conduct. As soon as the pulse stops, the cutoff bias automatically restores the tube to a non-conducting state. In scope-type engine analyzers, this system is used. The thyratrons are triggered through a voltage divider network by voltage spikes, of very short duration, from the ignition system. The high voltage from the ignition system is reduced through a voltage divider network usually located in the pick-up that is attached to the spark plugs and distributor of the engine under test.

14-10 Purpose of Voltage Regulators

All types of automotive tune-up and test equipment that is accurately calibrated requires some means of maintaining calibration over a wide range of supply voltage, whether it be from batteries or the 115 V ac power line.

14-11 Zener Diode Voltage Regulators

On vehicle battery-operated devices such as dwell meters, tachometers, and the like, the voltage is held constant by zener diodes which have a constant voltage drop over a wide range of currents. Some even use one zener diode to hold the instrument voltage at approximately 5.5 V when used on either 6 or 12 V. These use a voltage dropping resistor VDR1 in a circuit as shown in Figure 14-34a. A voltage regulator is not necessary when internal batteries are used to provide the power. Alternating current rectified power supplies providing low voltage and low current bias voltages use the same circuit.

14-12 Zener Diode and Transistor Voltage Regulators

Where more power is to be handled, a circuit similar to that shown in Figure 14-34b is used. In this case, a transistor handles the heavier current, with the zener diode acting only as a low current voltage reference. A circuit similar to this is used in transistorized voltage

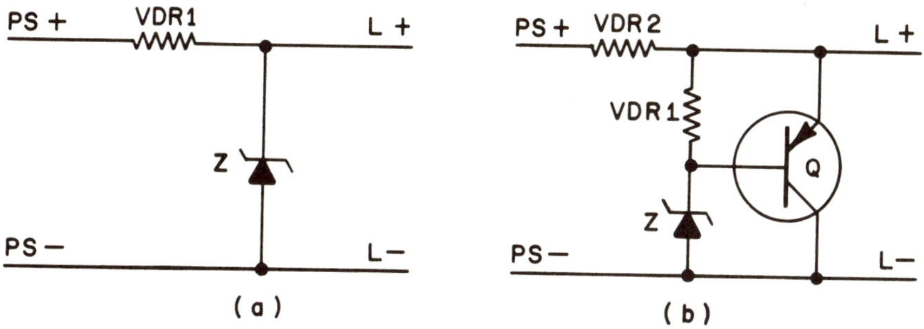

Figure 14-34 Zener diode and transistor voltage regulators

regulators to control the output of generators and alternators. In this application they may have an additional heavy-duty power transistor to control the field current.

In Figure 14-34a, the power supply PS must be higher than the zener voltage of the zener diode Z across the load L. The zener diode maintains its zener voltage over a limited range of current through it. Voltage dropping resistor VDR1 has a varying voltage drop across it, depending on the power supply PS voltage and the current drawn by the load L plus the zener diode current. Resistor VDR1 is calculated to supply enough current through the zener diode Z for it to maintain control, and yet, at the other extreme, limit the current through the zener diode Z to a safe value even at no load.

In Figure 14-34b, resistor VDR1 and zener diode Z are the same as in Figure 14-34a but the zener voltage is maintained at the base of transistor Q at very low current. The transistor Q and load voltage dropping resistor VDR2 now take over the control of heavier currents and maintain a constant load voltage regardless of a limited change in supply voltage PS, or change in the load L current.

14-13 Gas-Type Tube Voltage Regulators

Figure 14-35 shows the use of a gas-type tube in voltage regulator circuits for higher voltage power supplies. The gas glow-discharge (cold cathode) type of tube has a constant voltage drop over a limited range of currents, just as in the zener diode, and regulates in the same manner. About the only difference is that the gas tube requires a little higher starting voltage than its operating voltage, whereas the zener diode starting voltage is the same as its control voltage.

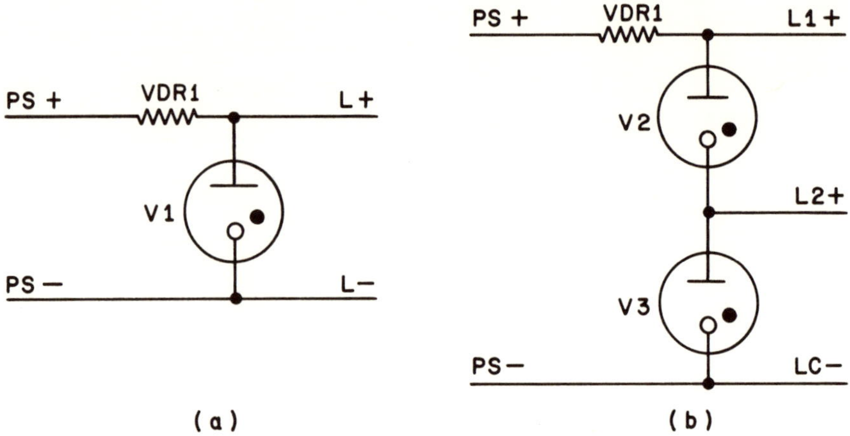

Figure 14-35 Gas-type tube voltage regulators

In Figure 14-35a is shown a single tube V1 which maintains a constant voltage across the load L. Figure 14-35b shows a circuit using two tubes, V2 and V3, in series to regulate higher voltages than the rating of one tube. A second regulated voltage can be tapped off at L2. Any number can be used having the same or different voltage drops by adding the voltage drop of all tubes, but they must all have approximately the same current rating for series connection. Zener diodes operate in series in the same manner.

Table 14-1 shows the important ratings of the most common voltage

Voltage Regulator Tube Ratings

Type	*Minimum dc Supply (volts)*	*Minimum dc Start (volts)*	*Operating Voltage*	*Operating Current (ma) Min.*	*Max.*	*Max. Reg. (volts)*
OA2	185	156	150–151	5	30	2–6
OA3	105	100	75	5	40	5
OB2	133	115	108	5	30	1–4
OB3	125	100	90	10	30	8
OC3	133	115	105–108	5	40	2–4
OD3	185	160	150–153	5	40	4–5.5

Table 14-1

regulator tubes most often encountered in automotive tune-up and test equipment.

14-14 Voltage Regulator Circuit Calculations

To calculate the value in ohms and the wattage rating of the voltage dropping resistor VDR1 in Figure 14-35, it is necessary to consider several factors. These are; the power supply full load voltage PSFLV, the power supply no load voltage PSNLV, the power supply voltage drop PSVD due to internal resistance, regulated load voltage RLV, the voltage dropping resistor voltage VDRV, the voltage dropping resistor current in milliamperes VDRMA, the voltage dropping resistor resistance in ohms VDRO, and the voltage dropping resistor wattage rating VDRW. These values can be calculated using the formulas below, but it is quicker to use the design nomograms shown in Figures 14-36 and 14-37.

Figure 14-36 determines the characteristics of the power supply PS. This nomogram is based on the two formulas:

$$(1)\quad \text{Percentage Regulation} = \frac{\text{PSNLV} - \text{PSFLV}}{\text{PSFLV}} \times 100$$

$$(2)\quad \text{PSVD} = \text{PSNLV} - \text{PSFLV}$$

Where PSNLV is the power supply no-load voltage, PSFLV is the power supply full-load voltage, and PSVD is the power supply internal voltage drop at full load.

The first three scales of Figure 14-36, starting at the bottom and reading up, solve equation (2). The third scale and the top horizontal scale are joined to determine the percentage regulation on the diagonal or "Z" scale to solve equation (1).

The following example will illustrate the use of the nomogram shown in Figure 14-36. Assume a power supply having a full-load voltage (PSFLV) of 200 V, and a no-load voltage (PSNLV) of 250 V resulting in a power supply voltage drop (PSVD) of 50 V. The 200-V point on the first (bottom) scale (PSFLV) is joined by a straight (dotted) line (1) to the 250-V point on the second (from the bottom) scale (PSNLV). This line is extended to the 50-V point on the third scale (PSVD). This point is joined to the 200-V point on the top horizontal scale (PSFLV) by a

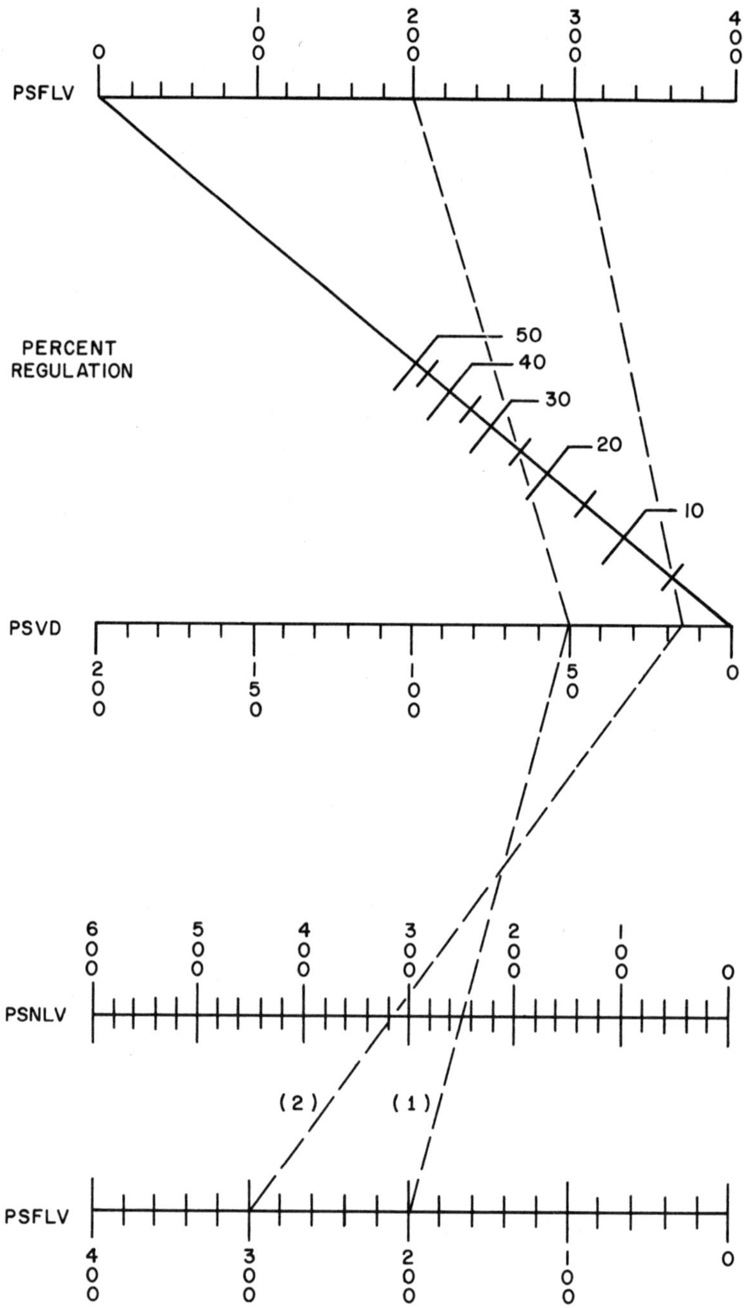

Figure 14-36 Design nomogram to determine percent regulation for various supply voltages

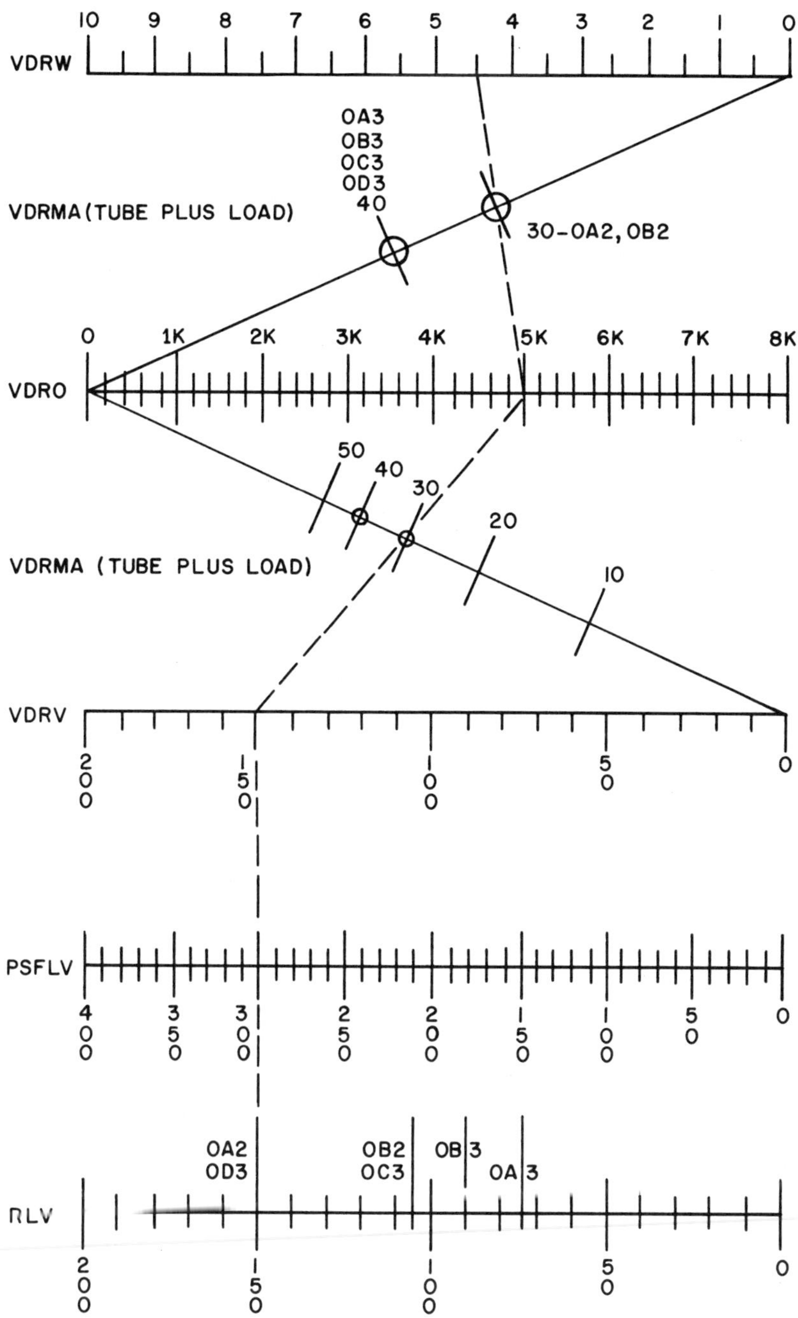

Figure 14-37 Design nomogram to determine resistance and wattage for series-limiting resistor

straight (dotted) line which intersects the percentage regulation diagonal line "Z" indicating 25% regulation of the power supply feeding the voltage regulator circuit.

A second example of a better regulated power supply is one that has a 300 V full-load voltage, and a 315 V no-load voltage having a 15-V drop. When this 15-V drop is divided by the 300 V full-load voltage the voltage regulation is 5% as indicated on the nomogram, Figure 14-36, dotted line (2).

The nomogram shown in Figure 14-37 is used to determine the resistance and wattage of the voltage dropping resistor VDR. The nomogram comprises a 3-scale horizontal alignment chart and two 3-scale "Z" nomograms. The first three scales (from bottom to top) solve the equation:

$$(3) \quad \mathrm{VDRV} = \mathrm{PSFLV} - \mathrm{RLV}$$

Where VDRV is the voltage drop across the series voltage dropping resistor VDR, PSFLV is the power supply full-load voltage, and RLV is the regulated load voltage maintained by the voltage regulator tube.

The six common voltage regulator tubes OA2, OA3, OB2, OB3, OC3, and OD3 are marked on the bottom scale at their regulating or operating voltage points (see Table 14-1). The first "Z" chart, from the bottom, solves the equation:

$$(4) \quad \mathrm{VDRO} = \frac{\mathrm{VDRV}}{\mathrm{I}}$$

Where VDRO is the required resistance in ohms of the series voltage dropping resistor VDR, VDRV is the voltage drop across the series voltage dropping resistor VDR and I is the total current in amperes to be drawn by the load current plus the current drawn by the regulator tube. Multiply the current I by 1000 to get VDRMA in milliamperes for chart use.

The second, or top, "Z" chart solves the equation:

$$(5) \quad \mathrm{VDRW} = \mathrm{I}^2 \times \mathrm{VDRO}$$

Where VDRW is the power dissipation of series voltage dropping resistor VDR, I is the total current through the resistor in amperes (multiply by 1000 to get VDRMA), and VDRO is the resistance in ohms of the resistor.

To continue with the previous example (2) of the nomogram in Figure 14-36, assume an OA2 tube is being used. Select the 150-V point on the first scale in Figure 14-37 and join it by a straight line (dotted) to the 300-V point on the adjacent scale PSFLV; continue to the third scale VDRV to 150 V drop across the series resistor. With no current drawn by the load, the voltage regulator tube must carry all of the current through the voltage dropping series resistor. As the load current is added, the tube draws less of the total current, therefore, the resistor must be selected to limit the tube current at no load. Since the OA2 tube is a 30 ma (maximum) tube, continue on a straight line from the 150-V point through the 30 ma point of the first diagonal scale VDRMA to the 5 Kohm point on the scale VDRO. Drawing a third straight line from this point through the second diagonal scale VDRMA at the 30 ma point to the scale VDRW and read 4.5 W. That is the minimum wattage rating. As a factor of safety, at least a 10-W resistor should be used.

The two nomograms shown in Figures 14-36 and 14-37 can be used for lower voltages, such as those used for zener diodes, by moving the decimal point one zero or digit to the left on all voltage, resistance, and watts scales, but not the VDRMA milliampere scale nor the percentage regulation scale.

The regulated power supplies are often used as an internal calibrating reference for dc voltmeters and other meters or circuits requiring adjustment or calibration.

15
Engine Component Testers

15-1 Single-Range Ohmmeter

A single-range ohmmeter of 0–200 ohms is satisfactory for all automotive service low-resistance measurements, except for measuring ignition coil secondary resistance which requires a second range of 7500 to 12,000 ohms. A single range meter is comprised of a meter calibrated in ohms, a fixed resistor, a battery, and a zero adjusting rheostat all connected in series. A multiple range ohmmeter requires only the addition of a fixed resistor for each range and a selector switch.

15-2 Ohmmeter Having Three Ranges and a Test Light

An ohmmeter having a built-in test lamp is shown in Figure 15-1 for measuring resistance in ohms and for continuity tests.

In Figure 15-1 selector switch SW1 is ganged with switch SW2. Position 1 connects the test light TL in series with battery BAT (2 dry cells–3 volts) and the test prods TP positive and negative. The test light TL will burn when the two test prods TP are touched together or connected to a very low resistance circuit component such as forward diode conductivity. Position 2 selects the ohmmeter range for zero to 500 Kohms, position 3 selects the ohmmeter range of zero to 5 Kohms, and position 4 selects the ohmmeter range of zero to 180 ohms. Meter M is calibrated to read ohms. Calibrating rheostat R1 is for zero ohms adjust on all ranges, and R3 is for 500 Kohm-range calibration; rheostat R2 is for 5 Kohm calibration; rheostat R4 is 180 ohm zero adjust; R5 is 180

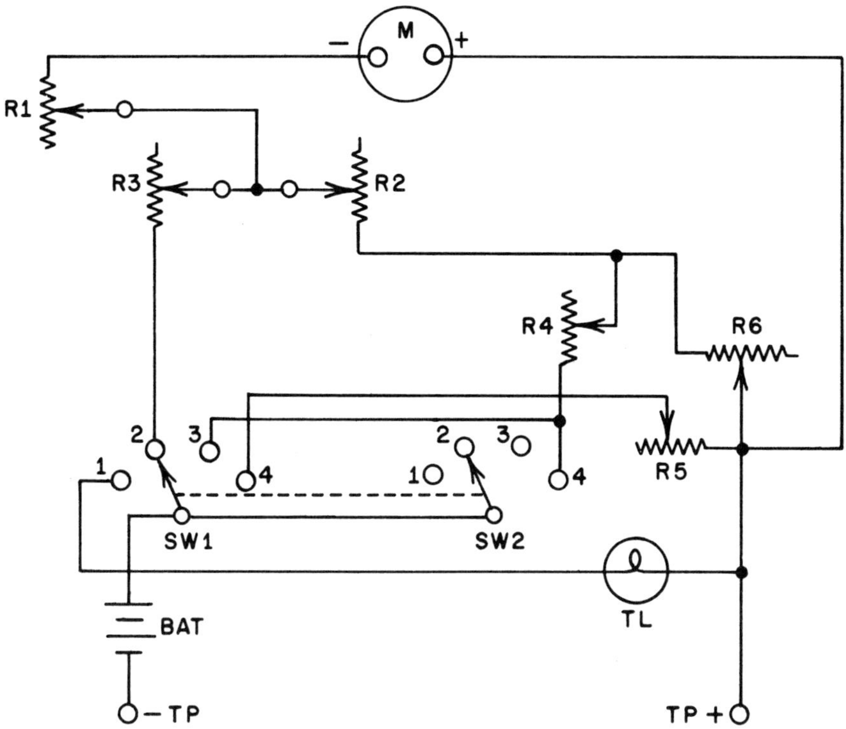

Figure 15-1 Three-range ohmmeter with test light

ohms zero adjust; and R6 is for 5 Kohms calibration adjust. Rheostat R1 is the main zero adjust located on the front panel, whereas the other rheostats are located inside the cabinet. The test lamp is convenient for testing diodes, transistors, and other solid states.

15-3 Volt–Ohm–Milliammeters

Multiple range volt-ohm-milliammeters (VOM), as the name implies, are versatile instruments in that they measure ohms, volts, and milliamperes with some having an ammeter range of 10 amperes, both ac and dc. They all operate fundamentally the same, differing only in the ranges and sensitivity of the meter which may have an ohms-per-volt sensitivity of 5000 or 20,000 ohms per volt dc and 1000 ohms per volt on ac.

For automotive electrical testing (except for radios and other electronic devices such as automotive tune-up and test equipment), the more rugged and less expensive 5000 ohms per volt dc and 1000 ohms per volt ac VOM is entirely satisfactory. Figure 15-2 shows the schematic wiring diagram of such a VOM suitable for automotive electrical measurements.

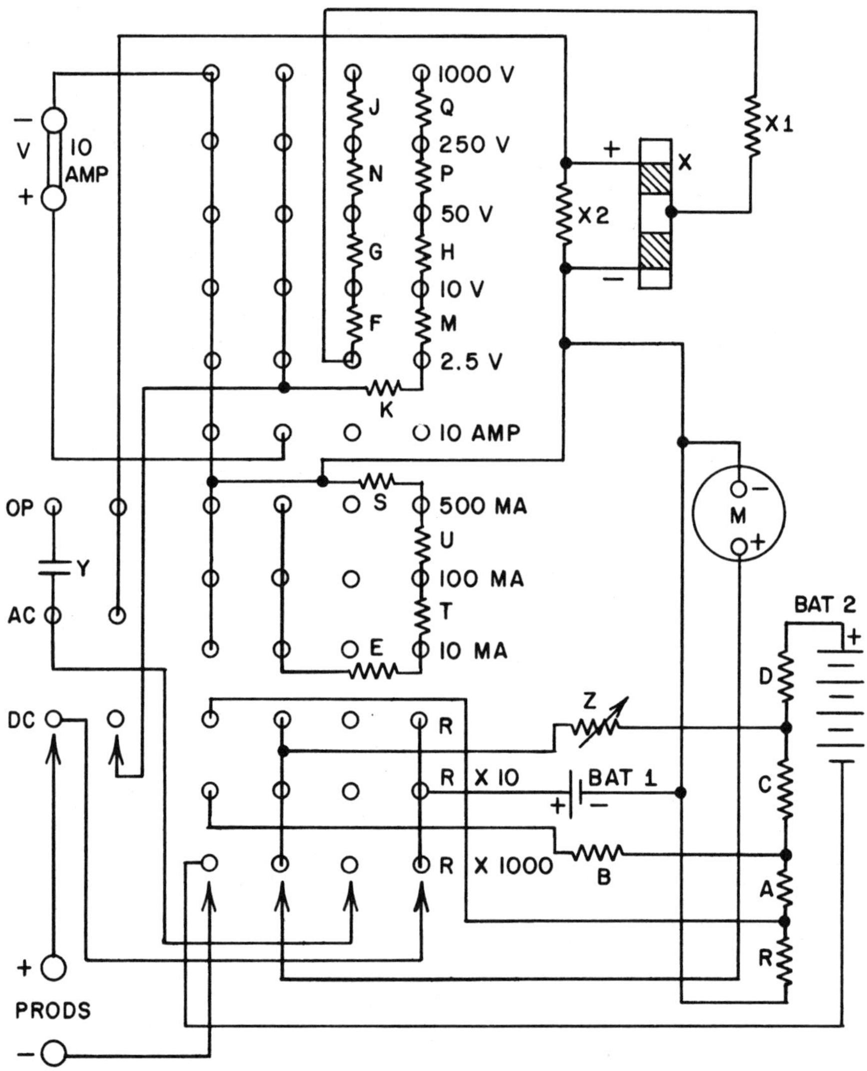

Figure 15-2 Multirange volt-ohm-milliammeter

In Figure 15-2, the meter M is a 200 micro-ampere, 1000 ohm, 200 millivolt movement; battery BAT 1 is 1-½-volt dry cell and battery BAT 2 is a 6-volt battery (four 1-½-volt dry cells). Resistors in ohms are A (2752), B (1050), C (2780), D (26,250), E (225), F (7500), G (40 K), H (200 K), J (750 K), K (11.5 meg) M (37,500), N (200 K), P (1 meg), Q (3750 K), R (27.8), S (0.5), T (22.5), U (2), V (10A shunt), Z (2 K pot.). Other components are; X (meter rectifier), X1 (rectifier series resistor), X2 (rectifier shunt resistor), Y (0.1 mfd capacitor).

The main selector switch is a four-deck 12-position rotary switch with the 12 ranges shown in Figure 15-2. The ac, dc, and output (OP) switch is a two-deck 3-position switch. Potentiometer Z is the zero ohms adjust for all ranges and is located on the front panel, all other resistors are precision resistors.

15-4 Transistorized Volt–Ohm–Milliammeter and Vacuum Tube Voltmeters

Where high input impedance is required in a test instrument, such as measuring ignition coil output peak volts, the vacuum tube voltmeter or its solid state equivalent is employed. The solid state equivalent of the VTVM was made possible by the development of the field-effect transistor (FET). The FET has an extremely high input impedance comparable to that of the vacuum tube. It is compatible with the power supply voltages and operating levels of bipolar transistors. The use of the FET makes it possible to use battery power for portable units. They also have the advantage of measuring currents and require no warm-up time as in the VTVM.

15-5 Typical Hi-Pot, Spark-and-Glow, Insulation Tester

To test for insulation leakage between any two electrical elements or windings, or any element and ground, a high potential (hi-pot) is applied to determine if the insulation is effective. The indicator may be a meter calibrated in megohms, or it may be a neon lamp that glows if the resistance value of the insulation is below an industry established acceptable standard of approximately 5 to 10 megohms.

A hi-pot insulation tester is shown schematically in Figure 15-3, where T1 is a step-up transformer intentionally having poor voltage regulation characteristics, to develop 500 V ac across the secondary S1 with 115 V, 60 Hz applied across the primary P terminals L1 and L2. The 500 V ac

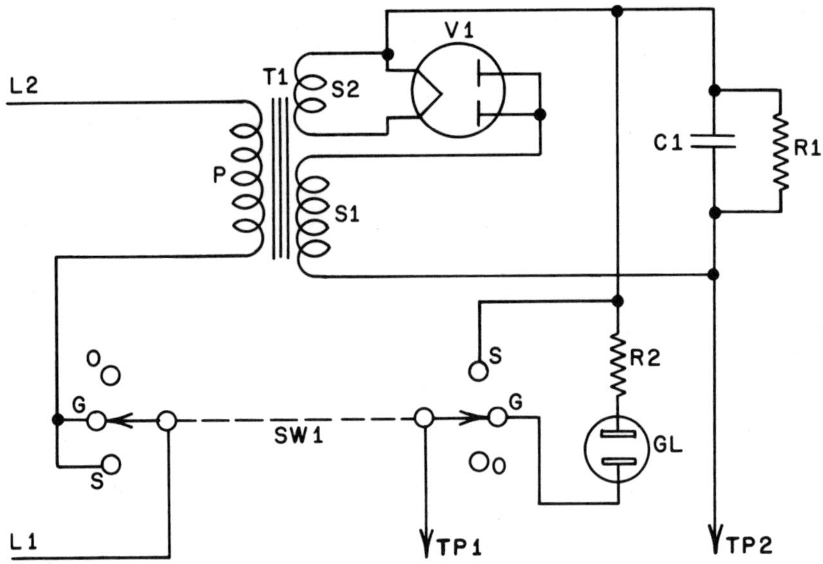

Figure 15-3 "Hi-pot" insulation tester circuit

is rectified by vacuum tube V1 (5Y3) to place a charge across capacitor C1 (0.47 mfd 600 V) which is shunted by resistor R1 (470 Kohms). Secondary S2 supplies filament voltage to tube V1.

The two-circuit, selector switch SW1 has three positions; position 1 is OFF O; position 2 is marked GLOW G, and will cause the neon glow lamp GL to glow if the resistance across the test prods TP1 and TP2 is below an acceptable value; and position 3 marked SPARK S which causes a spark at the test prods if the test elements are of very low resistance to each other or to ground. This spark is caused by the discharge of capacitor C1 which builds up to about 700 V open circuit voltage. R2 (100 Kohms) is a current limiting resistor.

15-6 Measuring Ignition Voltage by the Spark Gap Method

One of the simplest methods of measuring high voltages is to determine the length of the air gap which a given (minimum) voltage will cause a spark to pass between two electrodes. The voltage depends primarily on the size and shape of the electrodes, and the conditions of the air, such as temperature, pressure and humidity. The voltage is independent of the wave shape or frequency.

For normal conditions in open air, Table 15-1 shows the rms and peak voltages required to jump a gap between two number 00 double, long sewing needle points, supported axially at the end of linear conductors. The length of the needles must be at least twice the length of the air gap.

Spark Gap Voltages

		Needle Gap	
rms Volts	*Peak Volts*	*Millimeters*	*Inches*
10,000	14,140	11.9	.469
15,000	21,210	18.4	.724
20,000	28,280	25.4	1.00
25,000	35,350	33.0	1.30
30,000	42,420	41.0	1.61
35,000	49,490	51.0	2.00
40,000	56,560	62.0	2.44
45,000	63,630	75.0	2.95
50,000	70,600	90.0	3.54

Table 15-1

15-7 Calibrating High-Voltage Testers and Scopes

Calibrating high-voltage testers and spark plug voltages on an oscilloscope-type engine analyzer can be accomplished by using a high-voltage neon sign transformer, or other ac operated ignition transformers such as used on oil burners and the like.

Neon transformers have the secondary center-tapped at one-half of the rated output voltage. For example, a 15,000-V neon transformer will have 7500 V on each side of the center tap and ground. The rms voltage can be read on a high resistance ac voltmeter and the peak voltage can be calculated. If only a 5000-V range ac voltmeter is available, then reduce the primary voltage of the transformer to 60 V and read 3750 V.

To calculate peak voltage, multiply the rms voltage as read on an ac meter by 1.414 to get the peak voltage. For example, 7500 V rms × 1.414 equals 10,605 peak volts.

To calibrate the vertical spark plug voltage scale on the scope screen, connect the neon transformer to the secondary pick-up lead and ground in the same way as connected to the ignition coil. There should be no spark gap, and the neon transformer should be operated at no load. Operate the simulator, or engine, at a speed of 1500–1800 rpm until the patterns are stationary on the scope screen. The ac line voltage should be maintained at the same voltage that was used in measuring the secondary voltage, or allowance made for a higher or lower voltage. The secondary voltage will vary in proportion to the change in the primary voltage.

15-8 Typical Meter for Reading Ignition High Voltage

The usual ignition voltage meter for reading spark plug and coil output voltages consists of an ac voltmeter calibrated to read peak voltages instead of rms voltages. To prevent loading the ignition coil, and thereby reducing the voltage, a very high impedance voltmeter is employed, such as a vacuum tube voltmeter VTVM or the solid state high-input impedance equivalent. These meters usually have a single scale, calibrated zero to 50,000 V.

15-9 Single-Range Capacitor Tester

The only capacitor encountered in conventional Kettering ignition systems is the distributor capacitor having a value of 0.2 to 0.25 mfd which can be tested on a single-range capacitor tester, usually incorporated in other instruments.

A single-, or limited-range capacitor tester can be improvised by using a fixed resistor and an ac voltmeter. Connect the unknown capacitor in series with a 10 Kohm 5 to 10 W wirewound resistor. Connect this series circuit across 120 V, 60 Hz ac and read the voltage across the resistor. It should read approximately 80 V for a 0.2 mfd ignition capacitor. Check this voltage against several capacitors known to be good. Make up a curve or chart for other values desired in this range. If the capacitor is open, the voltmeter will read zero. If the capacitor is shorted, or has excessive internal leakage, the meter will read higher, or at full line voltage. Any ac voltmeter can be used, such as a VOM, 1000 ohms per volt ac meter.

15-10 Multi-Range Capacitor Tester and Ohmmeter

For servicing transistorized ignition systems and automotive tune-up and test equipment, a wide-range capacitor tester is required. Such a tester is shown in Figure 15-4.

This tester uses a bridge circuit which is tuned to balance by a "tuning eye" tube VT1 (1629) and the values read on a calibrated dial. This tester will also measure resistance. When the bridge circuit is balanced, the tuning eye tube will be open at its widest angle.

The power supply transformer T1 supplies 12 V ac across secondary S3 for the tube heaters, 55 V ac across secondary S2, and 500 V ac across secondary S1. Secondary S1 voltage is rectified by half-wave rectifier tube VT2 (1626) which supplies positive 150 V dc across capacitor C5 (8 mfd) and negative 410 V dc across capacitor C6 (8 mfd).

Switch SW1 is turned ON when switch SW2 (rear deck) and SW3 (front deck) are turned to all positions except OFF, as shown. Switch SW4 is the leakage test switch and is spring loaded to return to NORMAL, as shown. Switch SW5 is on the back of the power-factor control R10 (800 ohms). Other values are; resistors in ohms, R1 (10 meg), R2 (1 meg), R3 (220 K), R4 (470 K), R5 (1 K), R6 (200 K precision), R7 (2 K precision), R8 (10 K main control on calibrated dial), R9 (90 K precision), R11, R12, R13, R14, and R15 (22 K), R16 (10 K), R17 (47 K); capacitors in mfd, C1 (0.01), C2 (0.25), C3 (200 mmfd precision), C4 (0.02 precision), C5 (8), C6 (8), C7 (0.05), C8 (2 precision).

The range and function switches SW2 and SW3 are shown in the OFF position. Rotation is in the direction of the arrow and indexes in order as follows;

0. OFF, as shown.
1. Ohms RX1 (100–50 K)
2. Ohms RX100 (10 K–5 meg)
3. Capacitor (0.00001–0.005 mfd)
4. Capacitor (0.001–0.5 mfd)
5. Capacitor (0.1–50 mfd)
6. Capacitor (20–1000 mfd)
7. Direct current leakage test voltage 25 V
8. Direct current leakage test voltage 150 V
9. Direct current leakage test voltage 250 V

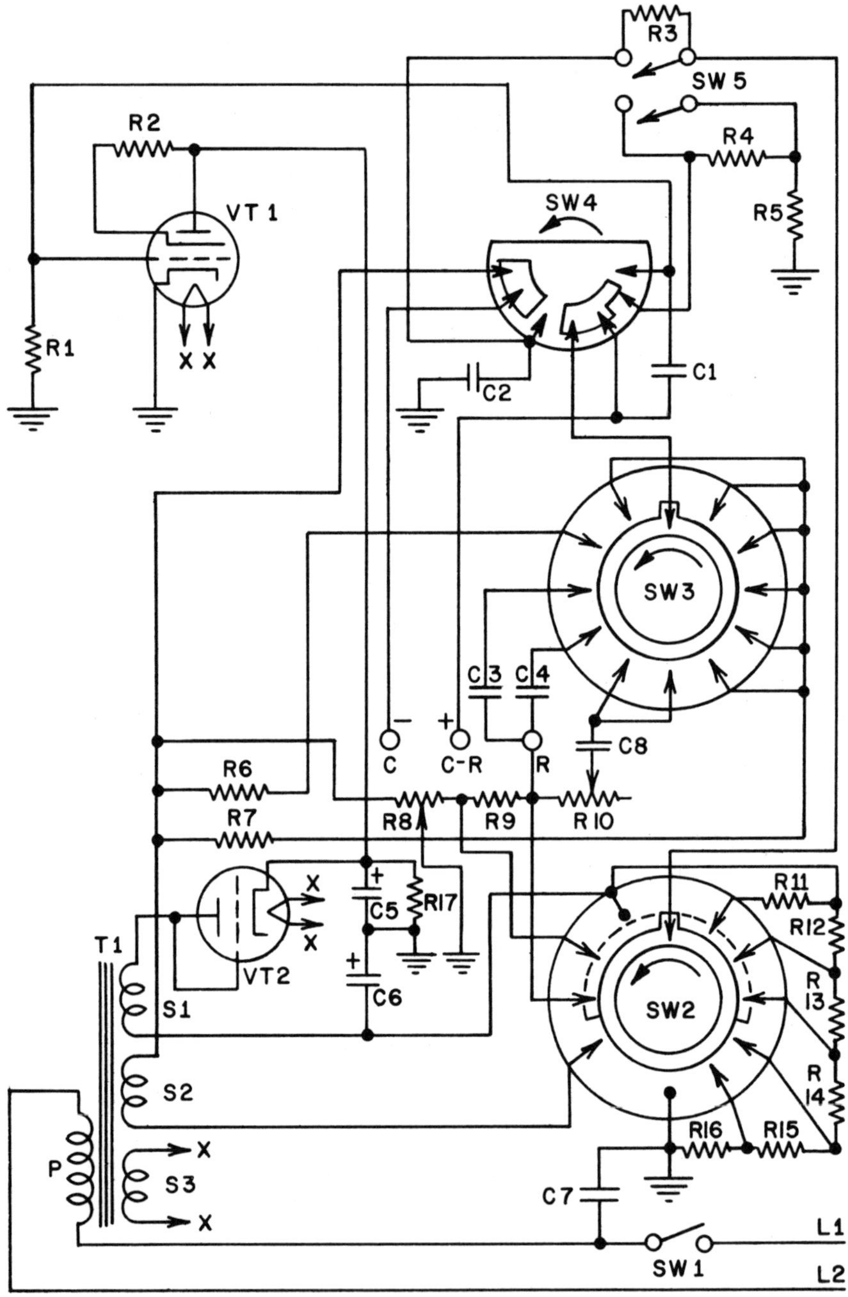

Figure 15-4 Multirange capacitor and resistor tester

10. Direct current leakage test voltage 350 V
11. Direct current leakage test voltage 450 V

Switch SW5 is on the rear of the power-factor control R10. In the position shown in Figure 15-4 the power-factor control is set for paper and mica capacitors. Switch SW5 closes when rheostat R10 is rotated for reading power-factor on electrolytic capacitors. Any capacitor showing a power-factor near or above 20 should be suspicioned, or replaced, depending on the circuit in which it is used. For example, a filter capacitor can show rated value of capacitance (mfd) but if it has a high power-factor its filtering action is hampered and *hum* can be evident.

15-11 Battery-Powered Capacitor Tester

Capacitor testers that operate from the vehicle battery, or a self-contained battery, are available usually in a single range of zero to 0.5 mfd.

These testers use a vibrator or transistorized oscillator as in timing lights, to change the battery dc voltage to alternating current. The remainder of the circuit is the same as in ac operated capacitor testers.

15-12 Testing Coils by Oscilloscope

The ignition coil of any engine can best be tested on the engine using an oscilloscope or a scope-type engine analyzer. The scope method of coil testing will show any of the following defects in the coil by the shape of the scope pattern.

A. High resistance, or open circuit, in coil primary.
B. High resistance, or open circuit, in coil secondary.
C. Shorted turns in coil primary.
D. Shorted turns in coil secondary.
E. High-voltage leakage to ground.
F. Reversed primary connection.

Any secondary or primary scope pattern that deviates from the normal pattern on all cylinder firings can usually be attributed to a defective coil, capacitor, or both. If replacing the capacitor does not correct it then replace coil.

15-13 Typical Coil Testers Using Spark Gap

The first coil testers, and low cost testers, use a vibrator, or chopper, in the coil primary circuit and measure the efficiency of the coil by the length and color of the spark across a spark gap that the secondary voltage can jump.

15-14 Typical Coil Testers Using a Calibrated Meter

A good coil tester should be capable of testing all types of ignition coils and measuring the output in peak volts and current (power) output on a calibrated meter. Such a tester should be capable of testing the following coil types.

A. Kettering-type ignition coil operating from distributor points at relatively low current.

B. Transistorized ignition coil systems using distributor points or pulse generator at relatively high current.

C. Capacitor-discharge (C-D) coil ignition systems using distributor points or pulse generator at relatively high peak current.

D. Magneto-type coils using distributor points, such as used on small engines and outboard motors.

Each of the above types of coils will require different settings on a selector switch.

The primary power is usually supplied by a vibrator, or transistorized oscillator, powered by a battery or ac. The output meter is usually of the high-impedance type of vacuum tube voltmeter (VTVM) or its solid state equivalent. The scale may be calibrated in peak volts and current, or simply a colored dial indicating GOOD or BAD. Some coil testers also include a capacitor test for the ignition capacitor.

15-15 Distributor Tester, Purpose and Construction

The distributor tester is a complete service instrument designed to check and calibrate all types of distributor assemblies accurately, after removal from the vehicle. The essential requirements for a complete distributor tester are as follows;

A. *Power supply.* The power supply is from the 115 V ac power line, but on earlier models a storage battery was also required.

B. *Drive motor.* The drive motor must be reversible and provide adjustable speeds of approximately 400 to 8000 crankshaft rpm (200 to 4000 distributor rpm). One type uses a ¼ hp constant-speed motor driving a rubber-faced drive pulley against a flat disc. By moving the motor in or out, the speed can be adjusted. A newer type uses a universal motor having an electronic speed control. The drive motor turns the distributor under test which turns in unison with a flash indicating dial.

C. *Distributor mounting.* The distributor housing is held in position by a stationary clamping device. The distributor drive shaft is clamped in a universal chuck or connected by adapters to the drive motor.

D. *Calibrated dial.* Figure 15-5 shows the distributor DIST mounted on the unit in the center of the dial assembly. The dial assembly consists of a stationary (but may be adjustable) dial SD calibrated in degrees and number of cylinders, and a rotating dial RD which turns with cam CAM of the distributor DIST.

E. *Flash tube indicator.* A xenon flash tube, as used in timing lights, is mounted on the back side of rotating dial RD behind a window in the shape of an arrow or pointer. When the tube flashes, it illuminates the flash indicator FI at every point on the stationary dial SD where the distributor points DP open. The power to the rotating flash tube must be conducted through brushes and collector rings. The trigger voltage to the flash tube consists of a high voltage ignition coil or pulse transformer. The coil may be mounted on the rotary shaft of the unit, in which case the primary current is fed through a low voltage brush and collector ring assembly. If the coil is mounted elsewhere, the high voltage is fed to the flash tube through a high voltage collector assembly.

F. *Tachometer.* The tachometer indicates the distributor and equivalent engine rpm and may be a mechanical tachometer driven from the rotating shaft, or it may be an electrical tachometer actuated from the distributor point pulses, or from a separate commutator. If a mechanical tachometer is used, or a separate commutator is used, it is not necessary to select the number of cylinders to get the correct tachometer reading.

G. *Dwell angle meter.* The dwell angle meter is usually a combination instrument serving as a dwell angle meter, a distributor capacitor tester, and a points resistance tester.

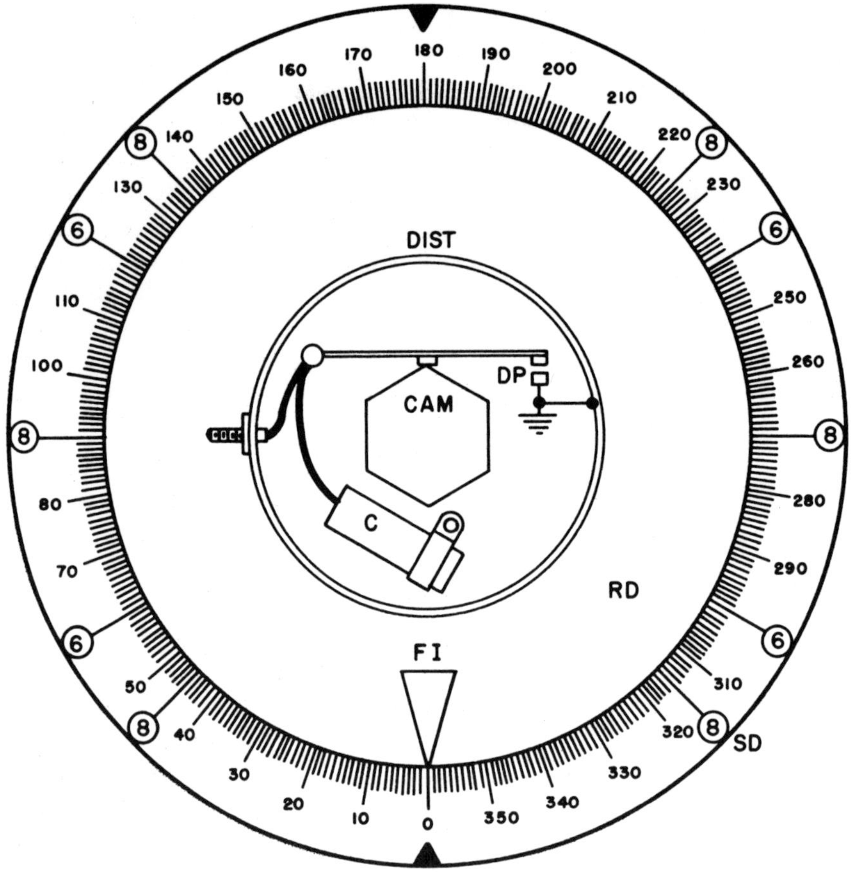

Figure 15-5 Distributor tester dial and distributor mounting

H. *Test selector switch.* The test selector switch turns the unit off and on and selects the various functions of the tester.

I. *Vacuum system.* The vacuum system consists of a motor-driven vacuum pump, a standard vacuum gauge, and a vacuum adjustment control for checking the distributor vacuum advance and retard system.

15-16 Distributor Tester, Operation

With the distributor mounted, and the direction of rotation set correctly for the particular distributor under test, the flash indicator will flash, showing the same number of arrows or pointers as there are number of cam lobes, or cylinders. With the speed set at low rpm, the distributor should be turned, or dial SD moved, until the nearest pointer FI is exactly at zero. All other pointers should then be spaced equally as follows; 4-cylinders—90° apart, 6-cylinders—60° apart, and 8-cylinders—45° apart. Any deviation from these angles indicates defects in the distributor.

The mechanical, or centrifugal, spark advance is tested by increasing the rpm. The pointer FI should move *against* rotation the number of degrees of advance, and toward the direction of rotation for degrees of retard.

The vacuum advance mechanism is checked by applying the specified vacuum and noting the spark advance or retard in degrees.

The distributor capacitor C is tested by turning the distributor until the points DP are open.

The distributor points DP resistance is measured with the points closed.

The dwell angle is measured on the dwell angle meter which indicates the *average* dwell angle of all lobes. The dwell angle for *each* lobe could be measured accurately on a distributor tester if the "Cantonwine timing light to measure dwell angle" (described elsewhere) is incorporated in the distributor tester. This would eliminate the dwell meter entirely.

To synchronize double breaker points, proceed as follows;

1. Move the protractor scale SD, or the distributor housing, so that one of the pointer flashes caused by the *stationary* points is at the zero degree mark.
2. If the distributor is the type that has *half* as many lobes on the cam as there are cylinders in the engine, adjust the *adjustable* breaker points plate assembly until the flashes caused by the *adjustable* points are in correct relationship, in accordance with the specifications.
3. If the distributor is the type having the same number of lobes on the cam as there are cylinders in the engine, then insert a piece of Bakelite or fibre insulation between the stationary contact points. Adjust the adjustable breaker plate assembly until flashes caused by the adjustable points occur at the same place on the protractor scale, or are in the correct relationship in accordance with the specifications.
4. If the distributor has *two* primary terminals, use the one con-

nected to the stationary points for setting one of the flashes to the zero degree mark on the protractor scale. Then move the lead to the terminal connected to the adjustable points. Adjust the adjustable breaker plate assembly until the flashes caused by the adjustable points occur at the same place on the protractor scale as the stationary points, or in accordance with specifications.

5. Adapters are usually available for adjusting older vehicles.

15-17 Spring Gauges

Many sizes and types of springs are used in automotive construction and some are more critical than others. Springs may lose their compression or tension in many ways and require testing to determine if they are within specifications.

Small spring gauges, measuring in ounces, are used for checking relay springs, armature brush springs, and the like.

The heavier spring gauges, measuring in pounds and ounces, are used to measure valve springs and other heavier springs.

The efficiency of a spring is measured by the pull or push in pounds or ounces that it takes to move the spring from an unloaded position to a specified new position in inches and fractions. The unloaded length should first be measured to find if it is within specifications.

Appendix A

1. Source Index

Source No.	*Manufacturer*	*Product*
1	Allen Test Products Division 2101 N. Pitcher Street Kalamazoo, Michigan 49007	*Battery chargers, testers, automotive tune-up and test equipment*
2	Allied Electronics 2400 W. Washington Boulevard Chicago, Illinois 60680	*Instruments and electronic parts*
3	American Bosch Arma Corporation American Bosch Sales Division Springfield, Massachusetts 01100	*Automotive electronic equipment*
4	Associated Equipment Corporation 1555 Salzman Avenue St. Louis, Missouri 63133	*Battery chargers, testers, and car-starting equipment*
5	Christie Electric Corporation Box 60020 or 3410 West 67th Street Los Angeles, California 90060	*Battery chargers and testers*
6	Delco-Remy Division General Motors Corporation Anderson, Indiana 46000	*Automotive ignition and electrical equipment*
7	Delta Products, Inc. Grand Junction, Colorado 81501	*Transistorized ignition systems and special equipment*

Source No.	*Manufacturer*	*Product*
8	Electro-Standards Laboratories 3650 N.W. 34th. Street Miami, Florida 33142	*Parts for battery chargers and automotive test equipment*
9	Ford Motor Company P.O. Box 2053 or 20000 Rotunda Drive Dearborn, Michigan 48121	*Automobiles and components*
10	Fox Products Company 4720 N. 18th. Street Philadelphia, Pennsylvania 19141	*Battery chargers and testers*
11	Industrial Instrument Works 3305 Tchoupitoulas Street New Orleans, Louisiana 70115	*Repair-calibration of precision meters and instruments*
12	Kal-Equip Company P.O. Box 188 or 411 Washington Street Otsego, Michigan 49078	*Battery chargers, testers, and automotive test equipment*
13	Lafayette Radio Electronics 111 Jericho Turnpike Syosset L.I., New York 11791	*Instruments and electronic parts*
14	P. R. Mallory and Company, Inc. 3029 E. Washington Street Indianapolis, Indiana 46201	*Vibrators, capacitors*
15	Marquette Corporation 3800 North Dunlap Street St. Paul, Minnesota 55112	*Battery chargers, testers, and automotive tune-up and test equipment*
16	Prestolite Division of Electra Corp. Toledo, Ohio 43601	*Automotive ignition systems and electrical components*
17	Stewart-Warner Corporation 1840 Diversey Parkway Chicago, Illinois 60614	*Wheel balancers and tachometers*
18	Sun Electric Corporation Harlem & Avondale Chicago, Illinois 60631	*Battery chargers, testers, and automotive tune-up and test equipment*
19	J. C. Whitney and Company 1917-19 Archer Avenue Chicago, Illinois 60616	*Automotive accessories and parts*

2. Brand Names and Sources

Brand Name	*Source No.*
Alemite	17
Atlas	15
Penny, J. C.	15
Rotunda (Ford)	1 and 5

Appendix B

1. Abbreviations

A Ammeter, Armature
ac Alternating Current
AFR Air-Fuel Ratio
AM Ammeter
amp Amperes

B Bulb, Black
BAT Battery
BL Blue
BR Brown, Bridge Rectifier
BZ Buzzer, Horn

C Capacitor, Cathode, Collector
°C Degrees Centigrade
CB Circuit Breaker
CH Choke
CCW Counterclockwise
CR Diode
CRT Cathode Ray Tube
CT Center Tap
CW Clockwise

D Diode
dc Direct Current
DPDT Double-Pole Double-Throw
DPST Double-Pole Single-Throw
DV Delay Voltage

E Voltage

F Fuse
°F Degrees Fahrenheit
FET Field-Effect Transistor
FM Fan Motor
FT Flash Tube
FW Full-Wave

G Green, Generator

H High
HP Horizontal Plate
hp Horsepower
HS Heat Sink
HW Half-Wave
HZ Horizontal
Hz Hertz-Cycles Per Second

I Input

K Thousand
kc Kilocycle

L Line, Lead, Low

M Motor, Meter
ma Milliamperes
MC Magnetic Clutch
mfd Microfarads
mmfd Micro-Microfarads

N.C. Normally Closed
NEG Negative
N.O. Normally Open
NPN Negative-Positive-Negative

O Orange
OC Open Circuit

P Primary, Pulley, Pressure
PC Piece
pF Picofarads
PIV Peak Inverse Voltage
PL Pilot Light
PNP Positive-Negative-Positive
POS Positive

Q Transistor

R Resistor, Thermistor, Red
R-C Resistor-Capacitor
REF Reference
REG Regulator
RF Radio Frequency
RFC Radio Frequency Choke
RL Relay, Solenoid
RLC Relay or Solenoid Contacts
rms Root Mean Square
rpm Revolutions Per Minute

S Switch, Secondary
SCR Silicon Controlled Rectifier
SH Shunt
SI Silicon
SL Slot
SPDT Single-Pole Double-Throw
SPST Single-Pole Single-Throw
SW Switch

T Transformer
TB Terminal Board
TH Thermostat, Thermometer, Thyratron
TI Timer
TM Timer Motor
TMC Timer Motor Contacts

V Volts, Voltmeter, Vacuum
VA Volt Amperes
VIB Vibrator
VM Voltmeter
VOM Volt-Ohms-Milliammeter
VP Vertical Plate
VT Vacuum Tube

W Watts, White
WBT Wheel Balancer Tester

Y Yellow

Z Zener Diode

Index

About the Author

The author, Charles R. Cantonwine B.S., E.E., is a graduate electrical engineer who majored in electronics. He is the author of the book titled *Battery Chargers and Testers,* also published by the Chilton Book Co. He has had many years' experience in all phases of electrical and mechanical designing, manufacturing, servicing, and merchandising, and is well qualified to write *Automotive Tune-Up and Test Equipment,* the first book on the subject.

During his long career he has functioned as an auto mechanic; a pioneer in police radio; an electric motor specialist for General Electric Co.; operator of his own electronics, radio, and TV shop; a holder of two amateur and one first-class commercial radio operator's license; an electronics engineer for the Naval Ordnance Laboratory in Washington, D.C., where he was the chief coordinator of, and helped develop, the famous Holmes antitorpedo device and other underwater ordnance during World War II; and as the national service manager and electrical buyer for Sears Roebuck & Co. Mr. Cantonwine holds many patents and patents pending, the latest of which is the "Cantonwine Timing Light to Measure Dwell Angle" (patents pending), that is 100% accurate and can never lose calibration. For the past several years he has been an operator of his own electric shop; building special equipment, repairing industrial electronics and motors. Mr. Cantonwine maintains an authorized service depot, operating state-wide, for all major brands of automotive tune-up and test equipment, battery chargers, and battery testers. He has developed the "Cantonwine Meter," designed especially for the peculiar needs of the automotive tune-up and test equipment industry.